Argument-Driven Inquiry in PHYSICS
VOLUME 2

ELECTRICITY AND MAGNETISM LAB INVESTIGATIONS for GRADES 9–12

Argument-Driven Inquiry in PHYSICS
VOLUME 2

ELECTRICITY AND MAGNETISM LAB INVESTIGATIONS for GRADES 9–12

Todd L. Hutner, Victor Sampson, Adam LaMee,
Daniel FitzPatrick, Austin Batson, and
Jesus Aguilar-Landaverde

Arlington, Virginia

Claire Reinburg, Director
Rachel Ledbetter, Managing Editor
Jennifer Merrill, Associate Editor
Andrea Silen, Associate Editor
Donna Yudkin, Book Acquisitions Manager

ART AND DESIGN
Will Thomas Jr., Director

PRINTING AND PRODUCTION
Catherine Lorrain, Director

NATIONAL SCIENCE TEACHING ASSOCIATION
1840 Wilson Blvd., Arlington, VA 22201
www.nsta.org/store
For customer service inquiries, please call 800-277-5300.

Copyright © 2020 by Argument-Driven Inquiry, LLC.
All rights reserved. Printed in the United States of America.
23 22 21 20   4 3 2 1

NSTA is committed to publishing material that promotes the best in inquiry-based science education. However, conditions of actual use may vary, and the safety procedures and practices described in this book are intended to serve only as a guide. Additional precautionary measures may be required. NSTA and the authors do not warrant or represent that the procedures and practices in this book meet any safety code or standard of federal, state, or local regulations. NSTA and the authors disclaim any liability for personal injury or damage to property arising out of or relating to the use of this book, including any of the recommendations, instructions, or materials contained therein.

PERMISSIONS
Book purchasers may photocopy, print, or e-mail up to five copies of an NSTA book chapter for personal use only; this does not include display or promotional use. Elementary, middle, and high school teachers may reproduce forms, sample documents, and single NSTA book chapters needed for classroom use only. E-book buyers may download files to multiple personal devices but are prohibited from posting the files to third-party servers or websites, or from passing files to non-buyers. For additional permission to photocopy or use material electronically from this NSTA Press book, please contact the Copyright Clearance Center (CCC) (*www.copyright.com*; 978-750-8400). Please access *www.nsta.org/permissions* for further information about NSTA's rights and permissions policies.

Library of Congress Cataloging-in-Publication Data
Names: Hutner, Todd, 1981- author. | Sampson, Victor, 1974- author. | FitzPatrick, Daniel (Clinical assistant professor of mathematics), author.
Title: Argument-driven inquiry in physics. volume 2, Electricity and magnetism lab investigations for grades 9-12 / by Todd L. Hutner, Victor Sampson, Daniel FitzPatrick, Austin Batson, Jesus Aguilar-Landaverde.
Other titles: Electricity and magnetism lab investigations for grades 9-12
Description: Arlington, VA : National Science Teaching Association, [2020] | Includes bibliographical references and index.
Identifiers: LCCN 2019051385 (print) | LCCN 2019051386 (ebook) | ISBN 9781681403779 (paperback) | ISBN 9781681403786 (pdf)
Subjects: LCSH: Electricity--Experiments. | Magnetism--Experiments. | Electricity--Study and teaching (Secondary) | Magnetism--Study and teaching (Secondary)
Classification: LCC QC533 .H88 2020 (print) | LCC QC533 (ebook) | DDC 537.078--dc23
LC record available at *https://lccn.loc.gov/2019051385*
LC ebook record available at *https://lccn.loc.gov/2019051386*

CONTENTS

Preface ix
About the Authors xv
Introduction xvii

SECTION 1
Using Argument-Driven Inquiry

Chapter 1. Argument-Driven Inquiry 3

Chapter 2. Lab Investigations 23

SECTION 2
Forces and Interactions: Electrostatics

INTRODUCTION LABS

Lab 1. Coulomb's Law: How Do the Amount of Charge on the Rod and the Mass of the Foil Used in an Electroscope Affect How Far Apart the Pieces of Foil Will Separate From Each Other?
    Teacher Notes 32
    Lab Handout 44
    Checkout Questions 50

Lab 2. Electric Fields and Electric Potential: How Does the Electric Potential Difference Change as You Move Away From the Positive Charge in an Electric Field?
    Teacher Notes 54
    Lab Handout 65
    Checkout Questions 72

APPLICATION LAB

Lab 3. Electric Fields in Biotechnology: How Does Gel Electrophoresis Work?
    Teacher Notes 76
    Lab Handout 87
    Checkout Questions 94

SECTION 3
Energy: Electric Current, Capacitors, Resistors, and Circuits

INTRODUCTION LABS

Lab 4. Capacitance, Potential Difference, and Charge: What Is the Mathematical Relationship Between the Potential Difference Used to Charge a Capacitor and the Amount of Charge Stored?
Teacher Notes 100
Lab Handout 110
Checkout Questions 116

Lab 5. Resistors in Series and Parallel: How Does the Arrangement of Four Lightbulbs in a Circuit Affect the Total Current of the System?
Teacher Notes 118
Lab Handout 128
Checkout Questions 134

Lab 6. Series and Parallel Circuits: How Does the Arrangement of Three Resistors Affect the Voltage Drop Across and the Current Through Each Resistor?
Teacher Notes 138
Lab Handout 150
Checkout Questions 156

Lab 7. Resistance of a Wire: What Factors Affect the Resistance of a Wire?
Teacher Notes 160
Lab Handout 172
Checkout Questions 178

APPLICATION LABS

Lab 8. Power, Voltage, and Resistance in a Circuit: What Is the Mathematical Relationship Between the Voltage of a Battery, the Total Resistance of a Circuit, and the Power Output of a Motor?
Teacher Notes 182
Lab Handout 193
Checkout Questions 199

Lab 9. Unknown Resistors in a Circuit: Given One Known Resistor and a Voltmeter, How Can You Determine the Resistance of the Unknown Resistors in a Circuit?
Teacher Notes 204
Lab Handout 217
Checkout Questions 223

SECTION 4

Forces and Interactions: Magnetic Fields and Electromagnetism

INTRODUCTION LABS

Lab 10. Magnetic Field Around a Permanent Magnet: How Does the Strength and Direction of a Magnetic Field Change as One Moves Around a Permanent Magnet?
Teacher Notes 230
Lab Handout 239
Checkout Questions 244

Lab 11. Magnetic Forces: What Is the Mathematical Relationship Between the Distance Between Two Magnets and the Strength of the Force Acting on Them?
Teacher Notes 248
Lab Handout 258
Checkout Questions 263

Lab 12. Magnetic Fields Around Current-Carrying Wires: How Does Changing the Magnitude and Direction of a Current in Two Parallel Wires Affect the Magnetic Field Around the Two-Wire System?
Teacher Notes 268
Lab Handout 280
Checkout Questions 286

Lab 13. Electromagnets: What Variables Affect the Strength of the Electromagnet?
Teacher Notes 290
Lab Handout 301
Checkout Questions 307

Lab 14. Wire in a Magnetic Field: What Variables Affect the Strength of the Force Acting on a Wire in the Magnetic Field?
Teacher Notes 310
Lab Handout 322
Checkout Questions 328

Lab 15. Electromagnetic Induction: What Factors Influence the Induced Voltage in a Loop of Wire Placed in a Changing Magnetic Field?
Teacher Notes 332
Lab Handout 344
Checkout Questions 350

APPLICATION LABS

Lab 16. Lenz's Law: Why Does the Magnet Fall Through the Metal Tube With an Acceleration That Is Not Equal to the Acceleration Due to Earth's Gravitational Field (−9.8 m/s^2)?
    Teacher Notes 354
    Lab Handout 367
    Checkout Questions 373

Lab 17. Electromagnetism: Why Does the Battery-and-Magnet "Car" Roll When It Is Placed on a Sheet of Aluminum Foil?
    Teacher Notes 378
    Lab Handout 390
    Checkout Questions 396

SECTION 5
Appendixes

Appendix 1. Standards Alignment Matrixes 403

Appendix 2. Overview of Crosscutting Concepts and Nature of Scientific Knowledge and Scientific Inquiry Concepts 417

Appendix 3. Timeline Options for Implementing ADI Lab Investigations 421

Appendix 4. Investigation Proposal Options 423

Appendix 5. Peer-Review Guides and Teacher Scoring Rubrics for Investigation Reports 433

    Image Credits 441
    Index 443

PREFACE

A Framework for K–12 Science Education (NRC 2012; henceforth referred to as the *Framework*) and the *Next Generation Science Standards* (NGSS Lead States 2013; henceforth referred to as the *NGSS*) call for a different way of thinking about why we teach science and what we expect students to know by the time they graduate high school. As to why we teach science, these documents emphasize that schools need to

> ensure by the end of 12th grade, *all* students have some appreciation of the beauty and wonder of science; possess sufficient knowledge of science and engineering to engage in public discussions on related issues; are careful consumers of scientific and technological information related to their everyday lives; are able to continue to learn about science outside school; and have the skills to enter careers of their choice, including (but not limited to) careers in science, engineering, and technology. (NRC 2012, p. 1)

The *Framework* and the *NGSS* are based on the idea that students need to learn science because it helps them understand how the natural world works, because citizens are required to use scientific ideas to inform both individual choices and collective choices as members of a modern democratic society, and because economic opportunity is increasingly tied to the ability to use scientific ideas, processes, and habits of mind. From this perspective, it is important to learn science because it enables people to figure things out or to solve problems.

These two documents also call for a reappraisal of what students need to know and be able to do by the time they graduate from high school. Instead of teaching with the goal of helping students remember facts, concepts, and terms, the *Framework* and *NGSS* now prioritize helping students to become *proficient* in science. To be considered proficient in science, the *Framework* suggests that students need to understand four disciplinary core ideas (DCIs) in the physical sciences,[1] be able to use seven crosscutting concepts (CCs) that span the various disciplines of science, and learn how to participate in eight fundamental scientific and engineering practices (SEPs; called science and engineering practices in the *NGSS*). The DCIs are key organizing principles that have broad explanatory power within a discipline. Scientists use these ideas to explain the natural world. The CCs are ideas that are used across disciplines. These concepts provide a framework or a lens that people can use to explore natural phenomena. As a result, these concepts often influence what people focus on or pay attention to when they attempt to understand how something works or why something happens. The SEPs are the different activities that scientists and engineers engage in as they attempt to generate new concepts, models, theories, or laws that are both valid and reliable. All three of these dimensions of science are

---

[1] Throughout this book, we use the term *physical sciences* when referring to the disciplinary core ideas of the *Framework* (in this context the term refers to a broad collection of scientific fields), but we use the term *physics* when referring to courses at the high school level (as in the title of the book).

PREFACE

important. Students not only need to know about the DCIs, CCs, and SEPs but also must be able to use all three dimensions at the same time to figure things out or to solve problems. These important DCIs, CCs, and SEPs are summarized in Figure 1.

FIGURE 1

The three dimensions of science in *A Framework for K–12 Science Education* and the *Next Generation Science Standards*

Science and engineering practices	Crosscutting concepts
1. Asking Questions and Defining Problems	1. Patterns
2. Developing and Using Models	2. Cause and Effect: Mechanism and Explanation
3. Planning and Carrying Out Investigations	3. Scale, Proportion, and Quantity
4. Analyzing and Interpreting Data	4. Systems and System Models
5. Using Mathematics and Computational Thinking	5. Energy and Matter: Flows, Cycles, and Conservation
6. Constructing Explanations and Designing Solutions	6. Structure and Function
7. Engaging in Argument From Evidence	7. Stability and Change
8. Obtaining, Evaluating, and Communicating Information	

Disciplinary core ideas for the physical sciences*
• PS1: Matter and Its Interactions
• PS2: Motion and Stability: Forces and Interactions
• PS3: Energy
• PS4: Waves and Their Applications in Technologies for Information Transfer

* These disciplinary core ideas represent one of the four subject areas in the *Framework* and the *NGSS*; the other subject areas are life sciences, earth and space sciences, and engineering, technology, and applications of science.

Source: Adapted from NRC 2012 and NGSS Lead States 2013.

To help students become proficient in science in ways described by the National Research Council in the *Framework*, teachers will need to use new instructional approaches that give students an opportunity to use the three dimensions of science to explain natural phenomena or develop novel solutions to problems. This is important because traditional instructional approaches, which were designed to help students "learn about" the concepts, theories, and laws of science rather than help them learn how to "figure out" how or why things work, were not created to foster the development of science proficiency inside the classroom. To help teachers make this instructional shift, this book provides 17 laboratory investigations designed using an innovative approach to lab instruction called argument-driven inquiry (ADI). This approach is designed to promote and support three-dimensional instruction inside classrooms by giving students an opportunity to use DCIs, CCs, and SEPs to construct and critique claims about how things work or why things happen. The lab activities described in this book will also enable students to develop the disciplinary-based literacy skills outlined in the *Common Core State Standards* for English language arts (NGAC and CCSSO 2010) because ADI gives students an opportunity to give presentations to their peers, respond to audience questions and critiques, and then write, evaluate, and revise reports as part of each lab. In addition, these investigations will help students learn many of the mathematical ideas and practices outlined in the *Common Core State Standards* for mathematics (NGAC and CCSSO 2010). Use of these labs, as a result, can help teachers align their teaching with current recommendations for improving classroom instruction in science and for making physics more meaningful for students.

The labs included in this book all focus on the topics of electricity and magnetism. Thus, these labs primarily focus on two of the four physical sciences DCIs from the *NGSS* that are outlined in Figure 1 (although some labs do align with other DCIs as well). These two DCIs are Motion and Stability: Forces and Interactions (PS2) and Energy (PS3). The other two DCIs for physical sciences from the *NGSS* are the focus of other books in the ADI series. All the labs, however, are well aligned with at least two of the seven CCs and seven of the eight SEPs. In addition, the labs in this book are well aligned with the big ideas and science practices for Advanced Placement (AP) Physics 1, 2, and C: Electricity and Magnetism (see Figure 2, p. xii). These labs, as a result, can be used in a wide range of physics courses, including, but not limited to, a conceptual physics course for 9th or 10th graders that is aligned with the *NGSS*, an introductory physics course for juniors or seniors, or even an AP Physics 1, 2, or C: Electricity and Magnetism course.

Finally, this book is the second volume in the ADI Physics series. The first volume has 23 labs focused on mechanics. The structure of the lab handouts is the same in both volumes, making the classroom use of both lab manuals seamless from a structural point of view.

FIGURE 2

Selected big ideas and science practices for AP Physics 1 and 2 and the content areas and science practices for AP Physics C: Electricity and Magnetism

AP Physics 1 and 2 big ideas	AP Physics 1 and 2 science practices
1. Systems: Objects and systems have properties such as mass and charge. Systems may have internal structure.	1. Modeling: Use representations and models to communicate scientific phenomena and solve scientific problems.
2. Fields: Fields existing in space can be used to explain interactions.	2. Mathematical Routines: Use mathematics appropriately.
3. Force Interactions: The interactions of an object with other objects can be described by forces.	3. Scientific Questioning: Engage in scientific questioning to extend thinking or to guide investigations.
4. Change: Interactions between systems can result in changes in those systems.	4. Experimental Methods: Plan and implement data collection strategies in relation to a particular scientific question.
5. Conservation: Changes that occur as a result of interactions are constrained by conservation laws.	5. Data Analysis: Perform data analysis and evaluation of evidence.
	6. Argumentation: Work with scientific explanations and theories.
	7. Making Connections: Connect and relate knowledge across various scales, concepts, and representations in and across domains.

AP Physics C: Electricity and Magnetism content areas

- Electrostatics
- Conductors, capacitors, dielectrics
- Electric circuits
- Magnetic fields
- Electromagnetism

AP Physics C: Electricity and Magnetism laboratory objectives

1. Visual representations
2. Question and method
3. Representing data and phenomena
4. Data analysis
5. Theoretical relationships
6. Mathematical routines
7. Argumentation

Source: Adapted from https://apcentral.collegeboard.org/pdf/ap-physics-1-course-and-exam-description.pdf?course=ap-physics-1-algebra-based (for AP Physics 1); https://apcentral.collegeboard.org/pdf/ap-physics-2-course-and-exam-description.pdf?course=ap-physics-2-algebra-based (for AP Physics 2); https://apcentral.collegeboard.org/pdf/ap-physics-c-electricity-and-magnetism-course-and-exam-description.pdf?course=ap-physics-c-electricity-and-magnetism (for AP Physics C: Electricity and Magnetism).

References

National Governors Association Center for Best Practices and Council of Chief State School Officers (NGAC and CCSSO). 2010. *Common core state standards*. Washington, DC: NGAC and CCSSO.

National Research Council (NRC). 2012. *A framework for K–12 science education: Practices, crosscutting concepts, and core ideas*. Washington, DC: National Academies Press.

NGSS Lead States. 2013. *Next Generation Science Standards: For states, by states*. Washington, DC: National Academies Press. *www.nextgenscience.org/next-generation-science-standards*.

ABOUT THE AUTHORS

Todd L. Hutner is an assistant professor of science education at the University of Alabama in Tuscaloosa. He received a BS and an MS in science education from Florida State University (FSU) and a PhD in curriculum and instruction from The University of Texas at Austin (UT-Austin). Todd's classroom experience includes teaching physics, Advanced Placement (AP) physics, and chemistry in Texas and teaching Earth science and astronomy in Florida. His current research focuses on the impact of both teacher education and education policy on the teaching practice of secondary science teachers.

Victor Sampson is an associate professor of STEM (science, technology, engineering, and mathematics) education at UT-Austin. He received a BA in zoology from the University of Washington, an MIT from Seattle University, and a PhD in curriculum and instruction with a specialization in science education from Arizona State University. Victor also taught high school biology and chemistry for nine years. He specializes in argumentation in science education, teacher learning, and assessment. To learn more about his work in science education, go to *www.vicsampson.com*.

Adam LaMee started teaching high school physics in Florida in 2003 and is currently the PhysTEC Teacher-in-Residence at the University of Central Florida (*http://sciences.ucf.edu/physics/phystec*). He received bachelor's degrees in physics and anthropology from FSU. Adam is a Quarknet Teaching and Learning Fellow, helped design Florida's 6–12 science courses and teacher certification exams, contributed to the Higgs boson discovery with CERN's CMS (Compact Muon Solenoid) experiment, and has researched game-based assessment and performance assessment alternatives to large-scale testing. Learn more about Adam's work at *adamlamee.com*.

Daniel FitzPatrick is a teacher, researcher, and developer of instructional methods and materials. He received a BS and MA in mathematics from UT-Austin and has taught both middle and high school mathematics in public and charter schools. Daniel previously served as a clinical assistant professor and master teacher in the UTeach program at UT-Austin before working full time on research in argumentation in mathematics.

Austin Batson is a physics teacher and assistant coach at Tom Glenn High School in Leander Independent School District, located in the Austin, Texas, area. He received a BS in biomedical engineering and a BA in the Plan II Honors Program from UT-Austin. After graduating, he spent a year in Poland teaching at the University of Bialystok through a Fulbright English Teaching Fellowship. Back in the United States, Austin helped open a new public high school, established the physics program, and helped lead a revamp of the district physics curriculum. His current instructional goals include implementing a blended learning classroom and increasing the amount of argument-driven inquiry practiced in his school.

Jesus Aguilar-Landaverde has been a physics, engineering, and astronomy teacher at all levels in high school for five years. During the same time, he has been a professional tutor in all secondary STEM subjects as well as three languages. During this time, he has designed dozens of inquiries, investigations, labs, and practicums to engage students at various stages in their development.

INTRODUCTION

The Importance of Helping Students Become Proficient in Science

The current aim of science education in the United States is for *all* students to become proficient in science by the time they finish high school. *Science proficiency*, as defined by Duschl, Schweingruber, and Shouse (2007), consists of four interrelated aspects. First, it requires an individual to know important scientific explanations about the natural world, to be able to use these explanations to solve problems, and to be able to understand new explanations when they are introduced to the individual. Second, it requires an individual to be able to generate and evaluate scientific explanations and scientific arguments. Third, it requires an individual to understand the nature of scientific knowledge and how scientific knowledge develops over time. Finally, and perhaps most important, an individual who is proficient in science should be able to participate in scientific practices (such as planning and carrying out investigations, analyzing and interpreting data, and arguing from evidence) and communicate in a manner that is consistent with the norms of the scientific community. These four aspects of science proficiency include the knowledge and skills that all people need to have in order to be able to purse a degree in science, be prepared for a science-related career, and participate in a democracy as an informed citizen.

This view of science proficiency serves as the foundation for the *Framework* (NRC 2012) and the *NGSS* (NGSS Lead States 2013). Unfortunately, our educational system was not designed to help students become proficient in science. As noted in the *Framework*,

> K–12 science education in the United States fails to [promote the development of science proficiency], in part because it is not organized systematically across multiple years of school, emphasizes discrete facts with a focus on breadth over depth, and does not provide students with engaging opportunities to experience how science is actually done. (p. 1)

Our current science education system, in other words, was not designed to give students an opportunity to learn how to use scientific explanations to solve problems, generate or evaluate scientific explanations and arguments, or participate in the practices of science. Our current system was designed to help students learn facts, vocabulary, and basic process skills because many people think that students need a strong foundation in the basics to be successful later in school or in a future career. This vision of science education defines *rigor* as covering more topics and *learning* as the simple acquisition of new ideas or skills.

Our views about what counts as rigor, therefore, must change to promote and support the development of science proficiency. Instead of using the number of different topics covered in a course as a way to measure rigor in our schools, we must

INTRODUCTION

start to measure rigor in terms of the number of opportunities students have to use the ideas of science as a way to make sense of the world around them. Students, in other words, should be expected to learn how to use the core ideas of science as conceptual tools to plan and carry out investigations, develop and evaluate explanations, and question how we know what we know. A rigorous course, would thus be one where students are expected to do science, not just learn about science.

Our views about what learning is and how it happens must also change to promote and support the development of science proficiency. Rather than viewing learning as a simple process where people accumulate more information over time, learning needs to be viewed as a personal and social process that involves "people entering into a different way of thinking about and explaining the natural world; becoming socialized to a greater or lesser extent into the practices of the scientific community with its particular purposes, ways of seeing, and ways of supporting its knowledge claims" (Driver et al. 1994, p. 8). Learning, from this perspective, requires a person to be exposed to the language, the concepts, and the practices of science that makes science different from other ways of knowing. This process requires input and guidance about "what counts" from people who are familiar with the goals of science, the norms of science, and the ways things are done in science. Thus, learning is dependent on supportive and informative interactions with others.

Over time, people will begin to appropriate and use the language, the concepts, and the practices of science as their own when they see how valuable they are as a way to accomplish their own goals. Learning therefore involves seeing new ideas and ways of doing things, trying out these new ideas and practices, and then adopting them when they are useful. This entire process, however, can only happen if teachers provide students with multiple opportunities to use scientific ideas to solve problems, to generate or evaluate scientific explanations and arguments, and to participate in the practices of science inside the classroom. This is important because students must have a supportive and educative environment to try out new ideas and practices, make mistakes, and refine what they know and what they do before they are able to adopt the language, the concepts, and the practices of science as their own.

A New Approach to Teaching Science

We need to use different instructional approaches to create a supportive and educative environment that will enable students to learn the knowledge and skills they need to become proficient in science. These new instructional approaches will need to give students an opportunity to learn how to "figure out" how things work or why things happen. Rather than simply encouraging students to learn about the facts, concepts, theories, and laws of science, we need to give students more opportunities to develop explanations for natural phenomena and design solutions to problems. This emphasis on "figuring things out" instead of "learning about things" represents a big change in the way we will need to teach science at all grade levels. To figure out how things work or why things happen in a way that is consistent with how science is actually done, students must do more than hands-on activities. Students must learn how to use disciplinary core ideas (DCIs), crosscutting concepts (CCs), and science and engineering practices (SEPs) to develop explanations and solve problems (NGSS Lead States 2013; NRC 2012).

A DCI is a scientific idea that is central to understanding a variety of natural phenomena. An example of a DCI in physics is Conservation of Energy and Energy Transfer. This DCI not only explains the relationship between the power dissipated by resistors in series and in parallel, but also explains why a magnet moving through a coil of wire feels a force resisting the motion.

CCs are those concepts that are important across the disciplines of science; there are similarities and differences in the treatment of the CC in each discipline. The CCs can be used as a lens to help people think about what to focus on or pay attention to during an investigation. For example, one of the CCs from the *Framework* is Energy and Matter: Flows, Cycles, and Conservation. This CC is important in many different fields of study. Physicists use this CC to study mechanics, thermodynamics, electricity, and magnetism. Biologists use this CC to study cells, growth and development, and ecosystems. It is important to highlight the centrality of this idea, and other CCs, for students as we teach the subject-specific DCIs.

SEPs describe what scientists do to investigate the natural world. The practices outlined in the *Framework* and the *NGSS* explain and extend what is meant by *inquiry* in science and the wide range of activities that scientists engage in as they attempt to generate and validate new ideas. Students engage in practices to build, deepen, and apply their knowledge of DCIs and CCs. The SEPs include familiar aspects of inquiry, such as Asking Questions and Defining Problems, Planning and Carrying Out Investigations, and Analyzing and Interpreting Data. More important, however, the SEPs include other activities that are at the core of doing science: Developing and Using Models, Constructing Explanations and Designing Solutions, Engaging in Argument From Evidence, and Obtaining, Evaluating, and Communicating Information. All of these SEPs are important to learn, because there is no single scientific method that all scientists must follow; scientists engage in different practices, at different times, and in different orders depending on what they are studying and what they are trying to accomplish at that point in time.

This focus on students using DCIs, CCs, and SEPs during a lesson is called *three-dimensional instruction* because students have an opportunity to use all three dimensions of science to understand how something works, to explain why

INTRODUCTION

something happens, or to develop a novel solution to a problem. When teachers use three-dimensional instruction inside their classrooms, they encourage students to develop or use conceptual models, design investigations, develop explanations, share and critique ideas, and argue from evidence, all of which allow students to develop the knowledge and skills they need to be proficient in science (NRC 2012). Current research suggests that all students benefit from three-dimensional instruction because it gives all students more voice and choice during a lesson and it makes the learning process inside the classroom more active and inclusive (NRC 2012).

We think the school science laboratory is the perfect place to integrate three-dimensional instruction into the science curriculum. Well-designed lab activities can provide opportunities for students to participate in an extended investigation where they can not only use one or more DCIs to understand how something works, to explain why something happens, or to develop a novel solution to a problem but also use several different CCs and SEPs during the same lesson. A teacher, for example, can give his or her students an opportunity to investigate the factors influencing the strength of an electromagnet. The teacher can then encourage them to use what they know about Types of Interactions (a DCI) and Ampère's law (an important idea in electricity and magnetism) and their understanding of Cause and Effect and of Scale, Proportion, and Quantity (two different CCs) to plan and carry out an investigation to figure out how the electromagnet works. During this investigation they must ask questions, analyze and interpret data, use mathematics, develop a model, argue from evidence, and obtain, evaluate, and communicate information (six different SEPs). Using multiple DCIs, CCs, and SEPs at the same time is important because it creates a classroom experience that parallels how science is done. This, in turn, gives all students who participate in a school science lab activity an opportunity to deepen their understanding of what it means to do science and to develop science-related identities. In the following section, we will describe how to promote and support the development of science proficiency during school science labs through three-dimensional instruction.

How School Science Labs Can Help Foster the Development of Science Proficiency Through Three-Dimensional Instruction

Science instruction in the 1980s and 1990s followed a similar sequence in most U.S. classrooms (Hofstein and Lunetta 2004; NRC 2005). This sequence began with the teacher introducing students to an important concept or principle through direct instruction, usually by giving a lecture about it or by assigning a chapter from a textbook to read. Next, the students were given a hands-on laboratory experience. As defined by the NRC (2005, p. 3), "[l]aboratory experiences provide opportunities

INTRODUCTION

for students to interact directly with the material world … using the tools, data collection techniques, models, and theories of science." The purpose of these laboratory experiences or "labs" was to help students understand the concept or principle that was introduced to them earlier by giving them a concrete experience with it. To ensure that students "got the right result" during the lab and that the lab actually illustrated, confirmed, or verified the target concept or principle, the teacher usually provided students with a step-by-step procedure to follow and a data table to fill out. Students were then asked to answer a set of analysis questions to ensure that everyone "reached the right conclusion" based on the data they collected during the lab. The lab experience would then end with the teacher going over what the students should have done during the hands-on activity, what they should have observed, and what answers they should have given in response to the analysis questions. This final review step was done to ensure that the students "learned what they were supposed to have learned" from the hands-on activity and was usually done, once again, through whole-class direct instruction.

Classroom-based research, however, suggests that this type of laboratory experience does little to help students learn key concepts. The National Research Council (2005, p. 5), for example, conducted a synthesis of several different studies that examined what students learn from this type of lab and found that "research focused on the goal of student mastery of subject matter indicates that typical laboratory experiences are no more or less effective than other forms of science instruction (such as reading, lectures, or discussion)." This finding was troubling when this report was released because, as noted earlier, the main goal of this type of lab was to help students understand an important concept or principle by giving them a hands-on and concrete experience with it. In addition, this type of lab does little to help students learn how to plan and carry out investigations or analyze and interpret data because students have no voice or choice during the activity. Students are expected to simply follow a set of directions rather than think about what data they will collect, how they will collect it, and what they will need to do to analyze it once they have it. This type of traditional or "cookbook" lab also can lead to misunderstanding about the nature of scientific knowledge and how this knowledge is developed over time due to the emphasis on following procedure and getting the right results.

Many science teachers started using more inquiry-based labs in the late 1990s and early 2000s to help address the shortcomings of more traditional cookbook lab activities. Inquiry-based lab experiences that are consistent with the definition of *inquiry* found in *National Science Education Standards* (NRC 1996) and *Inquiry and the National Science Education Standards* (NRC 2000) share five key features:

1. Students need to answer a scientifically oriented question.

INTRODUCTION

2. Students must collect data or use data collected by someone else.
3. Students formulate an answer to the question based on their analysis of the data.
4. Students connect their answer to some theory, model, or law.
5. Students communicate their answer to the question to someone else.

Many teachers also changed the traditional sequence of science instruction when they started using more inquiry-based labs. Rather than introducing students to an important concept or principle through direct instruction and then having students do a cookbook lab to demonstrate or confirm it, they used an inquiry-based lab as a way to introduce students to a new concept or principle and then gave them a formal definition of it (NRC 2012). This type of sequence is often described as an "activity before concept" approach to instruction because the activity provides a concrete and shared experience for students that a teacher can use to help explain a concept.

Although inquiry-based labs give students much more voice and choice, especially when compared with more traditional cookbook approaches, they do not do as much as they could do to promote the development of science proficiency. Teachers tend to use inquiry-based labs as a way to help students learn about a new idea rather than as a way to help students learn how to figure out how things work or why they happen. Students, as a result, rarely have an opportunity to learn how to use DCIs, CCs, and SEPs to develop explanations or solve problems. In addition, inquiry-based labs in the early 2000s rarely gave students an opportunity to participate in the full range of scientific practices. These inquiry-based labs were often designed so students had many opportunities to learn how to ask questions, plan and carry out investigations, and analyze and interpret data but few opportunities to learn how to participate in the practices that focus on how new ideas are developed, shared, refined, and eventually validated within the scientific community. These important practices include developing and using models, constructing explanations, arguing from evidence, and obtaining, evaluating, and communicating information (Duschl, Schweingruber, and Shouse 2007; NRC 2005). Most inquiry-based labs that were used in the 1990s and 2000s also did not give students an opportunity to improve their science-specific literacy skills. Students were rarely expected to read, write, and speak in a scientific manner because the focus of these labs was learning about content and how to collect and analyze data in science, not how to propose, critique, and revise ideas.

Changing the focus and nature of inquiry-based labs so they are more consistent with three-dimensional instruction can help address these issues. To implement such a change, teachers will not only have to focus on using DCIs, CCs, and SEPs during a lab but will also need to emphasize "how we know" in physics (i.e., how new knowledge is generated and validated) equally with "what we know" about electricity, magnetism, and conservation (i.e., the theories, laws, and unifying concepts). We have found that one way to make this shift in focus is to make the practice of arguing from evidence or scientific argumentation the central feature of all lab activities so this practice drives decision making during an investigation. We define *scientific argumentation* as the process of proposing, supporting, evaluating, and refining claims based on evidence (Sampson, Grooms, and Walker 2011). The *Framework* (NRC 2012) provides a good description of the role argumentation plays in science:

> Scientists and engineers use evidence-based argumentation to make the case for their ideas, whether involving new theories or designs, novel ways of collecting data, or interpretations of evidence. They and their peers then attempt to identify weaknesses and limitations in the argument, with the ultimate goal of refining and improving the explanation or design. (p. 46)

When teachers make the practice of arguing from evidence the central focus of lab activities, students have more opportunities to learn how to construct and support scientific knowledge claims through argument (NRC 2012). Students, as a result, have more opportunities to learn how scientific ideas are generated, shared, and refined over time. For example, when students know that they have to support a new idea through argument, they begin to think more about the goal of an investigation (e.g., identify a pattern, test a potential causal relationship, confirm a relationship) and the criteria that other people will use to determine if the new idea is valid or acceptable when they are ready to share it. Students are then able to make better decisions about what to do during an investigation (e.g., what data to collect, how to collect it, and how to analyze it given the goal of the investigation) based on what they are trying to do and their understanding of what will be convincing to others. They also have more opportunities to learn how to evaluate the ideas and arguments made by others. Students, as a result, learn how to read, write, and speak in scientific manner because they need to be able to propose and support their claims when they share them and evaluate, challenge, and refine the claims made by others.

We developed the argument-driven inquiry (ADI) instructional model (Sampson and Gleim 2009; Sampson, Grooms, and Walker 2009, 2011) as a way to change the focus and nature of labs so they are consistent with three-dimensional instruction. ADI gives students an opportunity to learn how to use DCIs, CCs, and SEPs to figure out how things work or why things happen. This instructional approach also places scientific argumentation at the center of all lab activities. ADI lab investigations, as a result, make lab activities more authentic and educative for students and thus help teachers promote and support the development of science proficiency. This instructional model reflects current theories about how people learn science (NRC 1999, 2005, 2008, 2012) and is also based on what is known about how to engage students in argumentation and other important scientific practices (Erduran and

Jimenez-Aleixandre 2008; McNeill and Krajcik 2008; Osborne, Erduran, and Simon 2004; Sampson and Clark 2008; Sampson, Enderle, and Grooms 2013). We explain the stages of ADI and how each stage works in Chapter 1.

How to Use This Book

The intended audience of the book is primarily practicing high school physics teachers. We recognize that physics teachers teach many different types of physics courses. Some courses are conceptual in nature, some are algebra based, and some are calculus based. We understand how teaching these different types of physics courses results in different challenges and needs. We have therefore designed the laboratory investigations included in this book to meet the needs of teachers who teach a wide range of courses. Some labs, for example, require students to determine a general relationship or trend and do not require a lot of mathematics. These labs can be used in a physics course that is more conceptual in nature. Other labs, in contrast, require students to develop a mathematical model that they can use to explain and predict changes in the energy of electromagnetic systems. These labs are intended for students in Advanced Placement (AP) Physics C: Electricity and Magnetism who are concurrently enrolled in or have successfully completed an introductory calculus course. The majority of the labs, however, were written for an algebra-based physics course. These labs require some algebra, such as determining a mathematical relationship between two variables (which is often, but not always, a linear relationship). All of the labs were designed to give students an opportunity to learn how to use DCIs, CCS, and SEPs to figure things out.

As we wrote the labs for this book, we kept in mind the fact that physics is often a two-year program of study in many school districts. Students usually take Physics I in 11th grade along with Algebra II and then in 12th grade take AP Physics 1, AP Physics 2, or AP Physics C (mechanics is a first-semester topic, and electricity and magnetism is a second-semester topic) along with either AP Statistics or AP Calculus. We have therefore aligned all the labs in this book with the *NGSS* performance expectations and the AP Physics 1 and 2 and AP Physics C: Electricity and Magnetism learning objectives so teachers can use these labs in either an introductory physics course or an AP physics course. We believe it is important to focus on three-dimensional instruction in both contexts because students need to learn how to use DCIs, CCs, and SEPs to figure out how things work or why things happen. Lab instruction is also a major component of the AP physics curriculum. In AP Physics 1 and 2 and AP Physics C: Electricity and Magnetism, the College Board recommends that at least 25% of instructional time be devoted to laboratory experiences. These experiences should therefore do more than demonstrate, illustrate, or verify a target concept; they should also promote and support the development of science proficiency.

INTRODUCTION

One of the recent advances in physics education has been the development of physics-specific equipment that students can use during investigations, such as probeware and video cameras for collecting data, and data analysis software, including video analysis software, which enables students to explore the data they collect during an investigation. We recognize that while some physics teachers work in settings where this equipment is readily available and funds are easily accessed to purchase additional equipment, many others do not work in such settings. Many of the labs included in this book can be conducted in lower-tech ways, by using batteries and wires that can be purchased at a local hardware store. Sometimes, however, a lab may not be worth doing if students do not have access to specific equipment. When equipment and/or materials are optional, the materials list table in the Teacher Notes indicates this, and we note in the Lab Handout that students "may also consider using" optional equipment. If the equipment and/or materials are not available to you, when introducing the lab just let students know they do not have the option to use them. We also recognize that the initial cost to purchase some equipment may be high, especially when compared with the equipment needed for a chemistry or biology course. However, the replacement costs for these labs are minimal because the equipment should last several years, particularly when compared with biology or chemistry courses, which require annual replacement of chemicals or specimens.

Finally, we want to make clear that we do not expect teachers to use every lab in this book over the course of an academic year. We wrote this book to support the teaching of electricity and magnetism, which is a topic found in the second-semester curriculum of an introductory physics course. Concepts included under the topic of electricity and magnetism include electrostatics; electric currents, capacitors, resistors and circuits; and magnetic fields and electromagnetism. We suggest that teachers who use this book choose two or three labs for each topic.

There are two types of labs included in the book. The first type of lab is called an introduction lab, and the second type of lab is called an application lab. Introduction labs should be used at the beginning of a unit. These labs often require little formal knowledge of the target concept before students begin the investigation. For example, the lab on Coulomb's law (Lab 1) is an introduction lab and does not require students to know the relationship between the charge on two objects and the forces between them, but students are still expected to use a DCI (Types of Interactions) and two CCs (Patterns and Structure and Function) to figure out the relationship between the amount of charge on an electroscope and the separation distance between the two pieces of foil. After students complete the lab, teachers can use other means of instructional to formalize the laws and formulas related to Coulomb's law. Application labs, on the other hand, are designed to come at the end of a unit. The intent of these labs is to give students an opportunity to apply their knowledge of a specific concept they learned about earlier in the course along with their knowledge of DCIs and CCs to a novel situation. For example, Lab 16 requires students to use their knowledge about Faraday's law, Lenz's law, and the Biot-Savart law to explain why a magnet dropped through a conducting tube will fall with a rate less than the acceleration due to gravity.

Organization of This Book

This book is divided into five sections. Section 1 includes two chapters: the first chapter describes the ADI instructional model, and the second chapter describes the development of the ADI lab investigations and provides an overview of what is included with each investigation. Sections 2–4 contain the 17 lab investigations. Each investigation includes three components:

- Teacher Notes, which provides information about the purpose of the lab and what teachers need to do to guide students through it.
- Lab Handout, which can be photocopied and given to students at the beginning of the lab. It provides the students with a phenomenon to investigate, a guiding question to answer, and an overview of the DCIs and CCs that students can use during the investigation.
- Checkout Questions, which can be photocopied and given to students at the conclusion of the lab activity as an optional assessment. The Checkout Questions consist of items that target students' understanding of the DCIs, the CCs, and the nature of scientific knowledge (NOSK) and the nature of scientific inquiry (NOSI) concepts addressed during the lab.

Section 5 consists of five appendixes:

- Appendix 1 contains several standards alignment matrixes that can be used to assist with curriculum or lesson planning.
- Appendix 2 provides an overview of the CCs and the NOSK and NOSI concepts that are a focus of the lab investigations. This information about the CCs and the NOSK and NOSI concepts is included as a reference for teachers.
- Appendix 3 provides several options (in tabular format) for implementing an ADI investigation over multiple 50-minute class periods.
- Appendix 4 provides options for investigation proposals, which students can use as graphic organizers to plan an investigation. The proposals can be photocopied and given to students during the lab.

- Appendix 5 provides two versions of a peer-review guide and teacher scoring rubric (one for high school and one for AP), which can also be photocopied and given to students.

Changes From ADI Physics Volume 1

We worked to make this volume in the ADI Physics series consistent in structure and tone with Volume 1, which covers mechanics. There are, however, some minor changes we want to point out. The first is the modified structure of the hints in the "Getting Started" section of the Lab Handout for students. Specifically, we have written our hints to emphasize the CCs related to the lab investigation. This should aid teachers in implementing the three-dimensional approach advocated by the NGSS.

Second, since the publication of Volume 1, the College Board has redesigned the AP Physics courses, including AP Physics 1 and 2 and AP Physics C: Electricity and Magnetism. This redesign and revision resulted in new standards, which we include in the standards alignment section of the Teacher Notes for each lab. As part of the revision, the College Board has also updated the scientific practices included in the courses. In the standards matrix in Appendix 1 (p. 403), we have aligned these labs with the new science practices for AP Physics 1 and 2 and AP Physics C: Electricity and Magnetism. As such, the science practices listed for the AP Physics courses are different from those listed in Volume 1.

Third, we have made some changes in how we address NOSK and NOSI. Researchers in science and science education who study the nature of science and nature of scientific inquiry have made updated recommendations for important NOSI concepts. In light of that, in this volume, we no longer include the NOSI concept of "the role of imagination and creativity in science." We also added the following NOSI concepts: (1) how scientists investigate questions about the natural or material world and (2) the assumptions made by scientists about order and consistency in nature.

Safety Practices in the Science Laboratory

It is important for all of us to do what we can to make school science laboratory experiences safer for everyone in the classroom. We recommend four important guidelines to follow. First, we need to have proper safety equipment such as, but not limited to, fume hoods, fire extinguishers, eye wash, and showers in the classroom or laboratory. Second, we need to ensure that students use appropriate personal protective equipment (PPE; e.g., sanitized indirectly vented chemical-splash goggles, chemical-resistant aprons and nonlatex gloves) during all components of lab activities (i.e., setup, hands-on investigation, and takedown). At a minimum, the PPE we provide for students to use must meet the ANSI/ISEA Z87.1D3 standard. Third, we must review and comply with all safety policies and procedures, including but not limited to appropriate chemical management, that have been established by our place of employment. Finally, and perhaps most important, we all need to adopt safety standards and better professional safety practices and enforce them inside the classroom or laboratory.

We provide safety precautions for each investigation and recommend that all teachers follow them to provide a safer learning experience inside the classroom. The safety precautions associated with each lab investigation are based, in part, on the use of the recommended materials and instructions, legal safety standards, and better professional safety practices. Selection of alternative materials or procedures for these activities may jeopardize the level of safety and therefore is at the user's own risk.

We also recommend that you encourage students to read the National Science Teaching Association's *High School Safety Acknowledgment Form* before allowing them to work in the laboratory for the first time. This document is available online at *http://static.nsta.org/pdfs/SafetyAcknowledgmentForm-HighSchool.pdf*. Your students and their parent(s) or guardian(s) should then sign the document to acknowledge that they understand the safety procedures that must be followed during a school science laboratory experience.

Remember that a lab includes three parts: (1) setup, which includes setting up the lab and preparing the materials; (2) the actual investigation; and (3) the cleanup, also called the *takedown*. The safety procedures and PPE we recommend for each investigation apply to all three parts.

References

Driver, R., H. Asoko, J. Leach, E. Mortimer, and P. Scott. 1994. Constructing scientific knowledge in the classroom. *Educational Researcher* 23: 5-12.

Duschl, R. A., H. A. Schweingruber, and A. W. Shouse, eds. 2007. *Taking science to school: Learning and teaching science in grades K–8*. Washington, DC: National Academies Press.

Erduran, S., and M. Jimenez-Aleixandre, eds. 2008. *Argumentation in science education: Perspectives from classroom-based research*. Dordrecht, The Netherlands: Springer.

Hofstein, A., and V. Lunetta. 2004. The laboratory in science education: Foundations for the twenty-first century. *Science Education* 88: 28–54.

McNeill, K., and J. Krajcik. 2008. Assessing middle school students' content knowledge and reasoning through written scientific explanations. In *Assessing science learning:*

Perspectives from research and practice, ed. J. Coffey, R. Douglas, and C. Stearns, 101–116. Arlington, VA: NSTA Press.

National Research Council (NRC). 1999. *How people learn: Brain, mind, experience, and school.* Washington, DC: National Academies Press.

National Research Council (NRC). 2000. *Inquiry and the National Science Education Standards.* Washington, DC: National Academies Press.

National Research Council (NRC). 2005. *America's lab report: Investigations in high school science.* Washington, DC: National Academies Press.

National Research Council (NRC). 2008. *Ready, set, science: Putting research to work in K–8 science classrooms.* Washington, DC: National Academies Press.

National Research Council (NRC). 2012. *A framework for K–12 science education: Practices, crosscutting concepts, and core ideas.* Washington, DC: National Academies Press.

NGSS Lead States. 2013. *Next Generation Science Standards: For states, by states.* Washington, DC: National Academies Press. *www.nextgenscience.org/next-generation-science-standards.*

Osborne, J., S. Erduran, and S. Simon. 2004. Enhancing the quality of argumentation in science classrooms. *Journal of Research in Science Teaching* 41 (10): 994–1020.

Sampson, V., and D. Clark. 2008. Assessment of the ways students generate arguments in science education: Current perspectives and recommendations for future directions. *Science Education* 92 (3): 447–472.

Sampson, V., and L. Gleim. 2009. Argument-driven inquiry to promote the understanding of important concepts and practices in biology. *American Biology Teacher* 71 (8): 471–477.

Sampson, V., P. Enderle, and J. Grooms. 2013. Argumentation in science and science education. *The Science Teacher* 80 (5): 30–33.

Sampson, V., J. Grooms, and J. Walker. 2009. Argument-driven inquiry: A way to promote learning during laboratory activities. *The Science Teacher* 76 (7): 42–47.

Sampson, V., J. Grooms, and J. Walker. 2011. Argument-driven inquiry as a way to help students learn how to participate in scientific argumentation and craft written arguments: An exploratory study. *Science Education* 95 (2): 217–257.

SECTION 1
Using Argument-Driven Inquiry

CHAPTER 1
Argument-Driven Inquiry

Stages of Argument-Driven Inquiry

The argument-driven inquiry (ADI) instructional model was designed to change the focus and nature of labs so they are consistent with the three-dimensional instructional approach. ADI therefore gives students an opportunity to learn how to use disciplinary core ideas (DCIs), crosscutting concepts (CCs), and science and engineering practices (SEPs) (NGSS Lead States 2013; NRC 2012) to figure out how things work or why things happen. This instructional approach also places scientific argumentation as the central feature of all laboratory activities. ADI lab investigations, as a result, make lab activities more authentic (students have an opportunity to engage in all the practices of science) *and* educative (students receive the feedback and explicit guidance that they need to improve on each aspect of science proficiency).

In this chapter, we will explain what happens during each of the eight stages of the ADI instructional model. These eight stages are the same for every ADI laboratory experience. As a result, students quickly learn what is expected of them during each stage of an ADI lab and can focus on learning how to use DCIs, CCs, and SEPs to develop explanations or solve problems. Figure 3 (p. 4) summarizes the eight stages of the ADI instructional model.

Stage 1: Identify the Task and the Guiding Question

An ADI lab activity begins with the teacher identifying a phenomenon to investigate and offering a guiding question for the students to answer. The goal of the teacher at this stage of the model is to capture the students' interest and provide them with a reason to complete the investigation. To aid in this, teachers should provide each student with a copy of the Lab Handout. This handout includes a brief introduction that provides a description of the puzzling phenomenon or a problem to solve, the DCI and CCs that students can use during the investigation, a reason to investigate, and the task the students will need to complete. This handout also includes information about the nature of the argument they will need to produce, some helpful tips on how to get started, and criteria that will be used to judge argument quality (e.g., the sufficiency of the claim and the quality of the evidence).

Teachers often begin an ADI investigation by selecting a different student to read each section of the Lab Handout out loud while the other students follow along. As the students read, they can annotate the text to identify important or useful ideas and information or terms that may be unfamiliar or confusing. After each section is read, the teacher can pause to clarify expectations, answer questions, and provide additional information as needed.

Teachers can also spark student interest by giving a demonstration or showing a video of the phenomenon.

It is also important for the teacher to hold a "tool talk" during this stage, taking a few minutes to explain how to use the available lab equipment, how to use a computer simulation, or even how to use software to analyze data. Teachers need to hold a tool talk because students are often unfamiliar with specialized lab equipment, simulations, or software. Even if the students are familiar with the available tools, they will often use them incorrectly or in an unsafe manner unless they are reminded about how the tools work and the proper way to use them. The teacher should therefore review specific safety protocols and precautions as part of the tool talk.

FIGURE 3

Stages of the argument-driven inquiry instructional model

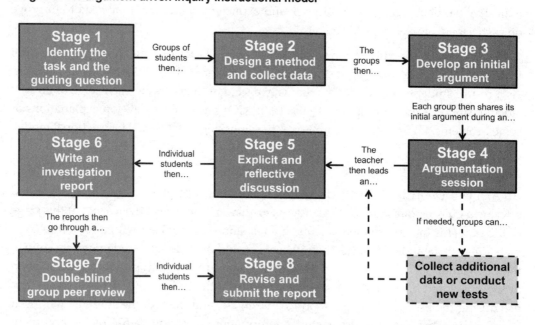

Including a tool talk during the first stage is useful because students often find it difficult to design a method to collect the data needed to answer the guiding question (the task of stage 2) when they do not understand how to use the available materials. We also recommend that teachers give students a few minutes to tinker with the equipment, simulation, or software they will be using to collect data as part of the tool talk. We have found that students can quickly figure out how the equipment, simulation, or software works and what they can and cannot do with it simply by tinkering with the available materials for 5–10 minutes. When students are given an opportunity to tinker with the equipment, simulation, or software as part of the tool talk, they end up designing much better investigations (the task of stage 2) because they understand what they can and cannot do with the tools they will use to collect data.

Once all the students understand the goal of the activity and how to use the available materials, the teacher should divide the students into small groups (we recommend three or four students per group) and move on to the second stage of the instructional model.

Stage 2: Design a Method and Collect Data

In stage 2, small groups of students develop a method to gather the data they need to answer the guiding question and carry out that method. How students complete this stage depends on the nature of the investigation. Some investigations call for groups to answer the guiding question by designing a controlled experiment, whereas others require students to analyze an existing data set (e.g., a database or information sheets). If students need assistance in designing their method, teachers can have them complete an investigation proposal. These proposals guide students through the process of developing a method by encouraging them to think about what type of data they will need to collect, how to collect it, and how to analyze it. We have included six different investigation proposals in Appendix 4 (p. 423) of this book that students can use to design their investigations. Investigation Proposal A (long or short version) can be used when students need to collect systematic observations for a descriptive investigation. Investigation Proposal B (long or short version) or Investigation Proposal C (long or short version) can be used when students need to design a comparative or experimental study to test potential explanations or relationships as part of their investigation. Investigation Proposal B requires students to design a test of two alternative hypotheses, and Investigation Proposal C requires students to design a test of three alternative hypotheses.

The overall intent of this stage is to provide students with an opportunity to interact directly with the natural world (or in some cases with data drawn from the natural world) using appropriate tools and data collection techniques and to learn how to deal with the uncertainties of empirical work. This stage of the model also gives students a chance to learn why some approaches to data collection or analysis work better than others and how the method used during a scientific investigation is based on the nature of the question and the phenomenon under investigation. At the end of this stage, students should have collected all the data they need to answer the guiding question.

Stage 3: Develop an Initial Argument

The next stage of the instructional model calls for students to develop an initial argument in response to the guiding question. To do this, each group needs to be encouraged to first analyze the measurements (e.g., temperature and mass) and/or observations (e.g., appearance and location) they collected during stage 2 of the model. Once the groups have analyzed and interpreted the results of their analysis, they can create an initial argument.

The argument consists of a claim, the evidence they are using to support their claim, and a justification of their evidence. The *claim* is their answer to the guiding question. The *evidence* consists of an analysis of the data they collected and an interpretation of the analysis. The *justification of the evidence* is a statement that defends their choice of evidence by explaining why it is important and relevant, making the concepts or assumptions underlying the analysis and interpretation explicit. The components of a scientific argument are illustrated in Figure 4.

FIGURE 4

The components of a scientific argument and criteria for evaluating its quality

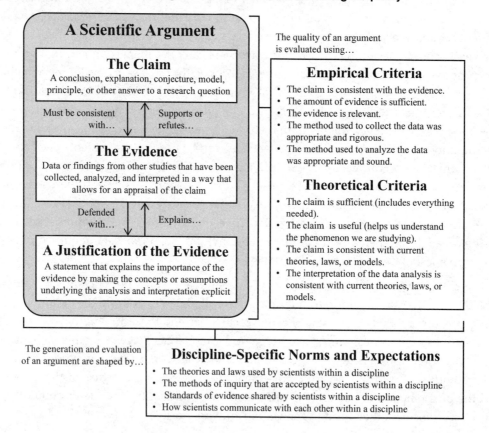

To illustrate each of the three structural components of a scientific argument, consider the following example. This argument was made in response to the guiding question, "What factors affect the resistance of a wire?"

Claim: As the length of the wire increases, the resistance of the wire also increases. The type of metal the wire is made from also affects the resistance.

Evidence: For both the aluminum and copper, as the length of the wire increased, the current decreased. This means the resistance increased. Also, the current for copper was always higher at each length than for aluminum. This leads us to conclude that the type of metal impacts the resistance.

Length (m)	Average current (ohms)	
	Copper wire	Aluminum wire
0.25	5.36	3.91
0.50	2.73	2.00
0.75	1.74	1.27
1.00	1.24	0.90
1.25	1.08	0.79
1.50	0.91	0.66

Justification of the evidence: We made the following assumptions when collecting our data:

1. Current is the movement of electrons through a wire.

2. We can't measure the resistance, but we can measure the current. Ohm's law states that V = IR, which means if the voltage is constant, then resistance is inversely proportional to the current.

3. The voltage of the battery was constant at 1.5 V for all trials.

4. The resistance from the wire is due to the collision between free electrons in the wire that can move when attached to a voltage source. Different metals have different properties, including the arrangement of their electrons. This may impact the resistance of the wire and is why we also tested an aluminum wire.

The claim in this argument provides an answer to the guiding question. The author then uses genuine evidence to support the claim by providing an analysis of the data collected (the table with the length of wire and average currents through each wire) and an interpretation of the analysis (that resistance increases with length). Finally, the author provides a justification of the evidence in the argument by making explicit the underlying concept and assumptions (the definition of a current, why wires have internal resistance, Ohm's law, and the voltage of the battery) guiding the analysis of the data and the interpretation of the analysis.

It is important for students to understand that, in science, some arguments are better than others. An important aspect of science and scientific argumentation involves the evaluation of the various components of the arguments put forward by others. Therefore, the framework provided in Figure 4 also highlights two types of criteria that students can and should be encouraged to use to evaluate an argument in science: empirical criteria and theoretical criteria. *Empirical criteria* include

- how well the claim fits with all available evidence,
- the sufficiency of the evidence,
- the relevance of the evidence,
- the appropriateness and rigor of the method used to collect the data, and
- the appropriateness and soundness of the method used to analyze the data.

Theoretical criteria, on the other hand, refer to standards that are important in science but are not empirical in nature; examples of these criteria are

- the sufficiency of the claim (i.e., Does it include everything needed?);
- the usefulness of the claim (i.e., Does it help us understand the phenomenon we are studying?);
- how consistent the claim is with accepted theories, laws, or models (i.e., Does it fit with our current understanding of force and interactions or stability and motion?); and
- how consistent the interpretation of the results of the analysis is with accepted theories, laws, or models (i.e., is the interpretation based on what is known about forces and interactions or stability and motion?).

What counts as quality within these different components, however, varies from discipline to discipline (e.g., physics, chemistry, geology, biology) and within the specific fields of each discipline (e.g., astrophysics, biophysics, nuclear physics, quantum mechanics, thermodynamics). This variation is due to differences in the types of phenomena investigated, what counts as an accepted mode of inquiry (e.g., descriptive studies, experimentation, computer modeling), and the theory-laden nature of scientific inquiry. It is important to keep in mind that "what counts" as a quality argument in science depends on the discipline and field.

To allow for the critique and refinement of the tentative argument during the next stage of ADI, each group of students should create their initial argument in a medium that can easily be viewed by the other groups. We recommend using a 2' × 3' whiteboard. Students should include the guiding question of the lab and the three main components of the argument on the board. Figure 5 shows the general layout for a presentation of an argument, and Figure 6 provides an example of an argument crafted by students. Students can also create their initial arguments using presentation software such as Microsoft's PowerPoint or Apple's Keynote and devote one slide to each component of an argument. The choice of medium is not important as long as students are able to easily modify the content of their argument as they work and others can easily view their argument.

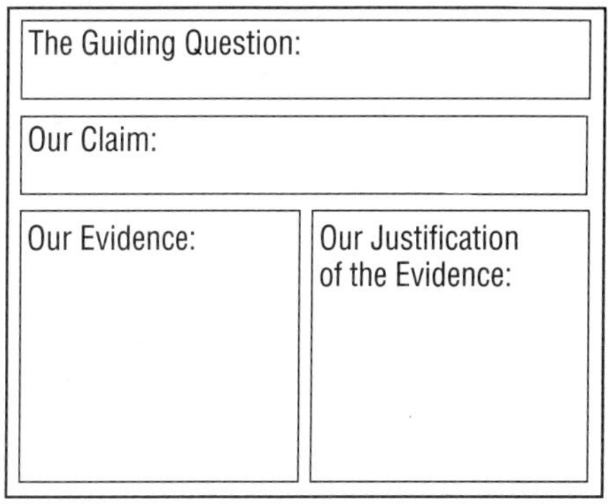

The intention of this stage of the model is to provide the student groups with an opportunity to make sense of what they are seeing or doing during the investigation. As students work together to create an initial argument, they must talk with each other and determine if their analysis is useful or not and how to best interpret the trends, differences, or relationships that they identify as a result of the analysis. They must also decide if the evidence (data that have been analyzed and interpreted) that they chose to include in their argument is relevant, sufficient, and an acceptable way to support their claim. This process, in turn, enables the groups of students to evaluate competing ideas and weed out any claim that is inaccurate, does not fit with all the available data, or contains contradictions.

This stage of the model is challenging for students because they are rarely asked to make sense of a phenomenon based on raw data, so it is important for teachers to actively work to support their "sense-making." In this stage, the teacher should circulate from group to group to act as a resource person for the students, asking questions urging them to think about what they are doing and why. To help students remember the goal of the activity, you can ask questions such as "What are you trying to figure out?" "Why is that information important?" or "Why is that analysis useful?" to encourage them to think about whether or not the data they are analyzing is relevant or the analysis is informative. To help them remember to use rigorous

FIGURE 5

The components of an argument that should be included on a whiteboard (outline)

The Guiding Question:	
Our Claim:	
Our Evidence:	Our Justification of the Evidence:

FIGURE 6

An example of a student-generated argument on a whiteboard

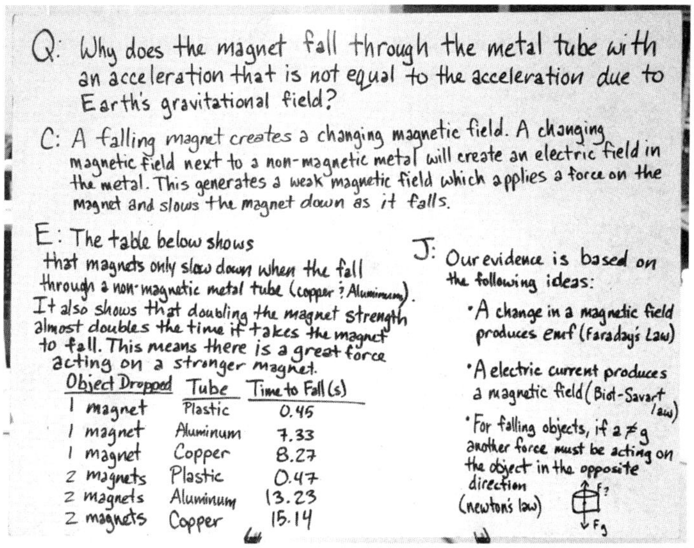

Q: Why does the magnet fall through the metal tube with an acceleration that is not equal to the acceleration due to Earth's gravitational field?

C: A falling magnet creates a changing magnetic field. A changing magnetic field next to a non-magnetic metal will create an electric field in the metal. This generates a weak magnetic field which applies a force on the magnet and slows the magnet down as it falls.

E: The table below shows that magnets only slow down when they fall through a non-magnetic metal tube (copper? Aluminum). It also shows that doubling the magnet strength almost doubles the time it takes the magnet to fall. This means there is a greater force acting on a stronger magnet.

Object Dropped	Tube	Time to Fall (s)
1 magnet	Plastic	0.45
1 magnet	Aluminum	7.33
1 magnet	Copper	8.27
2 magnets	Plastic	0.47
2 magnets	Aluminum	13.23
2 magnets	Copper	15.14

J: Our evidence is based on the following ideas:
- A change in a magnetic field produces emf (Faraday's Law)
- A electric current produces a magnetic field (Biot-Savart law)
- For falling objects, if $a \neq g$ another force must be acting on the object in the opposite direction (Newton's law)

criteria to evaluate the merits of a tentative claim, you can ask, "Does that fit with all the data?" or "Is that consistent with what we know about forces and motion?"

It is important to remember that at the beginning of the school year, students will struggle to develop arguments and will often rely on inappropriate criteria such as plausibility (e.g., "That sounds good to me") or fit with personal experience (e.g., "But that is what I saw on TV once") as they attempt to make sense of their data. However, as students learn why it is useful to use evidence in an argument, what makes evidence valid or acceptable in science, and why it is important to justify why they used a particular type of evidence through practice, *students will improve their ability to argue from evidence* (Grooms, Enderle, and Sampson 2015). This is an important principle underlying the ADI instructional model.

Stage 4: Argumentation Session

The fourth stage of ADI is the argumentation session. In this stage, each group is given an opportunity to share, evaluate, and revise their initial arguments by interacting with members from the other groups (see Figure 7). This stage is included in the model for three reasons:

FIGURE 7

A student presents her group's argument to students from other groups during the argumentation session.

1. Scientific argumentation (i.e., arguing from evidence) is an important practice in science because critique and revision lead to better outcomes.

2. Research indicates that students learn more about the content and develop better critical-thinking skills when they are exposed to alternative ideas, respond to the questions and challenges of other students, and evaluate the merits of competing ideas (Duschl, Schweingruber, and Shouse 2007; NRC 2012).

3. Students learn how to distinguish between ideas using rigorous scientific criteria and are able to develop scientific habits of mind (such as treating ideas with initial skepticism, insisting that the reasoning and assumptions be made explicit, and insisting that claims be supported by valid evidence) during the argumentation sessions.

FIGURE 8

A modified gallery walk format is used during the argumentation session to allow multiple groups to share their arguments at the same time.

This stage, as a result, provides the students with an opportunity to learn from and about scientific argumentation.

It is important to note, however, that supporting and promoting productive interactions between students in the classroom can be difficult because the practice of arguing from evidence is foreign to most students when they first begin participating in ADI. To aid these interactions, students are required to generate their arguments in a medium that can be seen by others. By looking at whiteboards, paper, or slides, students tend to focus their attention on evaluating evidence and the DCI that was used to justify the evidence rather than attacking the source of the ideas. As a result, this strategy often makes the discussion more productive and makes it easier for student to identify and weed out faulty ideas. It is also important for the students to view the argumentation session as an opportunity to learn. The teacher, therefore, should describe the argumentation session as an opportunity for students to collaborate with their peers and as a chance to give each other feedback so the quality of all the arguments can be improved, rather than as an opportunity to determine who is right or wrong.

To ensure that all students remain engaged during the argumentation session, we recommend that teachers use a modified "gallery walk" format rather than a whole-class presentation format. In the modified gallery walk format, one or two members of the group stay at their workstation to share their groups' ideas while the other group members go to different groups one at a time to listen to and critique the arguments developed by their classmates (see Figure 8 on previous page). This type of format ensures that all ideas are heard and more students are actively involved in the process. We recommend that the students who are responsible for critiquing arguments visit at least three different groups during the argumentation session. We also recommend that the presenters keep a record of the critiques made by their classmates and any suggestions for improvement. The students who are responsible for critiquing the arguments should also be encouraged to keep a record of good ideas or potential ways to improve their own arguments as they travel from group to group.

Just as is the case in earlier stages of ADI, it is important for the classroom teacher to be involved in (without leading) the discussions during the argumentation session. Once again, the teacher should move from group to group to keep students on task and model good scientific argumentation. The teacher can ask the presenter(s) questions such as "Why did you decide to analyze the available data like that?" or "Were there any data that did not fit with your claim?" to encourage students to use empirical criteria to evaluate the quality of the arguments. The teacher can also ask the presenter(s) to explain how the claim they are presenting fits with the theories, laws, or models of science or to explain why the evidence they used is important. In addition, the teacher can ask the students who are listening to the presentation questions such as "Do you think their analysis is accurate?" or "Do you think their interpretation is sound?" or even "Do you think their claim fits with what we know about forces and motion?" These questions can serve to remind students to use empirical and theoretical criteria to evaluate an argument during the discussions. Overall, it is the goal of the teacher at this stage of the lesson to encourage students to think about how they know what they know and why some claims are more valid or acceptable in science. This stage of the model, however, is not the time to tell the students that they are right or wrong.

At the end of the argumentation session, it is important to give students time to meet with their original group so they can discuss what they learned by interacting with individuals from the other groups and revise their initial arguments. This process can begin with the presenter(s) sharing the critiques and the suggestions for improvement that they heard during the argumentation session. The students who visited the other groups during the argumentation can then share their ideas for making the arguments better based on what they observed and discussed at other stations. Students often realize that the way they collected or analyzed data was flawed in some way at this point in the process. The teacher should therefore encourage students to collect new data or reanalyze the data they collected as needed. Teachers can also give students time to conduct additional tests of ideas or claims. At the end of this stage, each group should have a final argument that is much better than their initial one.

Stage 5: Explicit and Reflective Discussion

The teacher should lead a whole-class explicit and reflective discussion during stage 5 of ADI. The intent of this discussion is to give students an opportunity to think about and share what they know and how they know it. This stage enables the classroom teacher to ensure that all students understand the DCI and CCs they used during the investigation. It also encourages students to think about ways to improve their participation in scientific practices such as planning and carrying out investigations, analyzing and interpreting data, and arguing from evidence. At this point in the instructional sequence, the teacher should also encourage students to think about one or two nature of scientific knowledge (NOSK) or nature of scientific inquiry (NOSI) concepts. It is important to stress that an explicit and reflective discussion is not a lecture; it is an opportunity for students to think about important ideas and practices and to share what they know or do not understand. The more students talk during this stage, the more meaningful the experience will be for them and the more a teacher can learn about student thinking.

Teachers should begin the discussion by asking students to share what they know about the DCI and the CCs they used to figure things out during the lab (the DCI and CCs can be found in the "Your Task" section of the Lab Handout). The teacher can give several images as prompts and then ask students questions to encourage students to think about how these ideas or concepts helped them explain the phenomenon under investigation and how they used these ideas or concepts to provide a justification of the evidence in their arguments. The teacher should not tell the students what results they should have obtained or what information should be included in each argument. Instead, the teachers should focus on the students' thoughts about the DCI and CCs by providing a context for students to share their views and explain their thinking. Remember, this stage of ADI is a *discussion*, not a lecture. We provide recommendations about what teachers can do and the types of questions that teachers can ask to facilitate a productive discussion about the DCI and CCs during this stage as part of the teacher notes for each lab investigation.

Next, the teacher should encourage the students to think about what they learned about the practices of science and how to design better investigations in the future. This is important because students are expected to design their own investigations, decide how to analyze and interpret data, and support their claims with evidence in every ADI lab investigation. These practices are complex, and students cannot be expected to master them without being given opportunities to try, fail, and then learn from their mistakes. To encourage students to learn from their mistakes during a lab, students must have an opportunity to reflect on what went well and what went wrong during their investigation. The teacher should therefore encourage the students to think about what they did during their investigation, how they choose to analyze and interpret data, how they decide to argue from evidence, and what they could do better. The teacher can then use the students' ideas to highlight what does and does not count as quality or rigor in science and to offer advice about ways to improve in the future. Over time, students will gradually improve their abilities to participate in the practices of science as they learn what works and what does not. To help facilitate this process, in the Teacher Notes for each ADI lab investigation we provide questions that teachers can ask students to help elicit their ideas about the practices of science and set goals for future investigations.

The teacher should end this stage with an explicit discussion of one or two aspects of NOSK or nature of scientific inquiry NOSI, using what the students did during the investigation to help illustrate these important concepts (NGSS Lead States 2013). This stage provides a golden opportunity for explicit instruction about NOSK and how this knowledge develops over time in a context that is meaningful to the students. For example, teachers can use the lab as a way to illustrate the differences between

- observations and inferences,
- data and evidence, and
- theories and laws.

Teachers can also use the lab investigation as a way to illustrate NOSI. For example, teachers might discuss

- how the culture of science, societal needs, and current events influence the work of scientists;
- the wide range of methods that scientists can use to collect data;
- what does and does not count as an experiment in science; and
- the role that creativity and imagination play during an investigation.

Recent research suggests that students only develop an appropriate understanding of the nature of science when teachers discuss these concepts in an *explicit* fashion (Abd-El-Khalick and Lederman 2000; Lederman and Lederman 2004; Schwartz, Lederman, and Crawford 2004). In addition, by embedding a discussion of NOSK and NOSI into each lab investigation, teachers can highlight these important concepts over and over again throughout the school year rather than just focusing on them during a single unit. This type of approach makes it easier for students to learn these abstract and sometimes counterintuitive concepts. As part of the Teacher Notes for each lab investigation, we provide recommendations about which concepts to focus on and examples of questions that teachers can ask to facilitate a productive discussion about these concepts during this stage of the instructional sequence.

Stage 6: Write an Investigation Report
Stage 6 is included in the ADI model because writing is an important part of doing science. Scientists must be able to read and understand the writing of others as well as evaluate its worth. They also must be able to share the results of their own research through writing.

In addition, writing helps students learn how to articulate their thinking in a clear and concise manner, encourages metacognition, and improves student understanding of the content (Wallace, Hand, and Prain 2004). Finally, and perhaps most important, writing makes each student's thinking visible to the teacher (which facilitates assessment) and enables the teacher to provide students with the educative feedback they need to improve.

In stage 6, each student is required to write an individual investigation report using his or her group's argument. The report should be centered on three fundamental questions:

1. What question were you trying to answer and why?
2. What did you do to answer your question and why?
3. What is your argument?

Teachers should encourage students to use tables or graphs to help organize their evidence and require them to reference this information in the body of the report. Stage 6 is important because it allows them to learn how to construct an explanation, argue from evidence, and communicate information. It also enables students to master the disciplinary-based writing skills outlined in the *Common Core State Standards for English Language Arts* (*CCSS ELA*; NGAC and CCSSO 2010). The report can be written during class or can be assigned as homework.

The format of the report is designed to emphasize the persuasive nature of science writing and to help students learn how to communicate in multiple modes (words, figures, tables, and equations). The three-question format is well aligned with the components of a traditional laboratory report (i.e., introduction, procedure, results and discussion) but allows students to see the important role argument plays in science. We strongly recommend that teachers *limit the length of the investigation* report to two double-spaced pages or one single-spaced page. This limitation encourages students to write in a clear and concise manner by leaving little room for extraneous information. This limitation is less intimidating than a more lengthy report requirement, and it lessens the work required in the subsequent stages.

Stage 7: Double-Blind Group Peer Review

During stage 7, each student is required to submit to the teacher one or more copies of his or her investigation report. We recommend that students bring in multiple copies of their report to make it easier for a group of students to review at the same time; however, this is not a requirement if students are unable to bring in multiple copies of their reports. Instead of reading multiple copies of the same report as they review it, the group of reviewers can simply share a single copy of a report. Students should not place their names on the report before they turn it in to the teacher at the beginning of this stage; instead they should use an identification number to maintain anonymity and ensure that reviews are based on the

FIGURE 9

A group of students reviewing a report written by a classmate using the peer-review guide and teacher scoring rubric

ideas presented and not the person presenting the ideas.

We recommend that teachers place students into groups of three to review the reports (these groups can be different from the groups that students worked in during stages 1–4). The teacher then gives each group a report written by a single student (or the multiple copies of the report submitted by a single student) and a peer-review guide and teacher scoring rubric (PRG/TSR). We included two versions of the PRG/TSR in Appendix 5 (p. 433): one version is designed for a high school course, and one is designed for an Advanced Placement (AP) course. The students in the group are then asked to review the report (or copies of the report) as a team using the PRG/TSR (see Figure 9). The PRG/TSR contains specific criteria that are to be used by the group as they evaluate the quality of each section of investigation report as well as quality of the writing. There is also space for the reviewers to provide the author with feedback about how to improve the report. Once a group finishes reviewing a report as a team, they are given another report to review. When students are grouped together in threes, they only need to review three different reports. Be sure to give students only 15 minutes to review each set of reports (we recommend setting a timer to help manage time). When students are grouped into three and given 15 minutes to complete each review, the entire peer-review process can be completed in one 50-minute class period (3 different reports × 15 minutes = 45 minutes).

Reviewing each report as a group using the PRG/TSR is an important component of the peer-review process because it provides students with a forum to discuss "what counts" as high quality or acceptable and, in so doing, forces them to reach a consensus during the process. This method also helps prevent students from checking off "yes" for each criterion on the PRG/TSR without thorough consideration of the merits of the paper. It is also important for students to provide constructive and specific feedback to the author when areas of the paper are found to not meet the standards established by the PRG/TSR. The peer-review process provides students with an opportunity to read good and bad examples of the reports. This helps the students learn new ways to organize and present information, which in turn will help them write better on subsequent reports.

This stage of the model also gives students more opportunities to develop reading skills that are needed to be successful in science. Students must be able to determine the central ideas or conclusions of a text and determine the meaning of symbols, key terms, and other domain-specific words. In addition, students must be able to assess the reasoning and evidence that an author includes in a text to support his or her claim and compare or contrast findings presented in a text with those from other sources when they read a scientific text. Students can develop all these skills, as well as the other discipline-based reading standards found in the *CCSS ELA*, when they are required to read and critically review reports written by their classmates.

Stage 8: Revise and Submit the Report

The final stage in the ADI instructional model is to revise the report based on the suggestions given during the peer review. If the report meets all the criteria, the student may simply submit the paper to the teacher with the original peer-reviewed "rough draft" and PRG/TSR attached, ensuring that his or her name replaces the identification number. Students whose reports are found by the peer-review group to be acceptable can maintain the option to revise it if they so desire after reviewing the work of other students. If a report was found to be unacceptable by the reviewers during the peer-review stage, the author is required to rewrite his or her report using the reviewers' comments and suggestions as a guideline. The author is also required to explain what he or she did to improve each section of the report in response to the reviewers' suggestions (or explain why he or she decided to ignore the reviewers' suggestions) in the author response section of the PRG/TSR.

Once the report is revised, it is turned in to the teacher for evaluation with the original rough draft and the PRG/TSR attached. The teacher can then provide a score on the PRG/TSR in the column labeled "Teacher Score" and use these ratings to assign an overall grade for the report. This approach provides students with a chance to improve their writing mechanics and develop their reasoning and understanding of the content. This process also offers students the added benefit of reducing academic pressure by providing support in obtaining the highest possible grade for their final product.

The PRG/TSR is designed to be used with any ADI lab investigation, which allows teachers to use the same scoring rubric throughout the entire school year. This is beneficial for several reasons. First, the criteria for what counts as a high-quality report do not change from lab to lab. Students therefore quickly learn what is expected from them when they write a report and teachers do not have to spend valuable class time explaining the various components of the PRG/TSR each time they assign a report. Second, the PRG/TSR makes it clear what components of a report need to be improved next time, because the grade is not based on a holistic evaluation of the report. Students, as a result, can see which aspects of their writing are strong and which aspects need improvement. Finally, and perhaps most important, PRG/TSR provides teachers with a standardized measure of student performance that can be compared over multiple reports across semesters. Teachers can therefore track improvement over time.

The Role of the Teacher During Argument-Driven Inquiry

If the ADI instructional model is to be successful and student learning is to be optimized, the role of the teacher during a lab activity designed using this model must be different than the teacher's role during a more traditional lab. The teacher *must* act as a resource for the students, rather than as a director, as students work through each stage of the activity; the teacher must encourage students to think about *what they are doing* and *why they made that decision* throughout the process. This encouragement should take the form of probing questions that teachers ask as they walk around the classroom, such as "Why do you want to set up your equipment that way?" or "What type of data will you need to collect to be able to answer that question?"

Teachers must restrain themselves from telling or showing students how to "properly" conduct the investigation. However, they must emphasize the need to maintain high standards for a scientific investigation by requiring students to use rigorous standards for "what counts" as a good method or a strong argument in the context of science.

Finally, and perhaps most important for the success of an ADI activity, teachers must be willing to let students try and fail, and then help them learn from their mistakes. Teachers should not try to make the lab investigations included in this book "student-proof" by providing additional directions to ensure that students do everything right the first time. We have found that students often learn more from an ADI lab activity when they design a flawed method to collect data during stage 2 or analyze their results in an inappropriate manner during stage 3, because their classmates quickly point out these mistakes during the argumentation session (stage 4) and it leads to more teachable moments.

Because the teacher's role in an ADI lab is different from what typically happens in laboratories, we've provided a chart describing teacher behaviors that are consistent and inconsistent with each stage of the instructional model (see Table 1). This table is organized by stage because what the students and the teacher need to accomplish during each stage is different. It might be helpful to keep this table handy as a guide when you are first attempting to implement the lab activities found in the book.

TABLE 1

Teacher behaviors during the stages of the ADI instructional model

Stage	What the teacher does that is ...	
	Consistent with ADI model	**Inconsistent with ADI model**
1: Identify the task and the guiding question	• Sparks students' curiosity • "Creates a need" for students to design and carry out an investigation • Organizes students into collaborative groups of three or four • Supplies students with the materials they will need • Holds a "tool talk" to show students how to use equipment or to illustrate proper technique • Reviews relevant safety precautions and protocols • Provides students with hints • Allows students to tinker with the equipment they will be using later	• Provides students with possible answers to the research question • Tells students that there is one correct answer • Provides a list of vocabulary terms or explicitly describes the content addressed in the lab
2: Design a method and collect data	• Encourages students to ask questions as they design their investigations • Asks groups questions about their method (e.g., "Why did you do it this way?") and the type of data they expect from that design • Reminds students of the importance of specificity when completing their investigation proposals	• Gives students a procedure to follow • Does not question students about the method they design or the type of data they expect to collect • Approves vague or incomplete investigation proposals
3: Develop an initial argument	• Reminds students of the research question and what counts as appropriate evidence in science • Requires students to generate an argument that provides and supports a claim with genuine evidence • Asks students what opposing ideas or rebuttals they might anticipate • Provides related theories and reference materials as tools	• Requires only one student to be prepared to discuss the argument • Moves to groups to check on progress without asking students questions about why they are doing what they are doing • Does not interact with students (uses the time to catch up on other responsibilities) • Tells students the right answer
4: Argumentation session	• Reminds students of appropriate behaviors in the learning community • Encourages students to ask questions of peers • Keeps the discussion focused on the elements of the argument • Encourages students to use appropriate criteria for determining what does and does not count	• Allows students to negatively respond to others • Asks questions about students' claims before other students can ask • Allows students to discuss ideas that are not supported by evidence • Allows students to use inappropriate criteria for determining what does and does not count

Continued

TABLE 1 (*continued*)

5: Explicit and reflective discussion	• Encourages students to discuss what they learned about the content and how they know what they know • Encourages students to discuss what they learned about the nature of scientific knowledge and the nature of scientific inquiry • Encourages students to think of ways to be more productive next time	• Provides a lecture on the content • Skips over the discussion about the nature of scientific knowledge and the nature of scientific inquiry to save time • Tells students "what they should have learned" or "this is what you all should have figured out"
6: Write an investigation report	• Reminds students about the audience, topic, and purpose of the report • Provides the peer-review guide in advance • Provides an example of a good report and an example of a bad report	• Has students write only a portion of the report • Moves on to the next activity/topic without providing feedback
7: Double-blind group peer review	• Reminds students of appropriate behaviors for the review process • Ensures that all groups are giving a quality and fair peer review to the best of their ability • Encourages students to remember that while grammar and punctuation are important, the main goal is an acceptable scientific claim with supporting evidence and justification • Holds the reviewers accountable	• Allows students to make critical comments about the author (e.g., "This person is stupid") rather than their work (e.g., "This claim needs to be supported by evidence") • Allows students to just check off "Yes" on each item without providing a critical evaluation of the report
8: Revise and submit the report	• Requires students to edit their reports based on the reviewers' comments • Requires students to respond to the reviewers' ratings and comments • Has students complete the Checkout Questions after they have turned in their report	• Allows students to turn in a report without a completed peer-review guide • Allows students to turn in a report without revising it first

References

Abd-El-Khalick, F., and N. G. Lederman. 2000. Improving science teachers' conceptions of nature of science: A critical review of the literature. *International Journal of Science Education* 22: 665–701.

Duschl, R. A., H. A. Schweingruber, and A. W. Shouse, eds. 2007. *Taking science to school: Learning and teaching science in grades K–8*. Washington, DC: National Academies Press.

Grooms, J., P. Enderle, and V. Sampson. 2015. Coordinating scientific argumentation and the *Next Generation Science Standards* through argument driven inquiry. *Science Educator* 24 (1): 45–50.

Lederman, N. G., and J. S. Lederman. 2004. Revising instruction to teach the nature of science. *The Science Teacher* 71 (9): 36–39.

National Governors Association Center for Best Practices and Council of Chief State School Officers (NGAC and CCSSO). 2010. *Common core state standards*. Washington, DC: NGAC and CCSSO.

National Research Council (NRC). 2012. *A framework for K–12 science education: Practices, crosscutting concepts, and core ideas*. Washington, DC: National Academies Press.

NGSS Lead States. 2013. *Next Generation Science Standards: For states, by states*. Washington, DC: National Academies Press. *www.nextgenscience.org/next-generation-science-standards*.

Schwartz, R. S., N. Lederman, and B. Crawford. 2004. Developing views of nature of science in an authentic context: An explicit approach to bridging the gap between nature of science and scientific inquiry. *Science Education* 88: 610–645.

Wallace, C., B. Hand, and V. Prain, eds. 2004. *Writing and learning in the science classroom*. Boston: Kluwer Academic Publishers.

CHAPTER 2
Lab Investigations

This book includes 17 physics lab investigations designed around the argument-driven inquiry (ADI) instructional model and focusing on concepts related to electricity and magnetism. This is the second volume in the ADI Physics series—the first volume contains 23 lab investigations focusing on concepts related to mechanics. Please note that these investigations are not designed to replace an existing curriculum, but as a way to change the nature of the labs that are included in the curriculum. These investigations are designed to function as stand-alone lessons, which gives teachers the flexibility they need to decide which ones to use and when to use them during the academic year. We do not expect teachers to use every lab included in this book. We do, however, recommend that teachers attempt to incorporate between 8 and 12 labs from the combined set of 40 labs across the two volumes into their science curriculum to give students an opportunity to learn how to use disciplinary core ideas (DCIs), crosscutting concepts (CCs), and science and engineering practices (SEPs) to figure things out over time.

A teacher can use these investigations as a way to introduce students to a new concept related to a DCI at the beginning of a unit (introduction labs) or as a way to give students an opportunity to apply a specific concept related to a DCI that they learned about earlier in class in a novel situation (application labs). All of the labs, however, were designed to give students an opportunity to learn how to use at least one DCI and multiple CCs and SEPs to figure out how or why things happen during each investigation. Each lab is labeled as being either an introduction lab or an application lab, and labs are grouped together into sections based on traditional units in physics. The different physics content sections included in the book (such as Section 2: Forces and Interactions: Electrostatics) can be integrated into a curriculum in any order, but teachers should not assign an application lab as a way to introduce a concept.

The 17 lab investigations have been aligned with the following sources to facilitate curriculum and lesson planning (see Appendix 1 [p. 403] for the matrixes referenced in this list):

- *A Framework for K–12 Science Education* (see Standards Matrix A);
- The *Next Generation Science Standards* (see Standards Matrix B);
- Aspects of the nature of scientific knowledge (NOSK) and the nature of scientific inquiry (NOSI) (see Standards Matrix C);
- Advanced Placement (AP) Physics 1 and 2 big ideas and science practices (see Standards Matrix D)

- AP Physics C: Electricity and Magnetism course content and science practices and skills (see Standards Matrix E)
- The *Common Core State Standards for English Language Arts* (*CCSS ELA;* see Standards Matrix F)
- The *Common Core State Standards for Mathematics* (*CCSS Mathematics;* see Standards Matrix G)

We wrote many of the investigations included in this book to align with a specific performance expectation found in the *NGSS*. We also made sure to target standards included in both AP Physics 1 and 2 and AP Physics C: Electricity and Magnetism. Many of the ideas for the investigations in this book came from existing resources; however, we modified these existing activities to target the DCI and CCs found within a specific performance expectation and to fit with the ADI instructional model. Once we finished writing the labs, we had them reviewed for content accuracy. Several different science teachers then piloted the labs in their courses (including general and honors sections). We then revised each investigation based on their feedback. The revised version of each lab is included in this book.

Research that has been conducted on ADI in classrooms indicates that students have much better inquiry and writing skills after participating in at least eight ADI investigations over the course of an academic year and make substantial gains in their understanding of DCIs, CCs, SEPs, NOSK, and NOSI (Grooms, Enderle, and Sampson 2015; Sampson et al. 2013; Strimaitis et al. 2017). To learn more about the research on the ADI instructional model, visit *www.argumentdriveninquiry.com*.

Teacher Notes

Each teacher must decide when and how to use a laboratory experience to best support student learning. To help with this decision making, we have included Teacher Notes for each investigation. These notes include information about the purpose of the lab, the time needed to implement each stage of the model for that lab, the materials needed, and hints for implementation. We have also included a "Connections to Standards" section showing how each ADI lab activity is aligned with the standards documents listed earlier in this chapter. In the sections that follow, we will describe the information provided in each section of the Teacher Notes.

Purpose

This section describes the content of the lab and indicates whether the activity is designed to help students think about a new idea or think with a new idea. Labs that are designed to help students *think about* a new idea are called introduction labs. Introduction labs require students to explore new concepts. These labs are best used at the beginning of a unit of study. Labs that are designed to help students learn to *think with* a new idea are called application labs. Application labs require students to use an idea they are already familiar with to develop an explanation or to solve a problem. These labs are best used at the end of the unit of study.

Please note that because of the nature of the ADI approach, in both introduction labs and application labs very *little* emphasis needs to be placed on making sure the students "learn the vocabulary first" or "know the content" before the lab investigation begins. Instead, with the combination of the information provided in the Lab Handout and your students' evolving understanding of the DCIs, CCs, and SEPs, they will develop a better understanding of the content *as they work through the eight stages of ADI*. The "Purpose" section also highlights the NOSK or NOSI concepts that should be emphasized during the explicit and reflective discussion stage of the activity.

Underlying Physics Concepts

This section of the Teacher Notes provides a basic overview of the major concepts that the students will explore and or use during the investigation.

Timeline

Unlike most traditional labs, ADI labs typically take four or five days to complete. The amount of time it will take to complete each lab will vary depending on how long it takes to collect data and whether or not the students write in class or at home. The time associated with each ADI lab investigation may be longer in the first few labs your students conduct, but the time will be reduced as your students become familiar with the practices used in the model (argumentation, designing investigations, writing reports). We therefore provide suggestions about which stages of ADI you should be able to complete in a 50-minute class period (see Appendix 3 [p. 421]).

Materials and Preparation

This section describes the lab supplies (i.e., consumables and equipment) and instructional materials (e.g., Lab Handout, Investigation Proposal, and peer-review guide and teacher scoring rubric [PRG/TSR]) needed to implement the lab activity. The lab supplies listed are designed for one group; however, multiple groups can share if resources are scarce. We have also included specific suggestions for some lab supplies, based on our finding that these supplies worked best during the field tests. However, if needed, substitutions can be made. Always be sure to test all lab supplies before conducting the lab with the students, because using new materials often has unexpected consequences.

Safety Precautions and Laboratory Waste Disposal

This section provides an overview of potential safety hazards as well as safety protocols that should be followed to make the laboratory safer for students. These are based on legal safety standards and current better professional safety practices. Teachers should also review and follow all local polices and protocols used within their school district and/or school (e.g., the district chemical hygiene plan, Board of Education safety policies).

Topics for the Explicit and Reflective Discussion

This section begins with an overview of some of the DCIs and CCs that students use to figure things out during the lab. We provide advice about ways to encourage students to think about how these ideas or concepts helped them explain the phenomenon under investigation and how they used these ideas or concepts to provide a justification of the evidence in their arguments. The section also provides some advice for teachers about how to encourage students to reflect on ways to improve the design of their investigation in the future. This section concludes with an overview of the relevant NOSK and NOSI concepts to discuss during the explicit and reflective discussion and some sample questions that teachers can pose to help students be reflective about what they know about these concepts.

Hints for Implementing the Lab

Many teachers have tested these labs many different times. As a result, we have collected hints from the teachers for each stage of the ADI process. These hints are designed to help you avoid some of the pitfalls earlier teachers have experienced and make the investigation run smoothly. The section also includes tips for making the investigation safer.

Connections to Standards

This section is designed to inform curriculum and lesson planning by highlighting how the investigation can be used to address specific performance expectations from the *NGSS*, AP Physics 1 and 2, AP Physics C, *CCSS ELA*, and *CCSS Mathematics*.

Instructional Materials

The instructional materials included in this book are reproducible copy masters that are designed to support students as they participate in an ADI lab activity. The materials needed for each lab include a Lab Handout, the PRG/TSR, and a set of Checkout Questions. Some labs also require an investigation proposal and supplementary materials.

Lab Handout

At the beginning of each lab activity, each student should be given a copy of the Lab Handout. This handout provides information about the phenomenon that they will investigate and a guiding question for the students to answer. The handout also provides hints for students to help them design their investigation in the "Getting Started" section, information about what to include in their initial argument, and the requirements for the investigation report. The last part of the Lab Handout provides space for students to keep track of critiques, suggestions for improvement, and good ideas that arise during the argumentation session.

Peer-Review Guide and Teacher Scoring Rubric

The PRG/TSR is designed to make the criteria that are used to judge the quality of an investigation report explicit. Appendix 5 (p. 433) includes two versions of the PRG/TSR: a high school version and an undergraduate or AP version. We recommend that teachers make one copy of the appropriate version for each student and provide it to the students before they begin writing their investigation report. This will ensure that students understand how they will be evaluated. Then during the double-blind group peer-review stage of the model, each group should fill out the peer-review guide as they review the reports of their classmates. (Each group will need to review at least three reports.) The reviewers should rate the report on each criterion and then provide advice to the author about ways to improve. Once the review is complete, the author needs to revise his or her report and respond to the reviewers' rating and comments in the appropriate sections in the PRG/TSR.

The PRG/TSR should be submitted to the instructor along with the first draft and the final report for a final evaluation. To score the report, the teacher can simply fill out the "Teacher Score" column of the PRG/TSR and then total the scores.

Checkout Questions

To facilitate formative assessment inside the classroom, we have included a set of Checkout Questions for each lab investigation. The questions target the core physics concepts, the CCs, and the NOSK and NOSI concepts that are addressed in the lab. Students should complete the Checkout Questions on the same day they turn in their final report. One handout is needed for each student. The students should complete these questions on their own. The teacher can use the students' responses, along with the report, to determine if the students learned what they needed to during the lab, and then reteach as needed.

Investigation Proposal

To help students design better investigations, we have developed and included three different types of investigation proposals in this book, with a short version and a long version of each type (see Appendix 4 [p. 423]). These investigation proposals are optional, but we have found that students design and carry out much better investigations when they are required to fill out a proposal and then get teacher feedback about their method before they begin. We provide recommendations about which investigation proposal (A, B, or C) to use for a particular lab as part of the Teacher Notes. If a teacher decides to use an investigation proposal as part of a lab activity, we recommend providing one copy for each group. The Lab Handout for students also has a heading labeled "Investigation Proposal Required?" that is followed by "Yes" and "No" check boxes. The teacher should be sure to have students check the appropriate box on the Lab Handout when introducing the lab activity.

References

Grooms, J., P. Enderle, and V. Sampson. 2015. Coordinating scientific argumentation and the *Next Generation Science Standards* through argument driven inquiry. *Science Educator* 24 (1): 45–50.

National Governors Association Center for Best Practices and Council of Chief State School Officers (NGAC and CCSSO). 2010. *Common core state standards.* Washington, DC: NGAC and CCSSO.

National Research Council (NRC). 2012. *A framework for K–12 science education: Practices, crosscutting concepts, and core ideas.* Washington, DC: National Academies Press.

NGSS Lead States. 2013. *Next Generation Science Standards: For states, by states.* Washington, DC: National Academies Press. *www.nextgenscience.org/next-generation-science-standards.*

Sampson, V., P. Enderle, J. Grooms, and S. Witte. 2013. Writing to learn and learning to write during the school science laboratory: Helping middle and high school students develop argumentative writing skills as they learn core ideas. *Science Education* 97 (5): 643–670.

Strimaitis, A., S. Southerland, V. Sampson, P. Enderle, and J. Grooms. 2017. Promoting equitable biology lab instruction by engaging all students in a broad range of science practices: An exploratory study. *School Science and Mathematics* 117 (3–4): 92–103.

SECTION 2
Forces and Interactions

Electrostatics

Introduction Labs

LAB 1

Teacher Notes

Lab 1. Coulomb's Law: How Do the Amount of Charge on the Rod and the Mass of the Foil Used in an Electroscope Affect How Far Apart the Pieces of Foil Will Separate From Each Other?

Purpose

The purpose of this lab is to *introduce* students to the disciplinary core idea (DCI) of Types of Interactions (PS2.B) from the *NGSS* by having them explore the electric force between two charged pieces of metal in an electroscope. In addition, this lab can be used to help students understand two big ideas from AP Physics: (a) fields existing in space can be used to explain interactions and (b) the interactions of an object with other objects can be described by forces. This lab also gives students an opportunity to learn about the crosscutting concepts (CCs) of (a) Patterns and (b) Structure and Function from the *NGSS*. As part of the explicit and reflective discussion, students will also learn about (a) the difference between data and evidence in science and (b) how the culture of science, societal needs, and current events influence the work of scientists.

Underlying Physics Concepts

Coulomb's law states that the magnitude of the electric force between two point charges is directly proportional to the magnitude of each charge and inversely proportional to the square of the distance between the two charges. Coulomb's law is shown mathematically in Equation 1.1, where $\mathbf{F}_E$ is the electric force, k is Coulomb's constant, q_1 is the magnitude of charge at the first point, q_2 is the magnitude of charge at the second point, and $\mathbf{r}$ is the distance between the two points. In SI units, force is measured in newtons (N), charge is measured in coulombs (C), distance is measured in meters (m), and Coulomb's constant has a value of 8.99×10^9 N · m²/C².

$$\text{(Equation 1.1)} \quad \mathbf{F}_E = \frac{kq_1q_2}{\mathbf{r}^2}$$

Coulomb's constant can also be expressed in terms of other constants. Specifically, $k = 1/4\pi\varepsilon_o$, where ε_o is the permittivity of free space and is equal to 8.85×10^{-12} C²/N · m². Thus, Coulomb's law can be rewritten as shown in Equation 1.2. We point this out because some people prefer Equation 1.2 to Equation 1.1, but we prefer to use Coulomb's constant, as opposed to the permittivity of free space, in the equation for Coulomb's law.

$$\text{(Equation 1.2)} \quad \mathbf{F}_E = \frac{q_1q_2}{4\pi\varepsilon_o\mathbf{r}^2}$$

Coulomb's Law

How Do the Amount of Charge on the Rod and the Mass of the Foil Used in an Electroscope Affect How Far Apart the Pieces of Foil Will Separate From Each Other?

Similar to how objects with mass produce a gravitational field, objects with charge create an electric field. When modeling the electric field, the convention used is to show the electric field due to a positive charge as pointing out from the charge and the electric field due to a negative charge as pointing in toward the charge. When a charged object is in an electric field, the force on the object due to the field is proportional to the charge on the object. This relationship is shown mathematically in Equation 1.3, where F_E is the electric force acting on the charged object, E is the electric field strength, and q is the charge on the object in the electric field. In SI units, electric field strength is measured in newtons per coulomb (N/C) or volts per meter (V/m). (*Note:* In some labs, electric field is better analyzed as N/C, while in others it is easier to work with V/m. In this lab, it is best to use N/C).

(Equation 1.3) $F_E = Eq$

If we are to set Equation 1.3 equal to Equation 1.1 and solve for the field strength, the resulting relationship gives the electric field strength at a given point as directly proportional to the amount of charge at the source and inversely related to the square of the distance from the source of the electric field. This relationship is shown mathematically in Equation 1.4, where E is the electric field strength, q is the charge of the source of the field, and r is the distance from the field.

(Equation 1.4) $E = \dfrac{kq}{r^2}$

Students will begin by first charging an object external to the electroscope. They can either use a Van de Graaff generator or use a piece of fur on a charging rod. If they use a Van de Graaff generator, then they can approximate the charge by the time the generator is running—the longer the time, the more charge accumulated on the Van de Graaff generator. If they use the charging rod, then students can approximate the charge by the number of times they charge the rod. In either case, the charged object will be left with an excess of electrons. (*Note:* We strongly recommend using a charging rod and fur rather than a Van de Graaff generator. See the "Hints for Implementing the Lab" section for further discussion.)

Once the students have built some static charge on the rod or the Van de Graaff generator, they can then touch the charged object to the top of the electroscope. Touching the rod to the electroscope will lead some of the excess electrons to move from the charged object onto the electroscope, because charge always distributes in a manner that minimizes the force on each individual charge. Thus, charge will distribute evenly down the electroscope and onto the two pieces of metal foil. As suggested by Coulomb's law, the two pieces of metal foil will be repelled because like charges repel. You can also think of this phenomenon this way: the force acting on a charged object in an electric field will be away from the source of the electric field if the charge on the object is the same as the charge at the source of the electric field. Each piece of foil will create an electric field that, when analyzed in isolation, will cause the other piece of foil to move away.

LAB 1

FIGURE 1.1

Force diagram showing the forces acting on the piece of foil on the right due to the charged foil on the left in a charged electroscope. Gray lines represent the foil pieces, and solid arrows represent vectors.

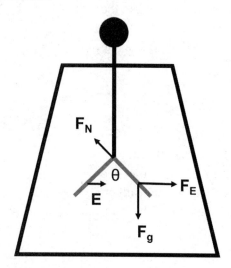

To understand the forces acting on the two pieces of foil, it helps to draw a force diagram (see Figure 1.1). In the diagram, we use gray lines to represent pieces of foil and solid arrows to represent vectors. Three forces act on each piece of foil on the electroscope: (1) the electric force from the other charged piece of metal foil, (2) the gravitational force acting on each piece of foil, and (3) the normal force acting between the hook of the electroscope and the metal foil. Figure 1.1 shows the forces acting on the foil piece on the right due to the electric field created by the foil piece on the left of the electroscope.

To analyze this situation, we can use Newton's second law, which states that the sum of the forces acting on an object causes that object to accelerate. Furthermore, the sum of the forces on the object is directly proportional to the acceleration of the object. Newton's second law is shown mathematically in Equation 1.5, where $\sum \mathbf{F}$ is the sum of the forces, or net force, acting on the object; m is the mass of the object; and $\mathbf{a}$ is the acceleration. In SI units, force is measured in newtons (N), mass is measured in kilograms (kg), and acceleration is measured in meters per second squared (m/s²).

$$\text{(Equation 1.5)} \quad \sum \mathbf{F} = m\mathbf{a}$$

We also know that the force due to gravity acting on an object is equal to the acceleration due to gravity times the object's mass, shown mathematically in Equation 1.6, where $\mathbf{F}_g$ is the force of gravity on an object and $\mathbf{g}$ is the acceleration due to gravity on Earth (9.8 m/s²).

$$\text{(Equation 1.6)} \quad \mathbf{F}_g = m\mathbf{g}$$

Once the electroscope is charged and the foil pieces come to rest, the sum of the forces acting on each piece of foil must be zero. Using vector addition, we can also identify the effects of different forces on the separation angle of the two pieces of foil. If more charge is added to the electroscope, $\mathbf{F}_E$ will increase in accordance with Coulomb's law. This will cause the angle of separation to also increase (θ in Figure 1.1). If the mass of the foil is increased, this will lead to an increase in the gravitational force acting on the foil, and the angle of separation will decrease.

Timeline

The instructional time needed to complete this lab investigation is 170–230 minutes. Appendix 3 (p. 421) provides options for implementing this lab investigation over several class periods. Option C (230 minutes) should be used if students are unfamiliar with

Coulomb's Law

How Do the Amount of Charge on the Rod and the Mass of the Foil Used in an Electroscope Affect How Far Apart the Pieces of Foil Will Separate From Each Other?

scientific writing, because this option provides extra instructional time for scaffolding the writing process. You can scaffold the writing process by modeling, providing examples, and providing hints as students write each section of the report. Option C can also be used if you are introducing students to the video analysis programs. Option D (170 minutes) should be used if students are familiar with scientific writing and have developed the skills needed to write an investigation report on their own. In option D, students complete stage 6 (writing the investigation report) and stage 8 (revising the investigation report) as homework.

Materials and Preparation

The materials needed to implement this investigation are listed in Table 1.1. The equipment can be purchased from a science supply company such as Flinn Scientific, PASCO, Vernier, or Ward's Science. Video analysis software can be purchased from Vernier (*Logger Pro*) or PASCO (SPARKvue or Capstone). These companies also have apps that can be used on Apple- or Android-based tablets and cell phones. We recommend consulting with your school's information technology coordinator to determine the best option for your students.

TABLE 1.1

Materials list for Lab 1

Item	Quantity
Consumables	
Aluminum foil	Several pieces per group
Cellophane tape	1 roll per group
Equipment and other materials	
Safety glasses with side shields or safety goggles	1 per student
Erlenmeyer flask	1 per group
Electroscope attachment	1 per group
Charging rod and fur OR Van de Graaff generator	1 per group
Copper wire	As needed
Scissors	1 per group
Ruler	1 per student
Electronic or triple beam balance	1 per group

Continued

LAB 1

Table 1.1 (continued)

Item	Quantity
Other equipment (see text below table)	
Investigation Proposal B (optional)	1 per group
Whiteboard, 2' × 3'*	1 per group
Lab Handout	1 per student
Peer-review guide and teacher scoring rubric	1 per student
Checkout Questions	1 per student
Equipment for video analysis (optional)	
Video camera	1 per group
Computer or tablet with video analysis software	1 per group

* As an alternative, students can use computer and presentation software such as Microsoft PowerPoint or Apple Keynote to create their arguments.

Science lab supply companies, such as those mentioned above, sell electroscopes that students can use to set up the equipment more quickly. Each supplier sells multiple electroscopes, with slight differences between them, that can be used to investigate the same concepts. To provide flexibility for teachers in purchasing, we have not committed to any specific set of equipment and instead have listed "other equipment" in Table 1.1. We also want to make it clear that specialized equipment is not necessary—the lab can be conducted with the materials purchased individually.

Use of video analysis software is optional, but using this software will allow students to more precisely measure the separation distance between the pieces of foil and, subsequently, the forces on the foil.

Be sure to use a set routine for distributing and collecting the materials during the lab investigation. One option is to set up the materials for each group at each group's lab station before class begins. This option works well when there is a dedicated section of the classroom for lab work and the materials are large and difficult to move. A second option is to have all the materials on a table or cart at a central location. You can then assign a member of each group to be the "materials manager." This individual is responsible for collecting all the materials his or her group needs from the table or cart during class and for returning all the materials at the end of the class. This option works well when the materials are small and easy to move (such as magnets, wire, and bulbs). It also makes it easy to inventory the materials at the end of the class before students leave for the day.

Coulomb's Law

How Do the Amount of Charge on the Rod and the Mass of the Foil Used in an Electroscope Affect How Far Apart the Pieces of Foil Will Separate From Each Other?

Safety Precautions and Laboratory Waste Disposal

Remind students to follow all normal lab safety rules. In addition, tell students to take the following safety precautions:

1. Wear sanitized safety glasses with side shields or goggles during lab setup, hands-on activity, and takedown.

2. Never put consumables in their mouth.

3. Handle all glassware with care.

4. If using a Van de Graaff generator, make sure that students do not cause the generator to discharge to them. This will cause a shock.

5. Use caution when working with sharp objects (e.g., scissors, wire) because they can cut or puncture skin.

6. Wash their hands with soap and water when they are done collecting the data.

There is no laboratory waste associated with this activity.

Topics for the Explicit and Reflective Discussion

Reflecting on the Use of Core Ideas and Crosscutting Concepts During the Investigation

Teachers should begin the explicit and reflective discussion by asking students to discuss what they know about the core ideas from the *NGSS* and other important ideas from physics that they used during the investigation. The following are some important concepts related to the DCI Types of Interactions that students need to use to determine how the charge and mass of the foil influence the angle of separation in the electroscope:

- Charge can be either positive or negative. Objects and systems with the same charge repel. Objects and systems with opposite charges attract. Coulomb's law provides a mathematical representation of this interaction.

- A field associates a value of some physical quantity with every point in space. Fields are a model that physicists use to describe interactions that occur over a distance. Fields permeate space, and objects experience forces due to their interaction with a field.

- The magnitude of the electric force exerted on a charged object in an electric field is expressed mathematically as $\mathbf{F}_E = \mathbf{E}q$. A positive charge creates a field out from the source of the charge, and a negative charge creates a field directed in toward the source. The direction of the force on a positive charge in an electric field is in the same direction as the field. The direction of the force on a negative charge in an electric field is in the opposite direction as the field.

- Electric charge is always conserved.

LAB 1

To help students reflect on what they know about electric forces and fields, we recommend showing them two or three images using presentation software that help illustrate these important ideas. You can then ask the students the following questions to encourage them to share how they are thinking about these important concepts:

1. What do we see going on in this image?
2. Does anyone have anything else to add?
3. What might be going on that we can't see?
4. What are some things that we are not sure about here?

You can then encourage students to think about how CCs played a role in their investigation. There are at least two CCs that students need to use to determine how the amount of charge and the mass of the foil affect the angle of separation between the two pieces of foil: (a) Patterns and (b) Structure and Function (see Appendix 2 [p. 417] for a brief description of these CCs). To help students reflect on what they know about these CCs, we recommend asking them the following questions:

1. Did you notice any patterns emerging between any variables in your investigation?
2. What predictions can you make using these patterns?
3. How does the structure of the electroscope influence how it functions?
4. Why is it important for scientists and engineers to know the relationship between structure and function for designing objects?

You can then encourage the students to think about how they used all these different concepts to help answer the guiding question and why it is important to use these ideas to help justify their evidence for their final arguments. Be sure to remind your students to explain why they included the evidence in their arguments and make the assumptions underlying their analysis and interpretation of the data explicit in order to provide an adequate justification of their evidence.

Reflecting on Ways to Design Better Investigations

It is important for students to reflect on the strengths and weaknesses of the investigation they designed during the explicit and reflective discussion. Students should therefore be encouraged to discuss ways to eliminate potential flaws, measurement errors, or sources of uncertainty in their investigations. To help students be more reflective about the design of their investigation and what they can do to make their investigations more rigorous in the future, you can ask them the following questions:

1. What were some of the strengths of the way you planned and carried out your investigation? In other words, what made it scientific?

2. What were some of the weaknesses of the way you planned and carried out your investigation? In other words, what made it less scientific?

3. What rules can we make, as a class, to ensure that our next investigation is more scientific?

Reflecting on the Nature of Scientific Knowledge and Scientific Inquiry

This investigation can be used to illustrate two important concepts related to the nature of scientific knowledge and the nature of scientific inquiry: (a) the difference between data and evidence in science and (b) how the culture of science, societal needs, and current events influence the work of scientists (see Appendix 2 [p. 417] for a brief description of these concepts). Be sure to review these concepts during and at the end of the explicit and reflective discussion. To help students think about these concepts in relation to what they did during the lab, you can ask them the following questions:

1. You had to talk about data and evidence during your investigation. Can you give me some examples of data and evidence from your investigation?

2. Can you work with your group to come up with a rule that you can use to decide if a piece of information is data or evidence? Be ready to share in a few minutes.

3. People view some types of research as being more important than other types of research because of cultural values and current events. Can you come up with some examples of how cultural values and current events have influenced the work of scientists?

4. Scientists share a set of values, norms, and commitments that shape what counts as knowing, how to represent or communicate information, and how to interact with other scientists. Can you work with your group to come up with a rule that you can use to decide if something is science or not science? Be ready to share in a few minutes.

You can also encourage the students to think about these concepts by showing examples of information from the lab using presentation software and then asking students to classify each one as data or evidence. You can also show images of recent scientific advances and ask them how cultural values or current events may have influenced the scientists in choosing to study that phenomena. Be sure to remind your students that it is important for them to understand what counts as scientific knowledge and how that knowledge develops over time in order to be proficient in science.

LAB 1

Hints for Implementing the Lab

- Allowing students to design their own procedures for collecting data gives students an opportunity to try, to fail, and to learn from their mistakes. However, you can scaffold students as they develop their own procedure by having them fill out an investigation proposal. The proposals provide a way for you to offer students hints and suggestions without telling them how to do it. You can also check the proposals quickly during a class period. For this lab we suggest using Investigation Proposal B.

- Learn how to use the electroscope (and the Van de Graaff generator, if used) before the lab begins. It is important for you to know how to use the equipment so you can help students when technical issues arise.

- Allow the students to become familiar with the electroscope (and the Van de Graaff generator, if used) as part of the tool talk before they begin to design their investigation. Give them 5–10 minutes to examine the equipment and materials before they begin designing their investigations. This gives students a chance to see what they can and cannot do with the equipment.

- We strongly suggest using the charging rod and fur instead of a Van de Graaff generator in this lab (in fact, we do not mention the generator option in the Lab Handout). This is for two reasons: (1) the charging rod and fur are much cheaper, and (2) accidental discharge of the charged rod will not cause a shock to students, whereas the Van de Graaff generator may. However, use of the generator allows for more precise measurements if students want to mathematically model the interactions, so we present this as an option in the Teacher Notes.

- For best results, this lab should be conducted when the air is relatively dry.

- Students will use different-size aluminum foil pieces during the course of this investigation. Although there is not a standard size to use, the size does not need to be more than a few centimeters on each side on the largest piece of foil. Balances may not be sensitive to changes in mass of the foil. Students can also use the area of the foil as a proxy for mass; assuming the density of the foil is uniform, they can calculate the mass from the area.

- This lab presents a good opportunity to talk about controlling variables. The magnitude of the charge established on the charging rod will be a function of both the number of times that students charge the charging rod and the length of motion the students run the fur over the rod each time they charge it. Students may not initially control the length of motion that they use to charge the rod. This will lead to error in their results.

- Be sure to allow students to go back and re-collect data at the end of the argumentation session. Students often realize that they made numerous mistakes when they were collecting data as a result of their discussions during the argumentation session. The students, as a result, will want a chance to re-collect data, and the re-collection of data should be encouraged when time allows. This also offers an opportunity to discuss what scientists do when they realize a mistake is made inside the lab.

If students use video analysis

- We suggest allowing students to familiarize themselves with the video analysis software before they finalize the procedure for the investigation, especially if they have not used such software previously. This gives students an opportunity to learn how to work with the software and to improve the quality of the video they take.
- Remind students to hold the video camera as still as possible. Any movement of the camera will introduce error into their analysis. If using actual camcorders, we recommend using a tripod to hold the camera steady. If students are using a camera on a cell phone or tablet, we recommend using a table to help steady the camera.
- Remind students to place a meterstick in the same field of view as the motion they are capturing with the video camera. Also, the meterstick should be approximately the same distance from the camera as the motion. Most video analysis software requires the user to define a scale in the video (this allows the software to establish distances and, subsequently, other variables dependent on distance and displacement).

Connections to Standards

Table 1.2 highlights how the investigation can be used to address specific performance expectations from the *NGSS;* learning objectives from AP Physics 1 and 2; learning objectives from AP Physics C: Electricity and Magnetism; *Common Core State Standards for English Language Arts* (*CCSS ELA*); and *Common Core State Standards for Mathematics* (*CCSS Mathematics*).

TABLE 1.2

Lab 1 alignment with standards

NGSS performance expectation	• HS-PS2-4: Use mathematical representations of Newton's Law of Gravitation and Coulomb's Law to describe and predict the gravitational and electrostatic forces between objects.

Continued

LAB 1

Table 1.2 (*continued*)

AP Physics 1 and AP Physics 2 learning objectives	• 1.B.1.1: Make claims about natural phenomena based on conservation of electric charge. • 2.C.2.1: Qualitatively and semiquantitatively apply the vector relationship between the electric field and the net electric charge creating that field. • 3.A.3.4: Make claims about the force on an object due to the presence of other objects with the same property: mass, electric charge. • 3.C.2.2: Connect the concepts of gravitational force and electric force to compare similarities and differences between the forces. • 3.G.2.1: Connect the strength of electromagnetic forces with the spatial scale of the situation, the magnitude of the electric charges, and the motion of the electrically charged objects involved. • 5.C.2.2: Design a plan to collect data on the electrical charging of objects and electric charge induction on neutral objects and qualitatively analyze that data. • 5.C.2.3: Justify the selection of data relevant to an investigation of the electrical charging of objects and electric charge induction on neutral objects.
AP Physics C: Electricity and Magnetism learning objectives	• ACT-1.A: Describe behavior of charges or system of charged objects interacting with each other.
Literacy connections (*CCSS ELA*)	• *Reading:* Key ideas and details, craft and structure, integration of knowledge and ideas • *Writing:* Text types and purposes, production and distribution of writing, research to build and present knowledge, range of writing • *Speaking and listening:* Comprehension and collaboration, presentation of knowledge and ideas

Continued

Table 1.2 (continued)

Mathematics connections (*CCSS Mathematics*)	• *Mathematical practices:* Make sense of problems and persevere in solving them, reason abstractly and quantitatively, construct viable arguments and critique the reasoning of others, model with mathematics, use appropriate tools strategically, attend to precision • *Number and quantity:* Reason quantitatively and use units to solve problems, represent and model with vector quantities, perform operations on vectors • *Algebra:* Interpret the structure of expressions, create expressions that describe numbers or relationships, understand solving equations as a process of reasoning and explain the reasoning, solve equations and inequalities in one variable, represent and solve equations and inequalities graphically • *Functions:* Understand the concept of a function and use function notation; interpret functions that arise in applications in terms of the context; analyze functions using different representations; build a function that models a relationship between two quantities; construct and compare linear, quadratic, and exponential models and solve problems; interpret expressions for functions in terms of the situation they model • *Statistics and probability:* Summarize, represent, and interpret data on two categorical and quantitative variables; interpret linear models; make inferences and justify conclusions from sample surveys, experiments, and observational studies

LAB 1

Lab Handout

Lab 1. Coulomb's Law: How Do the Amount of Charge on the Rod and the Mass of the Foil Used in an Electroscope Affect How Far Apart the Pieces of Foil Will Separate From Each Other?

Introduction

Current policy debates focus on our society's increasing need for electrical energy and how to produce such energy in ways that do not add greenhouse gases, such as carbon dioxide, to the atmosphere. One such proposal is through increased use of nuclear power. On the pro side, nuclear power is cheap to produce (once the power plant is built) and does not give off pollution. On the con side, however, is the possibility of nuclear contamination should the plant have any structural problems. When this occurs, radiation is able to "leak" out of the plant's containment mechanisms and affect the people who work in and around the power plant. In the past 50 years, there have been three catastrophic nuclear plant disasters: Three Mile Island in the United States in 1979, Chernobyl in the U.S.S.R. in 1986, and Fukushima in Japan in 2011.

Thus, many governments require strict safety protocols for nuclear power plants. One safety requirement is the use of a dosimeter, a device that is able to measure the amount of radiation passing through it. Power plants can have large, stationary dosimeters that measure the amount of radiation in a given area. People who work in the power plant often wear smaller, personal dosimeters that can measure the amount, or dose, of radiation they have been exposed to each day. If the dosimeters indicate too much radiation, then additional safety precautions are taken to ensure the health and safety of employees and to prevent further radiation leakage.

The earliest dosimeters worked by using an *electroscope,* which is a device that can help to measure the charge on an object. Dosimeters that use an electroscope work by charging the electroscope to a known value. When radiation passes by the electroscope, air molecules become ionized and are attracted to the charged pieces of foil in the electroscope. This will cause the metal foil to change the angle of separation between the two pieces. Figure L1.1 shows a permanent dosimeter near Fukushima, Japan. This dosimeter displays the radiation in units of microsieverts per

FIGURE L1.1

A permanent dosimeter located near Fukushima, Japan

Coulomb's Law

How Do the Amount of Charge on the Rod and the Mass of the Foil Used in an Electroscope Affect How Far Apart the Pieces of Foil Will Separate From Each Other?

hour (μSv/h). A sievert is an SI unit that describes the effect of radiation on the chance a person will develop cancer. One sievert corresponds to a 5.5% increase in the chance a person will develop cancer. A person standing near the dosimeter in Figure L1.1 for one hour would have a 1.1×10^{-6}% increase in developing cancer.

Charges come in two varieties—positive (carried by protons) and negative (carried by electrons). The *electric force* is the interaction due to two charged objects and is one of the fundamental forces of nature. The *magnitude* of the force between two objects depends on how many charged particles are on each object. The direction of the force between two charged objects depends on the type (i.e., positive or negative) of the charge. If the two objects have the same charge—they both are positively charged, for example—the force is *repulsive*. If the charges are opposite, the force between the two objects is *attractive*. Finally, charge obeys *conservation* laws—when two objects interact, the total charge in the system is the same before and after the interaction. This makes intuitive sense because charge is carried by protons and electrons, which both obey the law of conservation of mass.

When designing a permanent dosimeter, scientists must have an idea of how the electroscope will react to different charges, including both the type of charge (positive or negative) and the amount of charge. This information will allow scientists to design a dosimeter sensitive enough to respond to small amounts of radiation that are harmful to human health.

Your Task

Use what you know about the forces acting on charged objects, patterns, and how the structure of designed objects influences their function to design and carry out an investigation to determine the effect of charge on an electroscope.

The guiding question of this investigation is, *How does the amount of charge on the rod and the mass of the foil used in an electroscope affect how far apart the pieces of foil will separate from each other?*

Materials

You may use any of the following materials during your investigation:

Consumables
- Aluminum foil
- Cellophane tape

Equipment
- Safety glasses or goggles (required)
- Erlenmeyer flask
- Electroscope attachment
- Charging rod
- Fur
- Copper wire
- Scissors
- Ruler
- Electronic or triple beam balance

If you have access to the following equipment, you may also consider using a video camera and a computer or tablet with video analysis software.

LAB 1

Safety Precautions

Follow all normal lab safety rules. In addition, take the following safety precautions:

1. Wear sanitized safety glasses with side shields or goggles during lab setup, hands-on activity, and takedown.
2. Never put consumables in your mouth.
3. Handle all glassware with care.
4. If using a Van de Graaff generator, make sure that you do not cause the generator to discharge to yourself or others. This will cause a shock.
5. Use caution when working with sharp objects (e.g., scissors, wire) because they can cut or puncture skin.
6. Wash your hands with soap and water when you are done collecting the data.

Investigation Proposal Required?   ☐ Yes   ☐ No

Getting Started

To answer the guiding question, you will need to design and carry out two separate experiments to determine the effect of charge and the mass of the foil on the electroscope. Before you can design your investigation, however, you must determine what type of data you need to collect, how you will collect it, and how you will analyze it.

To determine *what type of data you need to collect,* think about the following questions:

- What are the boundaries and components of the system?
- How do the components of the system interact with each other?
- When is this system stable and under which conditions does it change?
- How could you keep track of changes in this system quantitatively?
- How might the structure of the electroscope relate to its function?

To determine *how you will collect the data,* think about the following questions:

- What is the independent variable and what is the dependent variable?
- What other factors will you need to control during each experiment?
- What scale or scales should you use when you take your measurements?
- How will you make sure that your data are of high quality (i.e., how will you reduce error)?
- How will you keep track of and organize the data you collect?

To determine *how you will analyze the data,* think about the following questions:

Coulomb's Law

How Do the Amount of Charge on the Rod and the Mass of the Foil Used in an Electroscope Affect How Far Apart the Pieces of Foil Will Separate From Each Other?

- What type of calculations, if any, will you need to make?
- What types of patterns might you look for as you analyze your data?
- What type of table or graph could you create to help make sense of your data?

Connections to the Nature of Scientific Knowledge and Scientific Inquiry

As you work through your investigation, you may want to consider

- the difference between data and evidence in science, and
- how the culture of science, societal needs, and current events influence the work of scientists.

Initial Argument

Once your group has finished collecting and analyzing your data, your group will need to develop an initial argument. Your initial argument needs to include a claim, evidence to support your claim, and a justification of the evidence. The *claim* is your group's answer to the guiding question. The *evidence* is an analysis and interpretation of your data. Finally, the *justification* of the evidence is why your group thinks the evidence matters. The justification of the evidence is important because scientists can use different kinds of evidence to support their claims. Your group will create your initial argument on a whiteboard. Your whiteboard should include all the information shown in Figure L1.2.

FIGURE L1.2
Argument presentation on a whiteboard

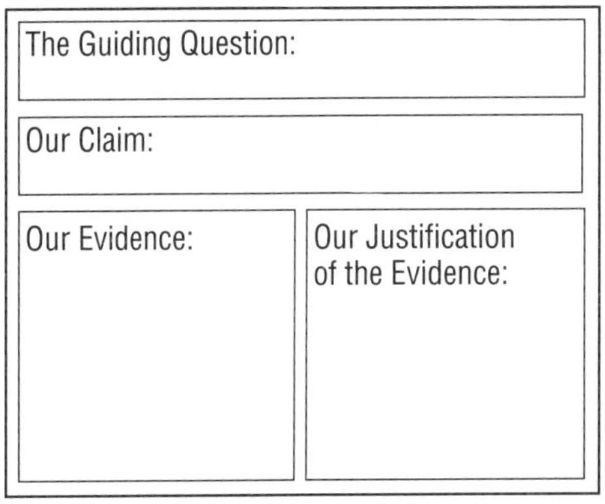

The Guiding Question:	
Our Claim:	
Our Evidence:	Our Justification of the Evidence:

Argumentation Session

The argumentation session allows all of the groups to share their arguments. One or two members of each group will stay at the lab station to share that group's argument while the other members of the group go to the other lab stations to listen to and critique the other arguments. This is similar to what scientists do when they propose, support, evaluate, and refine new ideas during a poster session at a conference. If you are presenting your group's argument, your goal is to share your ideas and answer questions. You should also keep a record of the critiques and suggestions made by your classmates so you can use this feedback to make your initial argument stronger. You can keep track of specific critiques and suggestions for improvement that your classmates mention in the space below.

LAB 1

Critiques about our initial argument and suggestions for improvement:

If you are critiquing your classmates' arguments, your goal is to look for mistakes in their arguments and offer suggestions for improvement so these mistakes can be fixed. You should look for ways to make your initial argument stronger by looking for things that the other groups did well. You can keep track of interesting ideas that you see and hear during the argumentation in the space below. You can also use this space to keep track of any questions that you will need to discuss with your team.

Interesting ideas from other groups or questions to take back to my group:

Once the argumentation session is complete, you will have a chance to meet with your group and revise your initial argument. Your group might need to gather more data or design a way to test one or more alternative claims as part of this process. Remember, your goal at this stage of the investigation is to develop the best argument possible.

Report

Once you have completed your research, you will need to prepare an *investigation report* that consists of three sections. Each section should provide an answer to the following questions:

1. What question were you trying to answer and why?
2. What did you do to answer your question and why?
3. What is your argument?

Your report should answer these questions in two pages or less. This report must be typed, and any diagrams, figures, or tables should be embedded into the document. Be sure to write in a persuasive style; you are trying to convince others that your claim is acceptable or valid!

LAB 1

Checkout Questions

Lab 1. Coulomb's Law: How Do the Amount of Charge on the Rod and the Mass of the Foil Used in an Electroscope Affect How Far Apart the Pieces of Foil Will Separate From Each Other?

1. The picture below shows a charged electroscope with the pieces of foil separated by an angle θ. Draw a free-body diagram showing all the forces acting on the piece of foil on the right.

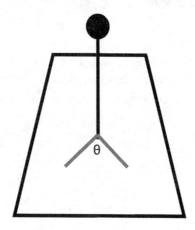

2. The mass of the foil on the right is 25 g, and the mass of the foil on the left is 40 g. When the electroscope is charged such that the angle of separation is 20°, the piece of foil on the right feels an electric force of 2 N.

   a. What force does the piece of foil on the left feel?

   b. How do you know?

Coulomb's Law

How Do the Amount of Charge on the Rod and the Mass of the Foil Used in an Electroscope Affect How Far Apart the Pieces of Foil Will Separate From Each Other?

3. Two electroscopes are charged. One electroscope has an angle of separation of 10° and the other electroscope has an angle of separation of 20°. What might have led to the angle of separation being different for the two electroscopes?

4. Scientists make choices on what investigations to conduct based on cultural values and current events.

    a. I agree with this statement.
    b. I disagree with this statement.

    Explain your answer, using examples from this investigation and at least one other investigation you have conducted.

LAB 1

5. *Data* and *evidence* are terms that have the same meaning in science.

    a. I agree with this statement.
    b. I disagree with this statement.

    Explain your answer, using examples from this investigation and at least one other investigation you have conducted.

6. Understanding the relationship between structure and function is an important part of scientific investigations. Why do scientists attend to this relationship during their investigations? In your answer, be sure to use examples from this investigation and at least one other investigation you have conducted.

7. Why is it useful to identify patterns during an investigation? In your answer, be sure to use examples from this investigation and at least one other investigation you have conducted.

LAB 2

Teacher Notes

Lab 2. Electric Fields and Electric Potential: How Does the Electric Potential Difference Change as You Move Away From the Positive Charge in an Electric Field?

Purpose

The purpose of this lab is to *introduce* students to the disciplinary core idea (DCI) of Types of Interactions (PS2.B) from the *NGSS* and the concept of electric potential difference by having them determine how the potential difference changes as one moves away from the positive source of an electric field. In addition, this lab can be used to help students understand two big ideas from AP Physics: (a) fields existing in space can be used to explain interactions and (b) the interactions of an object with other objects can be described by forces. This lab also gives students an opportunity to learn about the crosscutting concepts (CCs) of (a) Systems and System Models and (b) Energy and Matter: Flows, Cycles, and Conservation from the *NGSS*. As part of the explicit and reflective discussion, students will also learn about (a) how scientific knowledge changes over time and (b) how scientists use different methods to answer different types of questions.

Underlying Physics Concepts

Fields are an important concept in physics. All of the fundamental forces act on objects via a field. A field permeates space due to the presence of a body that interacts via a specific force. And, when an object feels a force, it is because the force is exerted on the object by the field. A gravitational field permeates space due to the presence of an object with mass. Other objects feel a gravitational force due to the field. For example, the gravitational force on the Moon from the Earth results from the Earth's gravitational field. The Earth, due to its mass, establishes a gravitational field. The gravitational field then exerts a force on the Moon. Similarly, an electric field permeates space due to the presence of a charged particle. A second charged particle will feel a force exerted by the electric field established by the first charged particle.

The electric field due to a point charge is directly proportional to the magnitude of the charge. The more charge at the point, the stronger the electric field. The strength of the electric field at a point in space is inversely proportional to the square of the distance of that point in space from the source of the charge. The strength of the electric field at a given point is shown mathematically in Equation 2.1, where $\mathbf{E}$ is the electric field, q is the electric charge creating the field, $\mathbf{r}$ is the distance from the charge, and k is Coulomb's constant. In SI units, electric field is measured in newtons per coulomb (N/C) or volts per meter (V/m), charge is measured in coulombs (C), distance is measured in meters (m), and Coulomb's constant has a value of $8.99 \times 10^9 \, \text{N} \cdot \text{m}^2/\text{C}^2$. The use of representing an electric field in volts per meter will become clear in the following paragraphs; 1 V/m = 1 N/C.

Electric Fields and Electric Potential

How Does the Electric Potential Difference Change as You Move Away From the Positive Charge in an Electric Field?

$$\text{(Equation 2.1)} \quad E = \frac{kq}{r^2}$$

When a second charge is placed into the electric field created by a first charge, it will experience a force due to the field, shown mathematically in Equation 2.2, where q_2 is the second charge placed in the electric field created by a first charge, **E** is the electric field established by the first charge, and **F** is the force acting on q_2 due to the electric field established by the first charge. In SI units, force is measured in newtons (N).

$$\text{(Equation 2.2)} \quad F = q_2 E$$

Combining Equations 2.1 and 2.2, we get Equation 2.3, which is Coulomb's law. Coulomb's law states that the magnitude of the electric force between two point charges is directly proportional to the magnitude of each charge and inversely proportional to the square of the distance between the two charges. From the perspective of the electric field, we can also say that each charge creates a field that the other charge interacts with. Furthermore, Newton's third law states that for every action, there is an equal and opposite reaction; in other words, if object 1 exerts a force on object 2, object 2 must exert an equal and opposite force on object 1. This holds true for the electric force. If charge 1 (q_1) exerts a force on charge 2 (q_2), then charge 2 must exert an equal and opposite force on charge 1.

$$\text{(Equation 2.3)} \quad F_E = \frac{kq_1 q_2}{r^2}$$

One advantage of the concept of fields is that it also makes it easy to think about the potential energy of a body in the field. Potential energy is the energy an object has due to its presence in a field. In mechanics, the gravitational potential energy (PE) of an object with mass (m) is related to the Earth's gravitational field (**g**) and its height above the ground (**h**, its position in the field), such that $PE = m\mathbf{gh}$. When an object at height **h** is released in Earth's gravitational field, work is done on the object, giving it kinetic energy. Similarly, the electric potential energy of a charged object in an electric field is directly proportional to the magnitude of the charged object, the strength of the electric field, and the position from the source of the electric field. This relationship is shown in Equation 2.4, where U_E is the electric potential energy, q is the magnitude of the charge in the electric field, **E** is the electric field strength at some distance from the source of the field, and **r** is the distance from the source of the field. In SI units, electric potential energy is measured in joules (J).

$$\text{(Equation 2.4)} \quad U_E = q\mathbf{Er}$$

As established in Equation 2.2, $q\mathbf{E}$ is equal to the force acting on charge q due to its presence in the electric field **E**. Thus, we can also express the electric potential energy in relationship to the force exerted on the charge by the field. This relationship is shown in Equation 2.5, where U_E is the potential energy, **F** is the electric force, and **r** is the distance from the source of the electric field. It is also important to point out that force times displacement is equal to work. And, according to the work-energy theorem, work is needed to change the type or amount of energy an object has. If we move a charge in an electric field from the source to a point **r** meters away using a force **F**, we will have done **Fr** joules of work, thereby giving the charge potential energy in the electric field.

$$\text{(Equation 2.5)} \quad U_E = \mathbf{Fr}$$

Because of the relationship between work and energy, it is also helpful to think about how charges move in electric fields. One question related to charges moving in electric fields is how the potential energy of a charge differs between two points in the field. This question is important because the answer allows us to make use of changes in the energy of the charge as it moves in the field (one example: a closed circuit connected to a battery is the movement of charge in an electric field). Thus, electric potential difference is a useful concept to physicists. The electric potential difference is the amount of work needed to move a test charge of 1 C between two points in an electric field. Equation 2.6 shows the formula to find the electric potential difference between two points in an electric field, where V is the electric potential difference, **E** is the electric field, and **r** is the distance from the source of the electric field. In SI units, electric potential difference is measured in volts (V), and 1 V = 1 J/C.

$$\text{(Equation 2.6)} \quad V = \mathbf{Er}$$

There are a few important things to note about the relationship expressed in Equation 2.6. First, voltage (V) is the difference in electric potential between two points separated by **r** meters, hence the name "electric potential difference." Second, we assume that the electric field is uniform between the two points in space. This is a valid assumption when talking about the space between two charged plates, such as in a capacitor, or the space between two point charges (as you move farther from one point, you move closer to the other point). Third, if we solve for the electric field, we will get voltage over meters, hence units for electric field can be expressed as volts per meter (V/m).

Finally, it is possible to also discuss electric potential difference in relationship to energy. Recall that 1 V = 1 J/C. Thus, we can think about the electric potential difference as the difference in potential energy per unit charge between two points in space. We can then represent the relationship between potential energy difference and electric potential difference using Equation 2.7, where U_{EA} is the potential energy of a charge at point A in an electric field, U_{EB} is the potential energy of a charge at point B in an electric field, q is the charge moving in the electric field, and V is the electric potential difference between point A and point B.

$$\text{(Equation 2.7)} \quad U_{EA} - U_{EB} = qV$$

Electric Fields and Electric Potential

How Does the Electric Potential Difference Change as You Move Away From the Positive Charge in an Electric Field?

The expression on the left side of Equation 2.7 ($U_{EA} - U_{EB}$) is also equal to the work done on the charge to move the charge from point A to point B. Hence, we can also think about electric potential difference as the work needed per unit charge to move a charge from point A to point B.

In this lab, students will set up an electric field using a 1.5V battery and conducting paper. You will attach the positive end of the battery to one metal thumbtack in the conducting paper and the negative end of the battery to the other metal thumbtack. The pins act as point charges with an electric field between them. Students will then use a voltmeter to measure the potential difference at various points on the paper. It is important to realize that this is the potential difference relative to the positive charge. In other words, the voltmeter will treat the negatively charged thumbtack as having no charge and the positively charged thumbtack as creating a field such that the potential difference as you move the voltage probe toward the positively charged thumbtack will tend toward the voltage of the battery. When you place the voltmeter on the positively charged thumbtack, you will get a reading of 1.5 V. If you move the voltage probe a few centimeters from the positively charged thumbtack to a point we will call point A, you will get a reading of less than 1.5 V. If we assume that the reading at point A was 1.25 V, then the electric potential difference between point A and a point infinitely far away from the positively charged thumbtack is 1.25 V (In reality, you will find that the voltage approaches zero as you move closer to the negatively charged thumbtack). This means it would require 1.25 J of work to move a charge of 1 C from infinitely far away to point A on the paper.

Timeline

The instructional time needed to complete this lab investigation is 200–280 minutes. Appendix 3 (p. 421) provides options for implementing this lab investigation over several class periods. Option E (280 minutes) should be used if students are unfamiliar with scientific writing, because this option provides extra instructional time for scaffolding the writing process. You can scaffold the writing process by modeling, providing examples, and providing hints as students write each section of the report. Option E can also be used if you are introducing students to the digital interface sensors and/or the data analysis software. Option F (200 minutes) should be used if students are familiar with scientific writing and have developed the skills needed to write an investigation report on their own. In option F, students complete stage 6 (writing the investigation report) and stage 8 (revising the investigation report) as homework.

Materials and Preparation

The materials needed to implement this investigation are listed in Table 2.1 (p. 58). The equipment can be purchased from a science supply company such as Flinn Scientific, PASCO, or Ward's Science.

LAB 2

TABLE 2.1
Materials list for Lab 2

Item	Quantity
Consumable	
D battery	1 per group
Equipment and other materials	
Safety glasses with side shields or safety goggles	1 per student
Battery holder	1 per group
Insulated copper wire	3 per group
Wire connectors	6 per group
Voltmeter or multimeter	1 per group
Metal thumbtacks	2 per group
Current conducting paper with conducting paint*	1 per group
Ruler	1 per group
Investigation Proposal C (optional)	1 per group
Whiteboard, 2' × 3'†	1 per group
Lab Handout	1 per student
Peer-review guide and teacher scoring rubric	1 per student
Checkout Questions	1 per student
Equipment for digital interface measurements (optional)	
Digital interface with USB or wireless connections	1 per group
Voltage measurement sensor	1 per group
Computer or tablet with appropriate data analysis software installed	1 per group

* In our pilot testing, we used the Electric Field Mapping kit from Flinn Scientific, which includes instructions on setting up the conducting paper and conducting paint. If you purchase equipment from other sources, we anticipate they will also include instructions for setting up the equivalent electric field.

† As an alternative, students can use computer and presentation software such as Microsoft PowerPoint or Apple Keynote to create their arguments.

We strongly suggest that you set up the electric field before students begin this lab. Most important, this includes applying the conducting paint to the conducting paper and letting it dry. Place the thumbtacks into the paint and paper in the appropriate location, and attach all the wires except the negative terminal of the battery to its corresponding thumbtack in the conducting paper. We also suggest trying to have the same color wires

Electric Fields and Electric Potential
How Does the Electric Potential Difference Change as You Move Away From the Positive Charge in an Electric Field?

for each component of the system across groups (e.g., use a green wire to connect the positive terminal of the battery to the positive thumbtack). This will allow you to give directions to the whole class in which students see you referencing the same color wire.

Be sure to use a set routine for distributing and collecting the materials during the lab investigation. One option is to set up the materials for each group at each group's lab station before class begins. This option works well when there is a dedicated section of the classroom for lab work and the materials are large and difficult to move. A second option is to have all the materials on a table or cart at a central location. You can then assign a member of each group to be the "materials manager." This individual is responsible for collecting all the materials his or her group needs from the table or cart during class and for returning all the materials at the end of the class. This option works well when the materials are small and easy to move (such as magnets, wire, and bulbs). It also makes it easy to inventory the materials at the end of the class before students leave for the day.

Safety Precautions and Laboratory Waste Disposal

Remind students to follow all normal lab safety rules. In addition, tell students to take the following safety precautions:

1. Wear sanitized safety glasses with side shields or goggles during lab setup, hands-on activity, and takedown.
2. Wire and other metals with electric current flowing through them may get hot. Use caution when handling components of a closed circuit.
3. Use caution when working with sharp objects (e.g., wires, tacks) because they can cut or puncture skin.
4. Never put consumables in their mouth.
5. Wash their hands with soap and water when they are done collecting the data.

Batteries and wire may be stored for future use. When batteries need replacing, dispose of old batteries according to manufacturer's recommendations.

Topics for the Explicit and Reflective Discussion

Reflecting on the Use of Core Ideas and Crosscutting Concepts During the Investigation

Teachers should begin the explicit and reflective discussion by asking students to discuss what they know about the core ideas from the *NGSS* and other ideas central to the study of physics that they used during the investigation. The following are some important concepts related to the DCI of Relationship Between Energy and Forces from the *NGSS* and to electric potential difference that students need to use to determine how the electric potential difference changes as you move from the positive charge:

LAB 2

- The magnitude of the electric force **F** exerted on an object with electric charge q by an electric field **E** is $F = qE$. The direction of the force is determined by the direction of the field (**E**) and the sign of the charge (q), with positively charged objects accelerating in the direction of the field and negatively charged objects accelerating in the direction opposite the field.
- Any object with a non-zero charge creates an electric field in space.
- The electric field is related to the electric potential difference via the equation $V =$ **Er.**
- Electric potential difference is the difference in electric potential energy between two points in space per unit charge due to an electric field. Electric potential difference is also equal to the work needed per unit charge to move a charge between two points.

To help students reflect on what they know about electric fields and electric potential difference, we recommend showing them two or three images using presentation software that help illustrate these important ideas. You can then ask the students the following questions to encourage them to share how they are thinking about these important concepts:

1. What do we see going on in this image?
2. Does anyone have anything else to add?
3. What might be going on that we can't see?
4. What are some things that we are not sure about here?

You can then encourage students to think about how CCs played a role in their investigation. There are at least two CCs that students need to use to determine how the electric potential difference changes as one moves away from a point charge: (a) Systems and System Models and (b) Energy and Matter: Flows, Cycles, and Conservation (see Appendix 2 [p. 417] for a brief description of these CCs). To help students reflect on what they know about these CCs, we recommend asking them the following questions:

1. How do scientists model electric fields?
2. How is the model of an electric field similar to other models you have used in physics?
3. Physicists use the term *electric potential difference* when talking about how charges move in space. What do we mean by *electric potential difference*?
4. Why is it important to know the value of the electric potential difference between two points?

You can then encourage the students to think about how they used all these different concepts to help answer the guiding question and why it is important to use these ideas to help justify their evidence for their final arguments. Be sure to remind your students

Electric Fields and Electric Potential
How Does the Electric Potential Difference Change as You Move Away From the Positive Charge in an Electric Field?

to explain why they included the evidence in their arguments and make the assumptions underlying their analysis and interpretation of the data explicit in order to provide an adequate justification of their evidence.

Reflecting on Ways to Design Better Investigations
It is important for students to reflect on the strengths and weaknesses of the investigation they designed during the explicit and reflective discussion. Students should therefore be encouraged to discuss ways to eliminate potential flaws, measurement errors, or sources of uncertainty in their investigations. To help students be more reflective about the design of their investigation and what they can do to make their investigations more rigorous in the future, you can ask them the following questions:

1. What were some of the strengths of the way you planned and carried out your investigation? In other words, what made it scientific?

2. What were some of the weaknesses of the way you planned and carried out your investigation? In other words, what made it less scientific?

3. What rules can we make, as a class, to ensure that our next investigation is more scientific?

Reflecting on the Nature of Scientific Knowledge and Scientific Inquiry
This investigation can be used to illustrate two important concepts related to the nature of scientific knowledge and the nature of scientific inquiry: (a) how scientific knowledge changes over time and (b) how scientists use different methods to answer different types of questions (see Appendix 2 [p. 417] for a brief description of these concepts). Be sure to review these concepts during and at the end of the explicit and reflective discussion. To help students think about these concepts in relation to what they did during the lab, you can ask them the following questions:

1. Scientific knowledge can and does change over time. Can you tell me why it changes?

2. Can you work with your group to come up with some examples of how scientific knowledge has changed over time? Be ready to share in a few minutes.

3. There is no universal step-by-step scientific method that all scientists follow. Why do you think there is no universal scientific method?

4. Think about what you did during this investigation. How would you describe the method you used to determine how the electric potential difference changes as one moves away from a point charge? Why would you call it that?

LAB 2

You can also use presentation software or other techniques to encourage your students to think about these concepts by showing examples of how our thinking about electricity and electric fields has changed over time and then asking students to discuss what they think led to those changes. You can also show one or more images of a "universal scientific method" that misrepresent the nature of scientific inquiry (see, e.g., *https://commons.wikimedia.org/wiki/File:The_Scientific_Method_as_an_Ongoing_Process.svg*) and ask students why each image is *not* a good representation of what scientists do to develop scientific knowledge. You can also ask students to suggest revisions to the image that would make it more consistent with the way scientists develop scientific knowledge. Be sure to remind your students that it is important for them to understand what counts as scientific knowledge and how that knowledge develops over time in order to be proficient in science.

Hints for Implementing the Lab

- Allowing students to design their own procedures for collecting data gives students an opportunity to try, to fail, and to learn from their mistakes. However, you can scaffold students as they develop their own procedure by having them fill out an investigation proposal. The proposals provide a way for you to offer students hints and suggestions without telling them how to do it. You can also check the proposals quickly during a class period. For this lab we suggest using Investigation Proposal C.

- Learn how to use the electric conducting paper and voltmeter (or multimeter) before the lab begins. It is important for you to know how to use the equipment so you can help students when technical issues arise.

- We strongly suggest that you set up the electric field before students begin this lab. This preparation includes applying the conducting paint to the conducting paper and letting it dry. See the "Materials and Preparation" section for more details.

- Allow the students to become familiar with the conducting paper and voltmeter (or multimeter) as part of the tool talk before they begin to design their investigation. Give them 5–10 minutes to examine the equipment and materials before they begin designing their investigations. This gives students a chance to see what they can and cannot do with the equipment.

- Because the resistance of the wire is minimal, the batteries will lose charge quickly. We recommend using fresh batteries each class period.

- Be sure to allow students to go back and re-collect data at the end of the argumentation session. Students often realize that they made numerous mistakes when they were collecting data as a result of their discussions during the argumentation session. The students, as a result, will want a chance to re-collect data, and the re-collection of data should be encouraged when time allows. This also offers an opportunity to discuss what scientists do when they realize a mistake is made inside the lab.

Electric Fields and Electric Potential
How Does the Electric Potential Difference Change as You Move Away From the Positive Charge in an Electric Field?

If students use digital interface measurement equipment and analysis

- We suggest allowing students to familiarize themselves with the data analysis software before they finalize the procedure for the investigation, especially if they have not used such software previously. This gives students an opportunity to learn how to work with the software and to improve the quality of the data they collect.
- Remind students to follow the user's guide to correctly connect any sensors to avoid damage to lab equipment.

Connections to Standards

Table 2.2 highlights how the investigation can be used to address specific performance expectations from the *NGSS;* learning objectives from AP Physics 1 and 2; learning objectives from AP Physics C: Electricity and Magnetism; *Common Core State Standards for English Language Arts (CCSS ELA);* and *Common Core State Standards for Mathematics (CCSS Mathematics).*

TABLE 2.2

Lab 2 alignment with standards

NGSS performance expectations	• HS-PS2-4. Use mathematical representations of Newton's Law of Gravitation and Coulomb's Law to describe and predict the gravitational and electrostatic forces between objects. • HS-PS3-5: Develop and use a model of two objects interacting through electric or magnetic fields to illustrate the forces between objects and the changes in energy of the objects due to the interaction.
AP Physics 1 and AP Physics 2 learning objectives	• 2.A.1: A vector field gives, as a function of position (and perhaps time), the value of a physical quantity that is described by a vector. • 2.C.1.2: Calculate any one of the variables—electric force, electric charge, and electric field—at a point given the values and sign or direction of the other two quantities. • 2.E.3.1: Apply mathematical routines to calculate the average value of the magnitude of the electric field in a region from a description of the electric potential in that region using the displacement along the line on which the difference in potential is evaluated. • 3.C.2.2: Connect the concepts of gravitational force and electric force to compare similarities and differences between the forces. • 5.A.2.1: Define open and closed systems for everyday situations and apply conservation concepts for energy, charge, and linear momentum to those situations. • 5.B.2.1: Calculate the expected behavior of a system using the object model (i.e., by ignoring changes in internal structure) to analyze a situation. Then, when the model fails, justify the use of conservation of energy principles to calculate the change in internal energy due to changes in internal structure because the object is actually a system.

Continued

Table 2.2. (continued)

AP Physics C: Electricity and Magnetism learning objectives	• CNV-1.A: Calculate the value of the electric potential in the vicinity of one or more point charges. • CNV-1.B: Mathematically represent the relationships between the electric charge, the difference in electric potential, and the work done (or electrostatic potential energy lost or gained) in moving a charge between two points in a known electric field. • CNV-1.D: Calculate the potential difference between two points in a uniform electric field and determine which point is at the higher potential. • CNV-1.G.a: Use the general relationship between electric field and electric potential to calculate the relationships between the magnitude of electric field or the potential difference as a function of position.
Literacy connections (*CCSS ELA*)	• *Reading:* Key ideas and details, craft and structure, integration of knowledge and ideas • *Writing:* Text types and purposes, production and distribution of writing, research to build and present knowledge, range of writing • *Speaking and listening:* Comprehension and collaboration, presentation of knowledge and ideas
Mathematics connections (*CCSS Mathematics*)	• *Mathematical practices:* Make sense of problems and persevere in solving them, reason abstractly and quantitatively, construct viable arguments and critique the reasoning of others, model with mathematics, use appropriate tools strategically, attend to precision • *Number and quantity:* Reason quantitatively and use units to solve problems, represent and model with vector quantities, perform operations on vectors • *Algebra:* Interpret the structure of expressions, create equations that describe numbers or relationships, understand solving equations as a process of reasoning and explain the reasoning, solve equations and inequalities in one variable, represent and solve equations and inequalities graphically. • *Functions:* Understand the concept of a function and use function notation; interpret functions that arise in applications in terms of the context; analyze functions using different representations; build a function that models a relationship between two quantities; construct and compare linear, quadratic, and exponential models and solve problems; interpret expressions for functions in terms of the situation they model • *Statistics and probability:* Summarize, represent, and interpret data on two categorical and quantitative variables; interpret linear models; make inferences and justify conclusions from sample surveys, experiments, and observational studies

Lab Handout

Lab 2. Electric Fields and Electric Potential: How Does the Electric Potential Difference Change as You Move Away From the Positive Charge in an Electric Field?

Introduction

The French philosopher and mathematician René Descartes (1596–1650) is widely known for his statement "Cogito, ergo sum" (translated from the Latin to English as "I think, therefore I am") and, within the field of mathematics, for the development of the Cartesian coordinate system (i.e., he invented the x-y coordinate system). He also did work in areas that we would now call physics, including on optics and forces. His ideas on forces included the notion that forces cannot act at a distance. Instead, according to Descartes, forces occur when two objects come in contact with one another (Hatfield 2018). While many of his ideas in philosophy and mathematics continue to be accepted today, his notion that forces cannot act at a distance was overturned by Isaac Newton (1642–1727). In Newton's seminal work, *Philosophiae Naturalis Principia Mathematica* (1687), one of the implications of the law of universal gravitation is that force can act at great distances. Subsequent work in physics showed that not only can the gravitational force act at a distance, but electric and magnetic forces can act at a distance as well.

To help understand how forces can act at a distance, physicists have introduced the concept of a field. All of the fundamental forces act on objects via a field. A field permeates space due to the presence of a body that interacts via a specific force. And, when an object feels a force, it is because the *force is exerted on the object by a field*. A gravitational field permeates space due to the presence of an object with mass. Other objects feel a gravitational force due to the field. As an example, the gravitational force on the Moon from the Earth results from the Earth's gravitational field. The Earth, due to its mass, establishes a gravitational field. The gravitational field then interacts with the Moon, producing the gravitational force. The Moon also creates its own gravitational field due to its mass, which the Earth interacts with, producing a gravitational force on the Earth from the Moon. Similarly, an electric field permeates space due to the presence of a charged particle. A second charged particle will feel a force exerted by the electric field established by a first charged particle.

Fields can also help us to understand the potential energy of objects placed in the field. Gravitational potential energy, for example, is the potential energy an object with mass has due to its position in a gravitational field. Similarly, electric potential energy is the potential energy a charged object has due to its position in an electric field. An electric field due to a positive point charge is directed out from the charge, and an electric field due to a negative point charge is directed in toward the charge. Physicists chose this convention because a positive field will cause a repulsive force on a positively charged object placed in the field. A negative field will cause an attractive force on a positively charged object in the field. Parts a and b of Figure L2.1 show an electric field due to a positive and negative charge, respectively.

When speaking about electric fields and electric potential energy, physicists often work with a quantity called electric potential difference, which is defined as the amount of work needed to move a charge of 1 coulomb (C) between two points. The units for electric potential difference are volts (V); 1 V is equal to 1 joule per coulomb (J/C). If there is an electric potential difference between two points equal to 1 V, then it takes 1 J of work to move a positive 1 C charge from the first point to the second point. We can find out how much work is done in total by multiplying the electric potential difference by the total amount of charge moved. Thus, electric potential difference is related to electric potential energy. Furthermore, we know that work is equal to force times displacement (**d**). If two points in an electric field are separated by a distance (*d*), we can determine the force needed to move a charge between the two points, because *W* = **Fd** (note that moving a charge between two points is a displacement with magnitude equal to the distance between the two points).

FIGURE L2.1

The electric field due to (a) a positive charge and (b) a negative charge

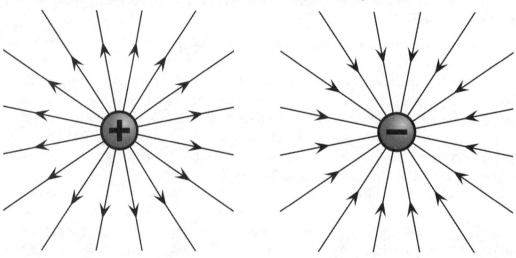

Many scientists are interested in understanding the movement of charges in an electric field because of our reliance on electric energy. To understand the movement of charges in an electric field, we must be able to explain and predict the electric potential difference between points in an electric field. In this investigation, you will have an opportunity to explore how the electric potential difference changes in an electric field.

Electric Fields and Electric Potential

How Does the Electric Potential Difference Change as You Move Away From the Positive Charge in an Electric Field?

Your Task

Use what you know about electric fields and electric potential difference, the relationship between work and energy, and systems and system models to design and carry out an investigation to determine how the electric potential difference changes due to the position in an electric field.

The guiding question of this investigation is, *How does the electric potential difference change as you move away from the positive charge in an electric field?*

Materials

You may use any of the following materials during your investigation:

Consumable
- D battery

Equipment
- Safety glasses with side shields or goggles (required)
- Battery holder
- Insulated copper wire
- Wire connectors
- Voltmeter or multimeter
- Metal thumbtacks
- Current conducting paper with conducting paint
- Ruler

If you have access to the following equipment, you may also consider using a digital voltage sensor with an accompanying interface and a computer or tablet.

Your teacher may have prepared the paper and placed the thumbtacks into the appropriate place on the paper.

Safety Precautions

Follow all normal lab safety rules. In addition, take the following safety precautions:

1. Wear sanitized safety glasses with side shields or goggles during lab setup, hands-on activity, and takedown.
2. Wire and other metals with electric current flowing through them may get hot. Use caution when handling components of a closed circuit.
3. Use caution when working with sharp objects (e.g., wires, tacks) because they can cut or puncture skin.
4. Never put consumables in your mouth.
5. Wash your hands with soap and water when you are done collecting the data.

LAB 2

Investigation Proposal Required?   ☐ Yes   ☐ No

Getting Started

To answer the guiding question, you will need to design and carry out an investigation to determine the relationship between the position in the electric field and the potential difference. Figure L2.2 illustrates how you can use the available equipment to set up the electric field and to measure the potential difference. The positively charged thumbtack will be considered the reference point. In other words, your measurement of the potential difference will be from any given point in the field to the positively charged thumbtack. Before you can design your investigation, however, you must determine what type of data you need to collect, how you will collect it, and how you will analyze it.

FIGURE L2.2
How to measure the electric potential difference in an electric field

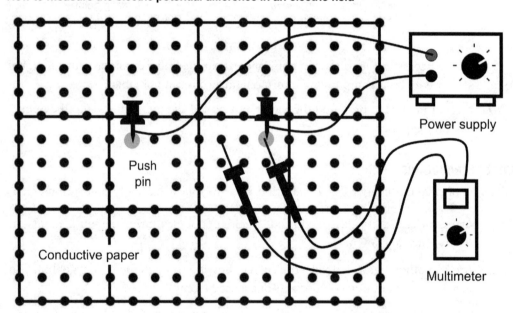

To determine *what type of data you need to collect,* think about the following questions:

- What are the boundaries and components of the system?
- How do the components of the system interact with each other?
- When is this system stable and under which conditions does it change?
- How could you keep track of changes in this system quantitatively?
- How will you determine the direction of the electric field?
- How useful is it to track how energy flows into, out of, or within this system?

Electric Fields and Electric Potential
How Does the Electric Potential Difference Change as You Move Away From the Positive Charge in an Electric Field?

- How do the components of the system interact with each other?

To determine *how you will collect the data,* think about the following questions:

- What is the independent variable and what is the dependent variable?
- What scale or scales should you use when you take your measurements?
- How will you make sure that your data are of high quality (i.e., how will you reduce error)?
- How will you keep track of and organize the data you collect?

To determine *how you will analyze the data,* think about the following questions:

- What type of calculations, if any, will you need to make?
- What types of patterns might you look for as you analyze your data?
- What type of table or graph could you create to help make sense of your data?
- Can you create a mathematical equation to describe the relationship between the electric potential difference and distance from the positive thumbtack?
- What units do any constants in your data have?

Connections to the Nature of Scientific Knowledge and Scientific Inquiry

As you work through your investigation, you may want to consider

- how scientific knowledge changes over time, and
- how scientists use different methods to answer different types of questions.

Initial Argument

Once your group has finished collecting and analyzing your data, your group will need to develop an initial argument. Your initial argument needs to include a claim, evidence to support your claim, and a justification of the evidence. The *claim* is your group's answer to the guiding question. The *evidence* is an analysis and interpretation of your data. Finally, the *justification* of the evidence is why your group thinks the evidence matters. The justification of the evidence is important because scientists can use different kinds of evidence to support their claims. Your group will create your initial argument on a whiteboard. Your whiteboard should include all the information shown in Figure L2.3.

FIGURE L2.3
Argument presentation on a whiteboard

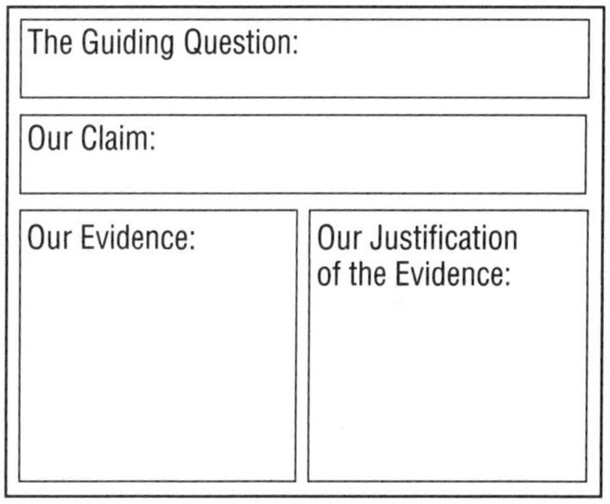

The Guiding Question:	
Our Claim:	
Our Evidence:	Our Justification of the Evidence:

LAB 2

Argumentation Session

The argumentation session allows all of the groups to share their arguments. One or two members of each group will stay at the lab station to share that group's argument, while the other members of the group go to the other lab stations to listen to and critique the other arguments. This is similar to what scientists do when they propose, support, evaluate, and refine new ideas during a poster session at a conference. If you are presenting your group's argument, your goal is to share your ideas and answer questions. You should also keep a record of the critiques and suggestions made by your classmates so you can use this feedback to make your initial argument stronger. You can keep track of specific critiques and suggestions for improvement that your classmates mention in the space below.

Critiques about our initial argument and suggestions for improvement:

If you are critiquing your classmates' arguments, your goal is to look for mistakes in their arguments and offer suggestions for improvement so these mistakes can be fixed. You should look for ways to make your initial argument stronger by looking for things that the other groups did well. You can keep track of interesting ideas that you see and hear during the argumentation in the space on the next page. You can also use this space to keep track of any questions that you will need to discuss with your team.

Electric Fields and Electric Potential
How Does the Electric Potential Difference Change as You Move Away From the Positive Charge in an Electric Field?

Interesting ideas from other groups or questions to take back to my group:

Once the argumentation session is complete, you will have a chance to meet with your group and revise your initial argument. Your group might need to gather more data or design a way to test one or more alternative claims as part of this process. Remember, your goal at this stage of the investigation is to develop the best argument possible.

Report

Once you have completed your research, you will need to prepare an *investigation report* that consists of three sections. Each section should provide an answer to the following questions:

1. What question were you trying to answer and why?
2. What did you do to answer your question and why?
3. What is your argument?

Your report should answer these questions in two pages or less. This report must be typed, and any diagrams, figures, or tables should be embedded into the document. Be sure to write in a persuasive style; you are trying to convince others that your claim is acceptable or valid!

References

Hatfield, G. 2018. René Descartes. In *The Stanford encyclopedia of philosophy*, ed. E. N. Zalta. *https://plato.stanford.edu/archives/sum2018/entries/descartes*.

Newton, I. 1687. *Philosophiae Naturalis Principia Mathematica* [Mathematical principles of natural philosophy]. London: S. Pepys.

LAB 2

Checkout Questions

Lab 2. Electric Fields and Electric Potential: How Does the Electric Potential Difference Change as You Move Away From the Positive Charge in an Electric Field?

Use the figure below to answer questions 1–3. The figure shows a point in space, A, that is a distance R from a positive charge.

1. The electric potential difference between the positive point charge and point A is 3 V. What is the potential difference when the distance from the positive point charge is 2R?

2. What is the potential difference when the distance from the positive charge to point A is R/2?

Electric Fields and Electric Potential
How Does the Electric Potential Difference Change as You Move Away From the Positive Charge in an Electric Field?

3. How much work is done moving a charge of $1.60217662 \times 10^{-19}$ C from point A to a point R/2 from the positive charge?

4. There is a single scientific method that all scientists follow when conducting investigations.

   a. I agree with this statement.
   b. I disagree with this statement.

   Explain your answer, using examples from this investigation and at least one other investigation you have conducted.

5. Scientists have always known about electric fields and electric potential.

   a. I agree with this statement.
   b. I disagree with this statement.

   Explain your answer, using an example from your investigation about electric potential.

LAB 2

6. Why do scientists track how the energy of objects within a system change over time? In your answer, be sure to use examples from this investigation and at least one other investigation you have conducted.

7. The concept of an electric field is a model of systems of charged particles. Why is it useful to use fields to model how charged particles interact? In your answer, be sure to use examples from this investigation and at least one other investigation you have conducted.

Application Lab

LAB 3

Teacher Notes

Lab 3. Electric Fields in Biotechnology: How Does Gel Electrophoresis Work?

Purpose

The purpose of this lab is for students to *apply* what they know about the disciplinary core ideas (DCIs) of Forces and Motion (PS2.A) and Types of Interactions (PS2.B) from the *NGSS* and about the concept of electric fields by having them generate a conceptual model to explain how gel electrophoresis works. In addition, this lab can be used to help students understand two big ideas from AP Physics: (a) fields existing in space can be used to explain interactions and (b) the interactions of an object with other objects can be described by forces. This lab also gives students an opportunity to learn about the crosscutting concepts (CCs) of (a) Cause and Effect: Mechanism and Explanation and (b) Stability and Change from the *NGSS*. As part of the explicit and reflective discussion, students will also learn about (a) how the culture of science, societal needs, and current events influence the work of scientists and (b) how scientists investigate questions about the natural or material world.

Underlying Physics Concepts

In this lab, students will run gel electrophoresis by applying an electric field to charged molecules, thereby causing the molecules to move toward either the cathode or the anode. When a charged object is placed in an electric field, the force on the object is directly proportional to both the charge of the object and the strength of the field at the location of the charged object. The mathematical relationship is shown in Equation 3.1, where **F** is the force on the charged object, q is the charge on the object, and **E** is the electric field strength at the location of the charged object. In SI units, force is measured in newtons (N), charge is measured in coulombs (C), and the electric field is measured in either newtons per coulomb or volts per meter (N/C or V/m). To see why the electric field can be measured in multiple units, see the Teacher Notes for Lab 2.

$$\text{(Equation 3.1)} \quad \mathbf{F} = q\mathbf{E}$$

The direction of the force depends on the direction of the electric field and if the object is positively or negatively charged. The direction of an electric field always points out or away from a positive charge and in or toward a negative charge. Thus, the direction of the force due to an electric field on a positive charge will be in the same direction as the electric field, and the direction on a negative charge will be in the opposite direction of the electric field. In this lab, the electric field direction is out from the cathode and in toward the anode.

Newton's second law states that the sum of the forces acting on an object causes that object to accelerate. Furthermore, the sum of the forces on the object is directly proportional to the acceleration of the object. Newton's second law is shown mathematically in Equation 3.2, where $\sum \mathbf{F}$ is the sum of the forces, or net force, acting on the object; m is the mass of the object; and $\mathbf{a}$ is the acceleration. In SI units, force is measured in newtons (N), mass is measured in kilograms (kg), and acceleration is measured in meters per second squared (m/s²).

(Equation 3.2) $\sum \mathbf{F} = m\mathbf{a}$

We can set the forces equal to each other from Equations 3.1 and 3.2 to get Equation 3.3.

(Equation 3.3) $m\mathbf{a} = q\mathbf{E}$

If we solve for the acceleration in Equation 3.3, we see that it is a function of the strength of the electric field, the charge on the object, and mass of the object. Acceleration is related to displacement via Equation 3.4, where $\mathbf{d}$ is the displacement of some object in a given unit of time, $\mathbf{v}_o$ is the initial velocity of the object, $\mathbf{a}$ is the acceleration of the object, and t is the time the object was moving.

(Equation 3.4) $\mathbf{d} = \mathbf{v}_o t + \frac{1}{2}\mathbf{a}t^2$

If the object starts from rest, the term $\mathbf{v}_o t$ is equal to zero, and we find that the displacement of the object is a function of the acceleration of the object and the time interval for which the force was acting on the object. We can also solve for the acceleration in Equation 3.4 and set it equal to the acceleration in Equation 3.3. Thus, we have Equation 3.5, which relates the displacement of the molecule in the gel to the charge on the molecule, the strength of the electric field, and the amount of time the field was applied to the gel.

(Equation 3.5) $\mathbf{d} = \dfrac{q\mathbf{E}t^2}{2m}$

In this lab, students should determine conceptually individual relationships between (1) the magnitude of the displacement and the electric field strength, (2) the magnitude of the displacement and the time the field is applied to the gel, and (3) the direction of the displacement and the direction of the electric field. That is, as they increase the electric field strength, specific molecules will move farther in any given time and for a specific electric field, the molecules will move farther the longer the field is applied. Finally, positive molecules will move toward the anode.

The last two factors that have an impact on the movement of the molecules are the magnitude of charge on the molecule and the mass of the molecule. It turns out that the relationship between the movement of the molecules and the magnitude of the charge and the mass, respectively, is a bit more complex, especially as it relates to macromolecules moving through the gel. This is because the gel exerts a drag force on the molecules in the opposite direction of the force due to the electric field. The drag force is a function

LAB 3

of the size of the molecule as well as the velocity of the molecule. Thus, as the molecule accelerates, the drag force will increase until the molecule reaches terminal velocity. For small molecules, such as the chemicals used in this lab and for DNA fragments used in forensic analysis, terminal velocity is reached rather quickly.

The size of the molecule is also a function of its mass (this is not necessarily true for all objects, but it is true for the chemicals in gel electrophoresis). Thus, the more massive the molecule, the larger the drag force. This means that the large molecules will reach terminal velocity before smaller molecules, and that the terminal velocity is inversely related to the mass and size of the molecule. That is, as the mass of the molecule increases, the terminal velocity of the molecule will decrease (because the drag force is larger). Thus, smaller molecules will travel farther in any given amount of time, because they will have a larger velocity.

As for the magnitude of the charge on the molecule, for DNA this is also a function of the size of the DNA fragment. DNA gets its charge from a negatively charged phosphate group that forms part of the backbone of the DNA molecule. The longer the fragments, the greater the charge on the molecule. All other things being equal, this would suggest that the larger molecules would experience a greater force. However, Equation 3.5 suggests that the mass and distance traveled have an inverse relationship, and the greater mass increases the drag force on the molecule.

If the mass and size of the molecule are held constant, then an increase in the charge on a molecule will lead to a greater displacement during a given time interval. However, because charge is a function of mass (for these chemicals), and because of the relationship of the mass to the displacement in Equation 3.5 and the increase drag force, the effect of increased charge is hard to experimentally verify in this lab.

Thus, in the lab, students may not determine the exact nature of the relationship between the mass of the molecules and displacement and the charge of the molecules and displacement. This is OK and is a good opportunity to reinforce that no single investigation can necessarily answer all of the questions we have about a physical system.

Timeline

The instructional time needed to complete this lab investigation is 220–280 minutes. Appendix 3 (p. 421) provides options for implementing this lab investigation over several class periods. Option A (280 minutes) should be used if students are unfamiliar with scientific writing, because this option provides extra instructional time for scaffolding the writing process. You can scaffold the writing process by modeling, providing examples, and providing hints as students write each section of the report. Option A can also be used if you think students need additional time to become familiar with the equipment. Option B (220 minutes) should be used if students are familiar with scientific writing and have developed the skills needed to write an investigation report on their own. In option B, students complete stage 6 (writing the investigation report) and stage 8 (revising the investigation report) as homework.

Electric Fields in Biotechnology
How Does Gel Electrophoresis Work?

Materials and Preparation

The materials needed to implement this investigation are listed in Table 3.1. The equipment can be purchased from a science supply company such as Carolina, Flinn Scientific, PASCO, or Ward's Science.

TABLE 3.1
Materials list for Lab 3

Item	Quantity
Consumables	
Agarose gel	Several per group
TBE running buffer	500 ml per group
Bromophenol blue (molar mass 669.96 g/mol; negative charge)	10 ml per group*
Crystal violet (molar mass 407.97 g/mol; positive charge)	10 ml per group*
Orange G (molar mass 452.38 g/mol; negative charge)	10 ml per group*
Methyl green (molar mass 458.47 g/mol; positive charge)	10 ml per group*
Xylene cyanol (molar mass 538.61 g/mol; negative charge)	10 ml per group*
Equipment and other materials	
Indirectly vented chemical-splash goggles, chemical-resistant nonlatex gloves and aprons	1 set per student
Gel electrophoresis chamber	1 per group
Power supply	1 per group
Micropipettes	1 per group
Ruler	1 per group
Other equipment (see text below table)	
Investigation Proposal A (optional)	1 per group
Whiteboard, 2' × 3'†	1 per group
Lab Handout	1 per student
Peer-review guide and teacher scoring rubric	1 per student
Checkout Questions	1 per student

* The students should put 10 µL of liquid in each chamber.

† As an alternative, students can use computer and presentation software such as Microsoft PowerPoint or Apple Keynote to create their arguments.

LAB 3

This is one of the more expensive labs in the book. We recommend consulting with the biology, anatomy and physiology, and/or forensic science teachers at your school to coordinate the use and/or purchase of the necessary equipment. Because these teachers are likely to have use for the gel electrophoresis equipment, they may already have the necessary equipment for you to use. If the equipment needs to be purchased, the ability to use the same equipment in multiple classes may allow you to split the cost between the physics course budget and the budgets of other science courses. Science lab supply companies, such as those mentioned at the beginning of this section, sell multiple types of gel electrophoresis equipment. Although they have slight differences, they can all be used to investigate the same concepts. To provide flexibility for teachers in purchasing, we have not committed to any specific set of equipment. The specific equipment does not matter; you should be able to do the lab with whatever the other teachers at your school have or want to purchase.

Be sure to use a set routine for distributing and collecting the materials during the lab investigation. One option is to set up the materials for each group at each group's lab station before class begins. This option works well when there is a dedicated section of the classroom for lab work and the materials are large and difficult to move. A second option is to have all the materials on a table or cart at a central location. You can then assign a member of each group to be the "materials manager." This individual is responsible for collecting all the materials his or her group needs from the table or cart during class and for returning all the materials at the end of the class. This option works well when the materials are small and easy to move (such as magnets, wire, and bulbs). It also makes it easy to inventory the materials at the end of the class before students leave for the day.

Safety Precautions and Laboratory Waste Disposal

Remind students to follow all normal lab safety rules. In addition, tell students to take the following safety precautions:

1. Review information about hazardous chemicals on safety data sheets and follow the safety precautions noted there.

2. Wear sanitized indirectly vented chemical-splash goggles and chemical-resistant nonlatex gloves and aprons during lab setup, hands-on activity, and takedown.

3. Never put consumables in their mouth.

4. Clean up any spilled liquids immediately to avoid a slip or fall hazard.

5. Handle all glassware with care.

6. Never return consumables to stock bottles.

7. Plug power cords only into ground fault circuit interrupter (GFCI)-protected circuits when using electrical equipment, and always keep away from water sources to prevent shock.

8. Immediately tell you if they spill hazardous chemicals on themselves, the work surface, or the floor.

9. Use safety equipment such as eyewash and shower stations if they spill hazardous chemicals on themselves.

10. Follow proper procedure for disposal of chemicals and solutions.

11. Wash their hands with soap and water when they are done collecting the data and disposing of chemicals and solutions.

Information about laboratory waste disposal methods is included in the Flinn catalog and reference manual; you can request a free copy at *www.flinnsci.com/flinn-freebies*.

Topics for the Explicit and Reflective Discussion

Reflecting on the Use of Core Ideas and Crosscutting Concepts During the Investigation

Teachers should begin the explicit and reflective discussion by asking students to discuss what they know about the core ideas from the *NGSS* and other ideas central to the study of physics that they used during the investigation. The following are some important concepts related to the core ideas of force and electric fields that students need to use to explain how gel electrophoresis works:

- A field associates a value of some physical quantity with every point in space. Fields are a model that physicists use to describe interactions that occur over a distance. Fields permeate space, and objects experience forces due to their interaction with a field.

- The relationship between the electric force on a charged object due to the object's location in an electric field is **F** = *q***E**. The magnitude of the force is determined by the charge on the object and the strength of the electric field. The direction of the force is determined by the direction of the field and the sign of the charge. The direction of the force due to an electric field on a positive charge will be in the same direction as the electric field, and the direction on a negative charge will be in the opposite direction of the electric field.

- Newton's second law states that the sum of the forces on an object cause the object to accelerate. Mathematically, Newton's second law is $\sum \mathbf{F} = m\mathbf{a}$.

- Other forces besides the force due to the electric field may be acting on an object in an electric field. In this lab, a drag force is exerted on the molecules by the gel.

LAB 3

To help students reflect on what they know about electric fields, we recommend showing them two or three images using presentation software that help illustrate these important ideas. You can then ask the students the following questions to encourage them to share how they are thinking about these important concepts:

1. What do we see going on in this image?
2. Does anyone have anything else to add?
3. What might be going on that we can't see?
4. What are some things that we are not sure about here?

You can then encourage students to think about how CCs played a role in their investigation. There are at least two CCs that students need to use to develop their explanatory model for how the gel electrophoresis works: (a) Cause and Effect: Mechanism and Explanation and (b) Stability and Change (see Appendix 2 [p. 417] for a brief description of these CCs). To help students reflect on what they know about these CCs, we recommend asking them the following questions:

1. Why is it important to identify cause-and-effect relationships in science?
2. What conditions need to be met for an investigation to show a cause-and-effect relationship?
3. Why is understanding factors that control rates of change important in designed systems?
4. What predictions are aided by understanding how systems change over time?

You can then encourage the students to think about how they used all these different concepts to help answer the guiding question and why it is important to use these ideas to help justify their evidence for their final arguments. Be sure to remind your students to explain why they included the evidence in their arguments and make the assumptions underlying their analysis and interpretation of the data explicit in order to provide an adequate justification of their evidence.

Reflecting on Ways to Design Better Investigations

It is important for students to reflect on the strengths and weaknesses of the investigation they designed during the explicit and reflective discussion. Students should therefore be encouraged to discuss ways to eliminate potential flaws, measurement errors, or sources of uncertainty in their investigations. To help students be more reflective about the design of their investigation and what they can do to make their investigations more rigorous in the future, you can ask them the following questions:

Electric Fields in Biotechnology
How Does Gel Electrophoresis Work?

1. What were some of the strengths of the way you planned and carried out your investigation? In other words, what made it scientific?

2. What were some of the weaknesses of the way you planned and carried out your investigation? In other words, what made it less scientific?

3. What rules can we make, as a class, to ensure that our next investigation is more scientific?

Reflecting on the Nature of Scientific Knowledge and Scientific Inquiry

This investigation can be used to illustrate two important concepts related to the nature of scientific knowledge and the nature of scientific inquiry: (a) how the culture of science, societal needs, and current events influence the work of scientists and (2) how scientists investigate questions about the natural or material world (see Appendix 2 [p. 417] for a brief description of these concepts). Be sure to review these concepts during and at the end of the explicit and reflective discussion. To help students think about these concepts in relation to what they did during the lab, you can ask them the following questions:

1. People view some types of research as being more important than other types of research because of cultural values and current events. Can you come up with some examples of how cultural values and current events have influenced the work of scientists?

2. Scientists share a set of values, norms, and commitments that shape what counts as knowing, how to represent or communicate information, and how to interact with other scientists. Can you work with your group to come up with a rule that you can use to decide if something is science or not science? Be ready to share in a few minutes.

3. Not all questions can be answered by science. Can you give me some examples of questions related to this investigation that can and cannot be answered by science?

4. Can you work with your group to come up with a rule that you can use to decide if a question can or cannot be answered by science?

You can also use presentation software or other techniques to encourage your students to think about these concepts. You can show examples of research projects that were influenced by cultural values and current events and ask students to think about what was going on in society when the research was conducted and why that research was viewed as being important for the greater good. You can show one or more examples of questions that can be answered by science (e.g., How does the mass of the molecule influence how far it moves in the gel?) and cannot be answered by science (e.g., Is a person guilty of a crime?) and then ask students why each example is or is not a question that can be answered by science. Be sure to remind your students that it is important for them to understand what counts as scientific knowledge and how that knowledge develops over time in order to be proficient in science.

Hints for Implementing the Lab

- Allowing students to design their own procedures for collecting data gives students an opportunity to try, to fail, and to learn from their mistakes. However, you can scaffold students as they develop their own procedure by having them fill out an investigation proposal. The proposals provide a way for you to offer students hints and suggestions without telling them how to do it. You can also check the proposals quickly during a class period. For this lab we suggest using Investigation Proposal A.

- Learn how to run gel electrophoresis before the lab begins. It is important for you to know how to use the equipment so you can help students when technical issues arise.

- Allow the students to become familiar with gel electrophoresis as part of the tool talk before they begin to design their investigation. Give them 5–10 minutes to examine the equipment and materials before they begin designing their investigations. This gives students a chance to see what they can and cannot do with the equipment.

- Students should use the same concentration of agarose in each gel they use because the concentration of agarose will affect the movement of the dyes. This is especially the case if students conduct multiple trials. As the gel exerts a drag force on the molecules, difference in gel type and consistency will change the drag force on the molecule.

- It can take extended time to run each gel. We suggest starting the gel at the beginning of class and then working on other assignments during class. Students from each group can periodically check on the gel to see if it has run to completion. If students want to conduct multiple trials, they may need to use multiple class periods.

- Be sure to allow students to go back and re-collect data at the end of the argumentation session. Students often realize that they made numerous mistakes when they were collecting data as a result of their discussions during the argumentation session. The students, as a result, will want a chance to re-collect data, and the re-collection of data should be encouraged when time allows. This also offers an opportunity to discuss what scientists do when they realize a mistake is made inside the lab.

Connections to Standards

Table 3.2 highlights how the investigation can be used to address specific performance expectations from the *NGSS*; learning objectives from AP Physics 1 and 2; learning objectives from AP Physics C: Electricity and Magnetism; *Common Core State Standards for English Language Arts* (*CCSS ELA*); and *Common Core State Standards for Mathematics* (*CCSS Mathematics*).

Electric Fields in Biotechnology
How Does Gel Electrophoresis Work?

TABLE 3.2

Lab 3 alignment with standards

NGSS performance expectation	• HS-PS3-5: Develop and use a model of two objects interacting through electric or magnetic fields to illustrate the forces between objects and the changes in energy of the objects due to the interaction.
AP Physics 1 and AP Physics 2 learning objectives	• 2.C.1.1: Predict the direction and the magnitude of the force exerted on an object with an electric charge q placed in an electric field E using the mathematical model of the relation between an electric force and an electric field: $\mathbf{F} = q\mathbf{E}$. • 2.C.5.3: Represent the motion of an electrically charged particle in the uniform field between two oppositely charged plates, and express the connection of this motion to projectile motion of an object with mass in Earth's gravitational field. • 3.A.1.1: Express the motion of an object using narrative, mathematical, and graphical representations. • 3.A.1.2: Design an experimental investigation of the motion of an object. • 3.A.1.3: Analyze experimental data describing the motion of an object and be able to express the results of the analysis using narrative, mathematical, and graphical representations. • 3.A.3.4: Make claims about the force on an object due to the presence of other objects with the same property: mass, electric charge. • 3.B.1.1: Predict the motion of an object subject to forces exerted by several objects using an application of Newton's second law in a variety of physical situations with acceleration in one dimension. • 3.B.1.4: Predict the motion of an object subject to forces exerted by several objects using an application of Newton's second law in a variety of physical situations.
AP Physics C: Electricity and Magnetism learning objectives	• ACT-1.A: Describe behavior of charges or system of charged objects interacting with each other. • ACT-1.D: Determine the motion of a charged object of specified charge and mass under the influence of an electrostatic force. • FIE-1.A: Using the definition of electric field, unknown quantities (such as charge, force, field, and direction of field) can be calculated in an electrostatic system of a point charge or an object with a charge in a specified electric field. • FIE-1.F: Determine the qualitative nature of the motion of a charged particle of specified charge and mass placed in a uniform electric field. • FIE-1.G: Sketch the trajectory of a known charged particle placed in a known uniform electric field.

Continued

LAB 3

Table 3.2 (*continued*)

Literacy connections (*CCSS ELA*)	• *Reading:* Key ideas and details, craft and structure, integration of knowledge and ideas • *Writing:* Text types and purposes, production and distribution of writing, research to build and present knowledge, range of writing • *Speaking and listening:* Comprehension and collaboration, presentation of knowledge and ideas
Mathematics connections (*CCSS Mathematics*)	• *Mathematical practices:* Make sense of problems and persevere in solving them, reason abstractly and quantitatively, construct viable arguments and critique the reasoning of others, model with mathematics, use appropriate tools strategically, attend to precision • *Number and quantity:* Reason quantitatively and use units to solve problems, represent and model with vector quantities, perform operations on vectors • *Algebra:* Interpret the structure of expressions, understand solving equations as a process of reasoning and explain the reasoning, solve equations and inequalities in one variable, represent and solve equations and inequalities graphically • *Functions:* Understand the concept of a function and use function notation; interpret functions that arise in applications in terms of the context; analyze functions using different representations; construct and compare linear, quadratic, and exponential models and solve problems; interpret expressions for functions in terms of the situation they model • *Statistics and probability:* Summarize, represent, and interpret data on two categorical and quantitative variables; interpret linear models; make inferences and justify conclusions from sample surveys, experiments, and observational studies

Lab Handout

Lab 3. Electric Fields in Biotechnology: How Does Gel Electrophoresis Work?

Introduction

The 20th century saw a proliferation of techniques, collectively referred to as forensic science, that apply scientific understandings to settle legal disputes. Often, these techniques are used in criminal proceedings by police officers, prosecutors, and defense attorneys to determine guilt or to exonerate a suspect. One prominent technique is to use DNA collected from a crime scene to determine if a suspect was at the scene of the crime. To do this, forensic scientists create a DNA profile from DNA collected at the scene of the crime and match it to a DNA profile of suspects. Figure L3.1 shows an example of a DNA profile for a forensic analysis. The strip on the left shows the profile for the DNA recovered from a crime scene. The strips on the right show DNA for three suspects. In the figure, it appears that suspect 2 was present at the crime scene. It is important to note that the DNA profile cannot prove that suspect 2 committed the crime. Instead, it rules out suspects 1 and 3. When combined with other evidence, the DNA profile can help build the case that suspect 2 committed a crime.

FIGURE L3.1

An example of a DNA profile in a forensic analysis during a criminal investigation

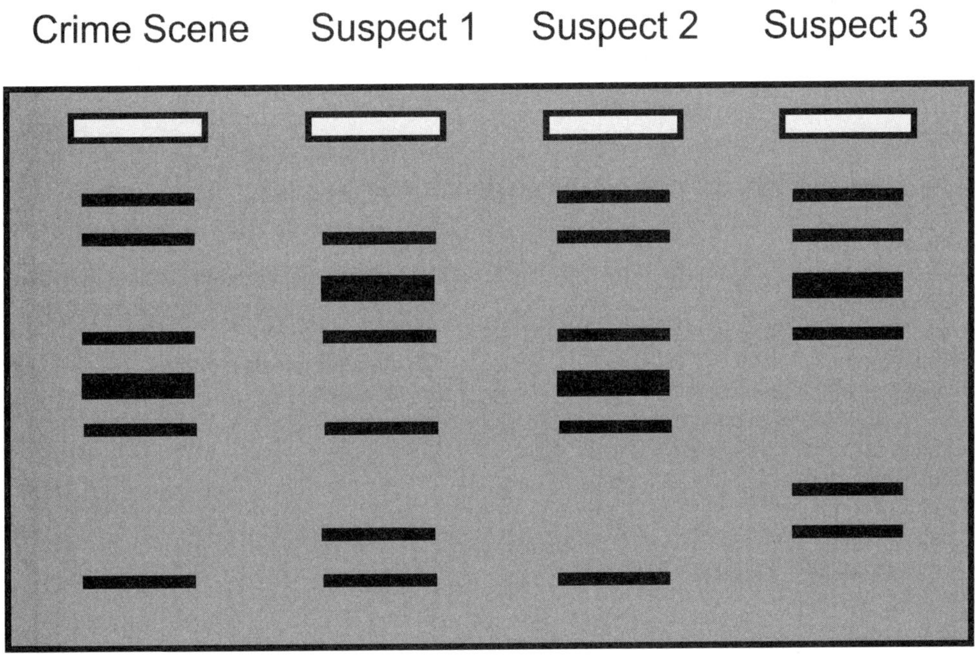

LAB 3

To create a DNA profile, forensic scientists use a technique called *gel electrophoresis*. In this process, scientists mix the DNA with an enzyme that will break the DNA molecule into smaller pieces, called DNA fragments. The DNA fragments are placed into a gel made of a specific type of polysaccharide (a complex carbohydrate molecule) that will allow the DNA fragments to move through the gel. Finally, the gel is placed into a gel chamber connected to a power supply. When the power supply is on, an *electric field* is established in the gel. The DNA molecules then move, and a DNA profile can be established. Each band in the DNA profile indicates the presence of DNA fragments at that point in the gel.

Gel electrophoresis is an important technique for analyzing other molecules besides DNA. Gel electrophoresis can also be used on RNA and proteins. In order to use a gel electrophoresis as part of the repertoire of forensic science techniques, it is important to know the precise mechanisms underlying this technique. Furthermore, different types of molecules behave differently when included in a gel electrophoresis. For example, all DNA will move toward the cathode. Proteins, however, can move toward either the cathode or the anode, depending on the specific protein. This knowledge is important for establishing the validity of any claims made using the results of the DNA profile.

Your Task

Use what you know about electric fields, forces, cause and effect, and rates of change in systems to design and carry out an investigation to develop a conceptual model that explains how a gel electrophoresis works. Your model should explain what factors influence the movement of DNA fragments and why you get bands of DNA fragments at specific points in the gel. For safety reasons, you will be using five chemicals as replacements for DNA. These chemicals are not reactive, so they can be mixed together if you choose to do so.

The guiding question of this investigation is, *How does gel electrophoresis work?*

Materials

You may use any of the following materials during your investigation:

Consumables
- Agarose gel
- TBE running buffer
- Bromophenol blue (molar mass 669.96 g/mol; negative charge)
- Crystal violet (molar mass 407.97 g/mol; positive charge)
- Orange G (molar mass 452.38 g/mol; negative charge)
- Methyl green (molar mass 458.47 g/mol; positive charge)
- Xylene cyanol (molar mass 538.61 g/mol; negative charge)

Equipment
- Indirectly vented chemical-splash goggles, chemical-resistant nonlatex gloves and aprons (required)
- Gel electrophoresis chamber
- Power supply
- Micropipettes
- Ruler

Electric Fields in Biotechnology
How Does Gel Electrophoresis Work?

Safety Precautions

Follow all normal lab safety rules. In addition, take the following safety precautions:

1. Review information about hazardous chemicals on safety data sheets and follow the safety precautions noted there.
2. Wear sanitized indirectly vented chemical-splash goggles and chemical-resistant nonlatex gloves and aprons during lab setup, hands-on activity, and takedown.
3. Never put consumables in your mouth.
4. Clean up any spilled liquids immediately to avoid a slip or fall hazard.
5. Handle all glassware with care.
6. Never return the consumables to stock bottles.
7. Plug power cords only into ground fault circuit interrupter (GFCI)–protected circuits when using electrical equipment, and always keep away from water sources to prevent shock.
8. Tell your teacher immediately if you spill chemicals on yourself, the work surface, or the floor.
9. Use safety equipment such as eyewash and shower stations if you spill hazardous chemicals on yourself.
10. Follow proper procedure for disposal of chemicals and solutions.
11. Wash your hands with soap and water when you are done collecting the data and disposing of chemicals and solutions.

Investigation Proposal Required?   ☐ Yes   ☐ No

Getting Started

To answer the guiding question and develop your conceptual model, you will need to design and carry out an investigation to determine how gel electrophoresis works. Figures L3.2 and L3.3 (p. 90) illustrate how you can use the available equipment to set up and run gel electrophoresis. First, lower the gel into the electrophoresis chamber. Then, fill the remainder of the chamber with the buffer solution by pouring the buffer into one side of the chamber until the solution is level with the top of the gel. Then, using a micropipette, add 10 µL of chemical into each well. Be careful not to overfill the well, and do not pierce the sides or bottom of the well with the micropipette. You can hook the positive voltage to the cathode and the negative voltage to the anode. Place the different chemicals in each well, and turn the power source on.

LAB 3

FIGURE L3.2
Top view of gel electrophoresis

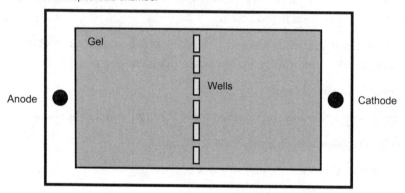

FIGURE L3.3
Side view of gel electrophoresis

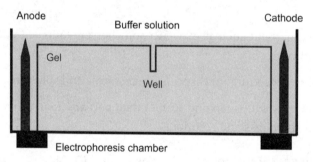

Before you can design your investigation, however, you must determine what type of data you need to collect, how you will collect it, and how you will analyze it.

To determine *what type of data you need to collect*, think about the following questions:

- What are the boundaries and components of the system?
- How do the components of the system interact with each other?
- When is this system stable and under which conditions does it change?
- How could you keep track of changes in this system quantitatively?
- What is going on at the unobservable level that could cause the things that you observe?
- Which factor(s) might control the rate of change in this system?

Electric Fields in Biotechnology
How Does Gel Electrophoresis Work?

To determine *how you will collect the data,* think about the following questions:

- What is the independent variable and what is the dependent variable?
- What other factors will you need to control during each experiment?
- What scale or scales should you use when you take your measurements?
- How will you make sure that your data are of high quality (i.e., how will you reduce error)?
- How will you keep track of and organize the data you collect?
- What type of research design needs to be used to establish a cause-and-effect relationship?
- How will you measure change over time during your investigation?

To determine *how you will analyze the data,* think about the following questions:

- What type of calculations, if any, will you need to make?
- What types of patterns might you look for as you analyze your data?
- What type of table or graph could you create to help make sense of your data?
- How could you use mathematics to describe a change over time?

Connections to the Nature of Scientific Knowledge and Scientific Inquiry

As you work through your investigation, you may want to consider

- how the culture of science, societal needs, and current events influence the work of scientists; and
- how scientists investigate questions about the natural or material world.

Initial Argument

Once your group has finished collecting and analyzing your data, your group will need to develop an initial argument. Your initial argument needs to include a claim, evidence to support your claim, and a justification of the evidence. The *claim* is your group's answer to the guiding question. The *evidence* is an analysis and interpretation of your data. Finally, the *justification* of the evidence is why your group thinks the evidence matters. The justification of the evidence is important because scientists can use different kinds of evidence to support their claims. Your group will create your initial argument on a whiteboard. Your whiteboard should include all the information shown in Figure L3.4.

FIGURE L3.4
Argument presentation on a whiteboard

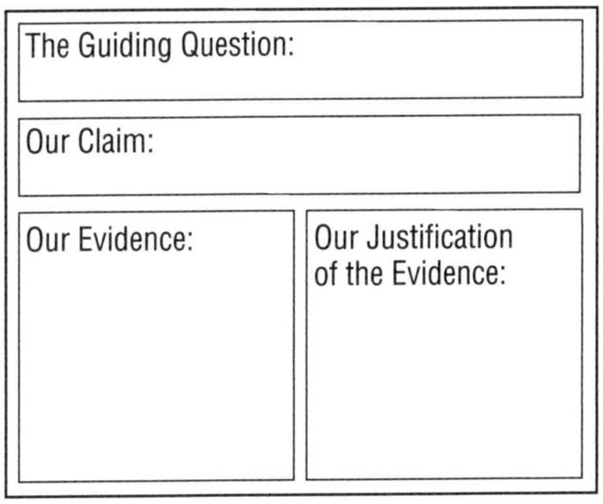

The Guiding Question:	
Our Claim:	
Our Evidence:	Our Justification of the Evidence:

LAB 3

Argumentation Session

The argumentation session allows all of the groups to share their arguments. One or two members of each group will stay at the lab station to share that group's argument, while the other members of the group go to the other lab stations to listen to and critique the other arguments. This is similar to what scientists do when they propose, support, evaluate, and refine new ideas during a poster session at a conference. If you are presenting your group's argument, your goal is to share your ideas and answer questions. You should also keep a record of the critiques and suggestions made by your classmates so you can use this feedback to make your initial argument stronger. You can keep track of specific critiques and suggestions for improvement that your classmates mention in the space below.

Critiques about our initial argument and suggestions for improvement:

If you are critiquing your classmates' arguments, your goal is to look for mistakes in their arguments and offer suggestions for improvement so these mistakes can be fixed. You should look for ways to make your initial argument stronger by looking for things that the other groups did well. You can keep track of interesting ideas that you see and hear during the argumentation in the space below. You can also use this space to keep track of any questions that you will need to discuss with your team.

Interesting ideas from other groups or questions to take back to my group:

Once the argumentation session is complete, you will have a chance to meet with your group and revise your initial argument. Your group might need to gather more data or design a way to test one or more alternative claims as part of this process. Remember, your goal at this stage of the investigation is to develop the best argument possible.

Report

Once you have completed your research, you will need to prepare an *investigation report* that consists of three sections. Each section should provide an answer to the following questions:

1. What question were you trying to answer and why?
2. What did you do to answer your question and why?
3. What is your argument?

Your report should answer these questions in two pages or less. This report must be typed, and any diagrams, figures, or tables should be embedded into the document. Be sure to write in a persuasive style; you are trying to convince others that your claim is acceptable or valid!

Checkout Questions

Lab 3. Electric Fields in Biotechnology: How Does Gel Electrophoresis Work?

1. A lab technician runs a gel electrophoresis for five minutes. After five minutes, two bands are separated by 4 cm.

    a. If the technician were to run the gel electrophoresis for an additional five minutes, how far apart would the bands be from each other?

    b. How do you know?

2. Why do some of the bands in the gel electrophoresis move toward the cathode while other bands move toward the anode?

3. There are no questions that science is unable to answer, provided the scientist has the correct set of equipment.

    a. I agree with this statement.
    b. I disagree with this statement.

    Explain your answer, using examples from this investigation and at least one other investigation you have conducted.

4. What questions are important to answer in science is sometimes influenced by cultural values and current events.

   a. I agree with this statement.
   b. I disagree with this statement.

   Explain your answer, using examples from this investigation and at least one other investigation you have conducted.

5. Why is it useful to understand the factors that control rates of change during an investigation? In your answer, be sure to use examples from this investigation and at least one other investigation you have conducted.

6. Why is it useful to identify cause-and-effect relationships during an investigation? In your answer, be sure to use examples from this investigation and at least one other investigation you have conducted.

SECTION 3

Energy

Electric Current, Capacitors, Resistors, and Circuits

Introduction Labs

LAB 4

Teacher Notes

Lab 4. Capacitance, Potential Difference, and Charge: What Is the Mathematical Relationship Between the Potential Difference Used to Charge a Capacitor and the Amount of Charge Stored?

Purpose

The purpose of this lab is to *introduce* students to the disciplinary core idea (DCI) of Conservation of Energy and Energy Transfer (PS3.B) from the *NGSS* and to capacitors by having them discover a mathematical relationship between the voltage source and the charge stored in a capacitor. In addition, this lab can be used to help students understand the following big ideas from AP Physics: (a) objects and systems have properties such as mass and charge, and systems may have internal structure; (b) interactions between systems can result in changes in those systems. This lab also gives students an opportunity to learn about the crosscutting concepts (CCs) of (a) Cause and Effect: Mechanism and Explanation and (b) Systems and System Models from the *NGSS*. As part of the explicit and reflective discussion, students will also learn about (a) the difference between observations and inferences in science and (b) how scientists use different methods to answer different types of questions.

Underlying Physics Concepts

Capacitors are ubiquitous elements in modern electronics. Their principal function is to store electrical charge. Alternatively, one can treat these devices as storage elements for electric potential energy. Whether working with a parallel-plate capacitor, a spherical capacitor, or a cylindrical capacitor, one can define the capacitance of a capacitor with Equation 4.1, where C is capacitance, Q is the magnitude of the charge on one of the plates (they are charged to equal and opposite charges), and ΔV is electric potential difference (voltage). In SI units, capacitance is measured in farads (F), electric charge is measured in coulombs (C), and electric potential difference is measured in volts (V).

$$\text{(Equation 4.1)} \quad C = \frac{Q}{\Delta V}$$

We can rearrange this relationship to find the total, maximum charge stored on a capacitor with Equation 4.2.

$$\text{(Equation 4.2)} \quad Q = C\Delta V$$

Capacitance, Potential Difference, and Charge
What Is the Mathematical Relationship Between the Potential Difference Used to Charge a Capacitor and the Amount of Charge Stored?

Since capacitance is a geometric quantity—that is, its value only depends on the dimensions of the "plates," their separation, and any dielectric material inserted between the plates—this value is constant for a given capacitor in Equation 4.2. One of the key ideas guiding this investigation, therefore, is the direct relationship between the maximum amount of charge stored in a capacitor and the applied potential difference.

Moreover, when a charged capacitor is allowed to discharge through a simple RC circuit (a circuit containing both a resistor and capacitor), it can be shown through standard ordinary differential equation (ODE) analysis that the current through the circuit as a function of time follows the exponential relationship in Equation 4.3. Here, I denotes electric current, ΔV_0 represents the initial potential difference across the capacitor plates at time zero, R is electrical resistance, C is capacitance, e is the base of the natural logarithm and has an approximate value of 2.71828 (note that e is irrational, and can only be approximated in a decimal representation), and t is an instant of time after the circuit is closed. In SI units, electric current is measured in amperes (A), electrical resistance is measured in ohms (Ω) and time is measured in seconds (s).

$$\text{(Equation 4.3)} \quad I(t) = \frac{\Delta V_0}{R} e^{-\frac{t}{RC}}$$

When this function is evaluated at $t = 0$, we find that the current is equal to the quantity $\frac{V_0}{R}$ which is the maximum current (I_{max}). Also note that at $t = 0$, the equation reduces to Ohm's law. The quantity RC is often referred to as the *RC time constant* or simply the *e-folding constant* of the circuit. This is because evaluating the exponential function when the time elapsed equals this quantity results in a current that is exactly 1 factor of e less than the maximum current. Likewise, when $t = 2RC$, the current drops by another factor of e and so on. So, we can rewrite Equation 4.3 in terms of these new quantities with Equation 4.4. The new variable, τ, is equal to the e-folding constant (RC). In SI units, τ has units of ohm-farads ($\Omega \cdot F$), which reduce simply to units of seconds (s).

$$\text{(Equation 4.4)} \quad I(t) = I_{max} e^{-\frac{t}{\tau}}$$

Next, assume that a capacitor is fully charged to a steady initial potential difference of ΔV_0. Then, we can use the fundamental relationship in Equation 4.2 to calculate the charge stored on each of the plates (Q_0).

$$\text{(Equation 4.5)} \quad \Delta V_0 = \frac{Q_0}{C}$$

A direct comparison of Equations 4.3, 4.4, and 4.5 reveals the proportionality that is at the heart of this inquiry.

$$\text{(Equation 4.6)} \quad I_{max} = \frac{1}{R} \Delta V_0 = \frac{1}{RC} Q_0$$

LAB 4

We also know that current is the flow of charge per unit time. The greater the current, the larger the amount of charge that is flowing in a given time interval. As can be seen in Equation 4.6, I_{max} is directly proportional to the total charge stored on a capacitor. Thus, the greater the I_{max} at time zero, the greater the initial charge on the capacitor. In this lab, students will charge the capacitor using different combinations of batteries to produce different voltage sources. After allowing the capacitor to fully charge, students will allow the capacitor to discharge and will measure the maximum current produced during the discharge of the capacitor. From there, they can infer the amount of charge stored on the capacitor, because of the relationship shown in Equation 4.6.

Timeline

The instructional time needed to complete this lab investigation is 220–280 minutes. Appendix 3 (p. 421) provides options for implementing this lab investigation over several class periods. Option A (280 minutes) should be used if students are unfamiliar with scientific writing, because this option provides extra instructional time for scaffolding the writing process. You can scaffold the writing process by modeling, providing examples, and providing hints as students write each section of the report. Option A can also be used if you are introducing students to the digital interface sensors and/or the data analysis software. Option B (220 minutes) should be used if students are familiar with scientific writing and have developed the skills needed to write an investigation report on their own. In option B, students complete stage 6 (writing the investigation report) and stage 8 (revising the investigation report) as homework.

Materials and Preparation

The materials needed to implement this investigation are listed in Table 4.1. The equipment can be purchased from a science supply company such as Flinn Scientific, PASCO, Vernier, or Ward's Science; some items are also available from Amazon.

Capacitance, Potential Difference, and Charge

What Is the Mathematical Relationship Between the Potential Difference Used to Charge a Capacitor and the Amount of Charge Stored?

TABLE 4.1

Materials list for Lab 4

Item	Quantity
Consumables	
AA alkaline batteries	4 per group
9V alkaline batteries	1 per group
Equipment and other materials	
Safety glasses with side shields or safety goggles	1 per student
AA four-battery holder	1 per group
Insulated alligator clip test leads	4 per group
Digital ammeter/multimeter with test leads	1 per group
Mini prototype breadboard (2.2" x 3.4") [5.5 cm x 8.5 cm]	1 per group
Male/male jumper wires	10 per group
Resistor	1 per group
Electrolytic capacitor (3300–6800 recommended)	1 per group
Investigation Proposal C (optional)	1 per group
Whiteboard, 2' × 3'*	1 per group
Lab Handout	1 per student
Peer-review guide and teacher scoring rubric	1 per student
Checkout Questions	1 per student
Equipment for digital interface measurements (optional)	
Digital interface with USB or wireless connections	1 per group
Current measurement sensor	1 per group
Computer or tablet with appropriate data analysis software installed	1 per group

* As an alternative, students can use computer and presentation software such as Microsoft PowerPoint or Apple Keynote to create their arguments.

Be sure to use a set routine for distributing and collecting the materials during the lab investigation. One option is to set up the materials for each group at each group's lab station before class begins. This option works well when there is a dedicated section of the classroom for lab work and the materials are large and difficult to move. A second option is to have all the materials on a table or cart at a central location. You can then assign a member of each group to be the "materials manager." This individual is responsible for

LAB 4

collecting all the materials his or her group needs from the table or cart during class and for returning all the materials at the end of the class. This option works well when the materials are small and easy to move (such as magnets, wire, and bulbs). It also makes it easy to inventory the materials at the end of the class before students leave for the day.

Safety Precautions and Laboratory Waste Disposal

Remind students to follow all normal lab safety rules. In addition, tell students to take the following safety precautions:

1. Wear sanitized safety glasses with side shields or goggles during lab setup, hands-on activity, and takedown.

2. Never put consumables in their mouth.

3. Wire and other metals with electric current flowing through them may get hot. Use caution when handling components of a closed circuit.

4. Handle electrical wires, including multimeter test leads, with caution. They have sharp ends, which can cut or puncture skin.

5. Never discharge capacitors by shorting them with the fingers or any other body part.

6. Never charge capacitors beyond the potential differences they are rated for (see label printed on capacitor), because they may explode.

7. Wash their hands with soap and water when they are done collecting the data.

Batteries and wire may be stored for future use. When batteries need replacing, dispose of old batteries according to manufacturer's recommendations.

Topics for the Explicit and Reflective Discussion

Reflecting on the Use of Core Ideas and Crosscutting Concepts During the Investigation

Teachers should begin the explicit and reflective discussion by asking students to discuss what they know about the core ideas they used during the investigation. The following are some important concepts related to the core idea of capacitors that students need to use to justify their evidence:

- Capacitance is a constant value determined by the material and geometric properties of a specific capacitor.

- Current is a measure of the flow of charge per unit time. The total charge is the integral of the current with respect to time.

- The current in a closed circuit is determined by the properties and arrangement of the individual circuit elements such as voltage sources, resistors, and capacitors.

Capacitance, Potential Difference, and Charge
What Is the Mathematical Relationship Between the Potential Difference Used to Charge a Capacitor and the Amount of Charge Stored?

To help students reflect on what they know about capacitors, we recommend showing them two or three images using presentation software that help illustrate these important ideas. You can then ask the students the following questions to encourage them to share how they are thinking about these important concepts:

1. What do we see going on in this image?
2. Does anyone have anything else to add?
3. What might be going on that we can't see?
4. What are some things that we are not sure about here?

You can then encourage students to think about how CCs played a role in their investigation. There are at least two CCs that students need to use to determine a mathematical relationship between capacitance and voltage: (a) Cause and Effect: Mechanism and Explanation and (b) Systems and System Models (see Appendix 2 [p. 417] for a brief description of these CCs). To help students reflect on what they know about these CCs, we recommend asking them the following questions:

1. How do scientists determine if two quantities have a cause-and-effect relationship?
2. What cause-and-effect relationships did you uncover in this investigation, if any?
3. What was the system under study and what components make up that system in this investigation?
4. In what ways did you model this system?

You can then encourage the students to think about how they used all these different concepts to help answer the guiding question and why it is important to use these ideas to help justify their evidence for their final arguments. Be sure to remind your students to explain why they included the evidence in their arguments and make the assumptions underlying their analysis and interpretation of the data explicit in order to provide an adequate justification of their evidence.

Reflecting on Ways to Design Better Investigations

It is important for students to reflect on the strengths and weaknesses of the investigation they designed during the explicit and reflective discussion. Students should therefore be encouraged to discuss ways to eliminate potential flaws, measurement errors, or sources of uncertainty in their investigations. To help students be more reflective about the design of their investigation and what they can do to make their investigations more rigorous in the future, you can ask them the following questions:

LAB 4

1. What were some of the strengths of the way you planned and carried out your investigation? In other words, what made it scientific?

2. What were some of the weaknesses of the way you planned and carried out your investigation? In other words, what made it less scientific?

3. What rules can we make, as a class, to ensure that our next investigation is more scientific?

Reflecting on the Nature of Scientific Knowledge and Scientific Inquiry

This investigation can be used to illustrate two important concepts related to the nature of scientific knowledge and the nature of scientific inquiry: (a) the difference between observations and inferences in science and (b) how scientists use different methods to answer different types of questions (see Appendix 2 [p. 417] for a brief description of these concepts). Be sure to review these concepts during and at the end of the explicit and reflective discussion. To help students think about these concepts in relation to what they did during the lab, you can ask them the following questions:

1. You had to make observations and inferences during your investigation. Can you give me some examples of these observations and inferences?

2. Can you work with your group to come up with a rule that you can use to decide if a piece of information is an observation or an inference? Be ready to share in a few minutes.

3. There is no universal step-by-step scientific method that all scientists follow. Why do you think there is no universal scientific method?

4. Think about what you did during this investigation. How would you describe the method you used to determine the relationship between the potential difference and charge stored on a capacitor? Why would you call it that?

You can encourage the students to think about these concepts by showing examples from the investigation that are either observations or inferences and ask students to classify each example and explain their thinking using presentation software. You can also show one or more images of a "universal scientific method" that misrepresent the nature of scientific inquiry (see, e.g., *https://commons.wikimedia.org/wiki/File:The_Scientific_Method_as_an_Ongoing_Process.svg*) and ask students why each image is *not* a good representation of what scientists do to develop scientific knowledge. You can also ask students to suggest revisions to the image that would make it more consistent with the way scientists develop scientific knowledge. Be sure to remind your students that it is important for them to understand what counts as scientific knowledge and how that knowledge develops over time in order to be proficient in science.

Capacitance, Potential Difference, and Charge
What Is the Mathematical Relationship Between the Potential Difference Used to Charge a Capacitor and the Amount of Charge Stored?

Hints for Implementing the Lab

- Allowing students to design their own procedures for collecting data gives them an opportunity to try, to fail, and to learn from their mistakes. However, you can scaffold students as they develop their own procedure by having them fill out an investigation proposal. The proposals provide a way for you to offer students hints and suggestions without telling them how to do it. You can also check the proposals quickly during a class period. For this lab we suggest using Investigation Proposal C.

- Learn how to use the digital ammeter/multimeter for the ranges of voltages (< 20 V) and currents (0–200 mA) used in this inquiry before the lab begins. It is important for you to know how to use the equipment so you can help students when technical issues arise.

- When charging/discharging, it is recommended that students wait 10 RC time constants before recharging or discharging these elements. If you do not wish to use this language at this stage, then 60 seconds is a good uniform time to wait for the ranges used in this investigation.

- This lab assumes students have had some prior experience with DC circuits (with resistors only) and wiring simple circuits with at most two elements and an ammeter. If this is the first time that students are working with a breadboard, it may be helpful to provide them with a wiring diagram.

- Allow the students to become familiar with the breadboard, resistor, and jumper cables as part of the tool talk before they begin to design their investigation. Give them 5–10 minutes to examine the equipment and materials before they begin designing their investigations. This gives students a chance to see what they can and cannot do with the equipment.

- Be sure to allow students to go back and re-collect data at the end of the argumentation session. Students often realize that they made numerous mistakes when they were collecting data as a result of their discussions during the argumentation session. The students, as a result, will want a chance to re-collect data, and the re-collection of data should be encouraged when time allows. This also offers an opportunity to discuss what scientists do when they realize a mistake is made inside the lab.

If students use digital interface measurement equipment and analysis

- We suggest allowing students to familiarize themselves with the data analysis software before they finalize the procedure for the investigation, especially if they have not used such software previously. This gives students an opportunity to learn how to work with the software and to improve the quality of the data they collect.

LAB 4

- Remind students to begin recording data just before they begin discharging their capacitors. Also, set the data collection time to last at least 30 seconds.
- Remind students to follow the user's guide to correctly connect any current probes to avoid damage to lab equipment. All currents in this lab are below 0.500 A, but short circuits may arise from improper wiring.

Connections to Standards

Table 4.2 highlights how the investigation can be used to address specific performance expectations from the *NGSS*; learning objectives from AP Physics 1 and 2; learning objectives from AP Physics C: Electricity and Magnetism; *Common Core State Standards for English Language Arts (CCSS ELA)*; and *Common Core State Standards for Mathematics (CCSS Mathematics)*.

TABLE 4.2

Lab 4 alignment with standards

NGSS performance expectation	• HS-PS3-1: Create a computational model to calculate the change in the energy of one component in a system when the change in energy of the other component(s) and energy flows in and out of the system are known.
AP Physics 1 and AP Physics 2 learning objectives	• 4.E.5.1. Make and justify a quantitative prediction of the effect of a change in values or arrangements of one or two circuit elements on the currents and potential differences in a circuit containing a small number of sources of emf, resistors, capacitors, and switches in series and/or parallel. • 4.E.5.2: Make and justify a qualitative prediction of the effect of a change in values or arrangements of one or two circuit elements on currents and potential differences in a circuit containing a small number of sources of emf, resistors, capacitors, and switches in series and/or parallel. • 4.E.5.3: Plan data collection strategies and perform data analysis to examine the values of currents and potential differences in an electric circuit that is modified by changing or rearranging circuit elements, including sources of emf, resistors, and capacitors.
AP Physics C: Electricity and Magnetism learning objectives	• CNV-4.A.a: Apply the general definition of capacitance to a capacitor attached to a charging source. • CNV-4.A.b: Calculate unknown quantities such as charge, potential difference, or capacitance for physical system with a charged capacitor.

Continued

Capacitance, Potential Difference, and Charge

What Is the Mathematical Relationship Between the Potential Difference Used to Charge a Capacitor and the Amount of Charge Stored?

Table 4.2 (*continued*)

Literacy connections (*CCSS ELA*)	• *Reading*: Key ideas and details, craft and structure, integration of knowledge and ideas • *Writing*: Text types and purposes, production and distribution of writing, research to build and present knowledge, range of writing • *Speaking and listening:* Comprehension and collaboration, presentation of knowledge and ideas
Mathematics connections (*CCSS Mathematics*)	• *Mathematical practices*: Make sense of problems and persevere in solving them, reason abstractly and quantitatively, construct viable arguments and critique the reasoning of others, model with mathematics, use appropriate tools strategically, attend to precision • *Number and quantity*: Reason quantitatively and use units to solve problems, represent and model with vector quantities, perform operations on vectors • *Algebra*: Interpret the structure of expressions, create equations that describe numbers or relationships, understand solving equations as a process of reasoning and explain the reasoning, solve equations and inequalities in one variable, represent and solve equations and inequalities graphically. • *Functions*: Understand the concept of a function and use function notation; interpret functions that arise in applications in terms of the context; analyze functions using different representations; build a function that models a relationship between two quantities; construct and compare linear, quadratic, and exponential models and solve problems; interpret expressions for functions in terms of the situation they model • *Statistics and probability*: Summarize, represent, and interpret data on two categorical and quantitative variables; interpret linear models; make inferences and justify conclusions from sample surveys, experiments, and observational studies

LAB 4

Lab Handout

Lab 4. Capacitance, Potential Difference, and Charge: What Is the Mathematical Relationship Between the Potential Difference Used to Charge a Capacitor and the Amount of Charge Stored?

Introduction

During the period of scientific and philosophical advancement known as the Enlightenment (approximately coinciding with the 18th century), two of the key discoveries by scientists conducting experiments on electrostatics were that (1) electric charge could be explained in terms of two states (positive and negative) and (2) some materials could better allow charges to move through them than others. As this apparent movement of charge began to be more fully understood, a new question arose: Is it possible to store charge for later use? The invention and discovery of the effects of the famous Leyden jar (see Figure L4.1) soon answered this question in the affirmative. A Leyden jar works by placing a thin conducting material, such as a metal foil, on the inside and outside of a glass jar. The metal rod exiting the top of the jar is connected to the conducting material on the inside. Charge can then be stored on the Leyden jar by connecting the metal ball to a voltage source.

FIGURE L4.1

A cross-section of an early Leyden jar. Component A and component B are the metal foil covering the inside and outside of the glass jar.

Today, scientists and engineers use devices known as capacitors as a means to store electric energy in circuits. Capacitors have become so ubiquitous in modern technology that, almost without exception, every electrical device in the room you are sitting in at this moment contains multiple capacitors. The size of capacitors can range from networks large enough to fill a warehouse for industrial applications to precise circuit elements smaller than a millimeter across. In fact, the property of capacitance is central to the physics of many models of touch screens used today in cell phones and laptop computers.

Physicists often derive relationships applicable to a number of different capacitor geometries from the model for the parallel-plate capacitor. When an uncharged capacitor is connected in series with a resistor and a source of potential difference (e.g., a voltage source), both the charge on each of the plates and the potential difference between

Capacitance, Potential Difference, and Charge

What Is the Mathematical Relationship Between the Potential Difference Used to Charge a Capacitor and the Amount of Charge Stored?

the plates increase to their maximum following exponential relationships. For instance, the potential difference relationship can be written as $\Delta V(t) = \Delta V_{max}(1-e^{-\frac{t}{RC}})$, where ΔV is the potential difference at some time after the circuit is closed with the voltage source, R is electrical resistance, C is capacitance, t is time, and ΔV_{max} is the voltage reached between the two plates as $t \to \infty$ (in practice, as $t \gg RC$). Similarly, when one connects a charged capacitor to a simple RC circuit, one can observe the current in the series circuit decreasing with the same rate as the capacitor discharges: $I(t) = I_{max}e^{-\frac{t}{RC}}$. Here, I is electric current at some time t and I_{max} is the maximum current measured in the circuit at time $t = 0$ s. Current is a measure of the flow of charge per unit time.

As mentioned earlier, the primary use of capacitors is to store energy. At the same time, inventors, researchers, and visionaries of today are faced with the challenge of finding sustainable, ethical, and efficient solutions to the energy demands of the world in the 21st century. Given that alternative electric power generation and storage methods are a core area of development, it is no surprise that understanding the physics of capacitors is essential for any working researcher. The quantities of electric charge, electric potential difference, and capacitance are closely related in calculating the amount of energy stored in capacitors. Your goal is to begin to arrive at these relationships by creating a conceptual-mathematical model relating two of these: electric charge and potential difference.

Your Task

Use what you know about capacitors and circuits, systems and system models, and cause and effect to design and carry out an investigation to determine how the potential difference used to charge a capacitor is related to the amount of charged stored on that capacitor.

The guiding question of this investigation is, *What is the mathematical relationship between the potential difference used to charge a capacitor and the amount of charge stored?*

Materials

You may use any of the following materials during your investigation:

Consumables
- AA batteries
- 9V battery

Equipment
- Safety glasses with side shields or goggles (required)
- AA battery holder
- Alligator clips

- Digital multimeter
- Breadboard
- Jumper wires
- Resistor
- Capacitor

If you have access to the following equipment, you may also consider using a digital current sensor with an accompanying interface and a computer or tablet.

LAB 4

Safety Precautions

Follow all normal lab safety rules. In addition, take the following safety precautions:

1. Wear sanitized safety glasses with side shields or goggles during lab setup, hands-on activity, and takedown.

2. Never put consumables in your mouth.

3. Wire and other metals with electric current flowing through them may get hot. Use caution when handling components of a closed circuit.

4. Handle electrical wires, including ammeter/multimeter test leads, with caution. They have sharp ends, which can cut or puncture skin.

5. Never discharge capacitors by shorting them with your fingers or any other body part.

6. Never charge capacitors beyond the potential differences (voltages) for which they are rated, because they may explode. Read the label printed on the capacitor. Always ask your teacher if you need help interpreting the label.

7. Wash your hands with soap and water when you are done collecting the data.

Investigation Proposal Required?   ☐ Yes   ☐ No

Getting Started

To answer the guiding question, you will need to design and carry out an investigation to determine how the potential difference used to charge a capacitor is related to the amount of charge stored on that capacitor. Figure L4.2 illustrates a sample circuit showing how you can use the available equipment to study the electrical properties of a capacitor in an RC circuit. Caution: Switch 1 and switch 2 should never be closed at the same time to avoid shorting your batteries. Before you can design your investigation, however, you must determine what type of data you need to collect, how you will collect it, and how you will analyze it.

FIGURE L4.2

A sample RC circuit with an unconnected ammeter

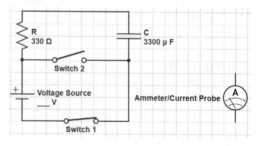

To determine *what type of data you need to collect*, think about the following questions:

- What are the boundaries and components of the RC circuit system?
- How do the components of the system interact with each other?

Capacitance, Potential Difference, and Charge
What Is the Mathematical Relationship Between the Potential Difference Used to Charge a Capacitor and the Amount of Charge Stored?

- When is this system stable and under which conditions does it change?
- How could you keep track of changes in this system quantitatively?
- How will you measure the voltage source?
- Where would be the most appropriate place to connect the ammeter/multimeter and in what configuration?
- When should you be collecting data: during charging or discharging?
- How will you create your circuit on a breadboard, and how can you verify that it matches your intended circuit diagram?

To determine *how you will collect the data*, think about the following questions:

- What is the independent variable and what is the dependent variable?
- What other factors will you need to control during each experiment?
- What scale or scales should you use when you take your measurements?
- How will you make sure that your data are of high quality (i.e., how will you reduce error)?
- How will you keep track of and organize the data you collect?
- What type of research design needs to be used to establish a cause-and-effect relationship?
- What conditions need to be satisfied to establish a cause-and-effect relationship?
- Capacitors charge and discharge within seconds: How will you design data collection around this?

To determine *how you will analyze the data*, think about the following questions:

- What type of calculations, if any, will you need to make?
- What types of patterns might you look for as you analyze your data?
- What type of table or graph could you create to help make sense of your data?
- How will you model the system to indicate how the potential difference across the capacitor is related to the maximum charge stored?

Connections to the Nature of Scientific Knowledge and Scientific Inquiry

As you work through your investigation, you may want to consider

- the difference between observations and inferences in science, and
- how scientists use different methods to answer different types of questions.

LAB 4

Initial Argument

Once your group has finished collecting and analyzing your data, your group will need to develop an initial argument. Your initial argument needs to include a claim, evidence to support your claim, and a justification of the evidence. The *claim* is your group's answer to the guiding question. The *evidence* is an analysis and interpretation of your data. Finally, the *justification* of the evidence is why your group thinks the evidence matters. The justification of the evidence is important because scientists can use different kinds of evidence to support their claims. Your group will create your initial argument on a whiteboard. Your whiteboard should include all the information shown in Figure L4.3.

FIGURE L4.3
Argument presentation on a whiteboard

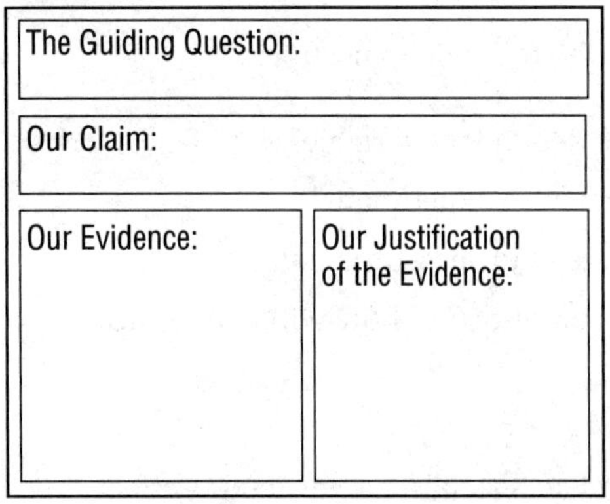

The Guiding Question:	
Our Claim:	
Our Evidence:	Our Justification of the Evidence:

Argumentation Session

The argumentation session allows all of the groups to share their arguments. One or two members of each group will stay at the lab station to share that group's argument, while the other members of the group go to the other lab stations to listen to and critique the other arguments. This is similar to what scientists do when they propose, support, evaluate, and refine new ideas during a poster session at a conference. If you are presenting your group's argument, your goal is to share your ideas and answer questions. You should also keep a record of the critiques and suggestions made by your classmates so you can use this feedback to make your initial argument stronger. You can keep track of specific critiques and suggestions for improvement that your classmates mention in the space below.

Critiques about our initial argument and suggestions for improvement:

Capacitance, Potential Difference, and Charge
What Is the Mathematical Relationship Between the Potential Difference Used to Charge a Capacitor and the Amount of Charge Stored?

If you are critiquing your classmates' arguments, your goal is to look for mistakes in their arguments and offer suggestions for improvement so these mistakes can be fixed. You should look for ways to make your initial argument stronger by looking for things that the other groups did well. You can keep track of interesting ideas that you see and hear during the argumentation in the space below. You can also use this space to keep track of any questions that you will need to discuss with your team.

Interesting ideas from other groups or questions to take back to my group:

Once the argumentation session is complete, you will have a chance to meet with your group and revise your initial argument. Your group might need to gather more data or design a way to test one or more alternative claims as part of this process. Remember, your goal at this stage of the investigation is to develop the best argument possible.

Report

Once you have completed your research, you will need to prepare an *investigation report* that consists of three sections. Each section should provide an answer to the following questions:

1. What question were you trying to answer and why?
2. What did you do to answer your question and why?
3. What is your argument?

Your report should answer these questions in two pages or less. This report must be typed, and any diagrams, figures, or tables should be embedded in the document. Be sure to write in a persuasive style; you are trying to convince others that your claim is acceptable or valid!

LAB 4

Checkout Questions

Lab 4. Capacitance, Potential Difference, and Charge: What Is the Mathematical Relationship Between the Potential Difference Used to Charge a Capacitor and the Amount of Charge Stored?

The picture below shows a simple RC circuit. A physics class is investigating the relationships between the various components of the circuit. Initially, the students use a voltage source with a potential difference of V and the capacitor is able to hold a charge of value Q. Use this information to answer questions 1–3.

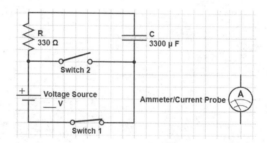

1. How much charge will the capacitor hold if the voltage source is increased to a potential difference of $4V$?

2. How do you know?

3. What would happen to the total charge stored by the capacitor if the capacitance is doubled?

Capacitance, Potential Difference, and Charge
What Is the Mathematical Relationship Between the Potential Difference Used to Charge a Capacitor and the Amount of Charge Stored?

4. Scientists use different methods to answer different types of questions.

    a. I agree with this statement.
    b. I disagree with this statement.

   Explain your answer, using examples from this investigation and at least one other investigation you have conducted.

5. *Observation* and *inference* are terms that have the same meaning in science.

    a. I agree with this statement.
    b. I disagree with this statement.

   Explain your answer, using examples from this investigation and at least one other investigation you have conducted.

6. Scientists often define a system under study and then make models of the system they are investigating. Why is it useful for scientists to make models of a system during an investigation? In your answer, be sure to use examples from this investigation and at least one other investigation you have conducted

7. Why is determining cause-and-effect relationships an important part of many scientific investigations? In your answer, be sure to use examples from this investigation and at least one other investigation you have conducted.

LAB 5

Teacher Notes

Lab 5. Resistors in Series and Parallel: How Does the Arrangement of Four Lightbulbs in a Circuit Affect the Total Current of the System?

Purpose

The purpose of this lab is to *introduce* students to the disciplinary core idea (DCI) of Conservation of Energy and Energy Transfer (PS3.B) from the *NGSS* by having them measure current in circuits of different resistor arrangements. In addition, this lab can be used to help students understand two big ideas from AP Physics: (a) objects and systems have properties such as mass and charge, and systems may have internal structure; and (b) changes that occur as a result of interactions are constrained by conservation laws. This lab also gives students an opportunity to learn about the crosscutting concepts (CCs) of (a) Patterns and (b) Energy and Matter: Flows, Cycles, and Conservation from the *NGSS*. As part of the explicit and reflective discussion, students will also learn about (a) the difference between data and evidence in science and (b) the assumptions made by scientists about order and consistency in nature.

Underlying Physics Concepts

Equation 5.1 shows how the total current in a multiresistor circuit can be predicted using Ohm's law, which states that the current through any closed circuit is directly proportional to the potential difference (i.e., voltage) applied to the circuit and inversely proportional to the resistance of the circuit. In Equation 5.1, I_{total} is the total current through the circuit, V is the voltage applied by the battery, and R_{eq} is the circuit's equivalent resistance. In SI units, current is measured in amperes (A), voltage in volts (V), and resistance in ohms (Ω). We define the *equivalent resistance* as the resistance of a single resistor that could replace the components of the circuit without changing the current drawn from the voltage source. Figure 5.1 shows what is meant by the equivalent resistance.

$$(\text{Equation 5.1}) \quad I_{total} = V/R_{eq}$$

FIGURE 5.1

The equivalent resistance (R_{eq}) of two resistors in series

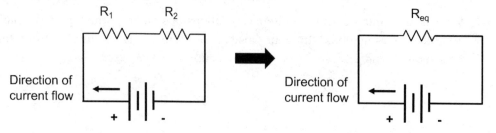

Resistors in Series and Parallel
How Does the Arrangement of Four Lightbulbs in a Circuit Affect the Total Current of the System?

This investigation sets the stage for students to then learn to calculate the equivalent resistance of a circuit. Resistors in series have an equivalent resistance equal to the sum of the individual resistances shown in Equation 5.2, where R_1 and R_2 represent the resistance of individual resistors in the series circuit and R_n is the resistance of the nth resistor in the circuit.

(Equation 5.2) $R_{eq} = R_1 + R_2 + \ldots R_n$

From a conceptual perspective, we can understand why the equivalent resistance for resistors in series is additive. If we take the point of view of a single electron moving through a circuit, then when resistors are in series, the electron must pass through the first resistor and then the second resistor. As resistors impede the flow of current (charge per unit time), the movement of the electron is impeded by both resistors in series. As the number of resistors in series is increased, the total resistance will also increase.

As shown in Equation 5.1, for a constant voltage source the total current is inversely proportional to the equivalent resistance. Thus, when resistors are wired in series, the equivalent resistance increases and the total current drawn from the voltage source decreases. Conceptually, this makes sense. Current is the flow of charge per unit time. The more resistors any single charge encounters, the slower the charge flows. Alternatively, this means that fewer charges are flowing into and out of the voltage source in a given time.

Determining the equivalent resistance for resistors in parallel is a bit more involved, as shown in Equation 5.3. Again, R_1 and R_2 are the resistance of the individual resistors in the parallel circuit and R_n is the resistance of the nth resistor in the parallel circuit. Figure 5.2 shows two resistors in parallel and the circuit with the equivalent resistance.

(Equation 5.3) $1/R_{eq} = 1/R_1 + 1/R_2 + \ldots 1/R_n$

FIGURE 5.2

Two resistors in parallel and the equivalent resistance

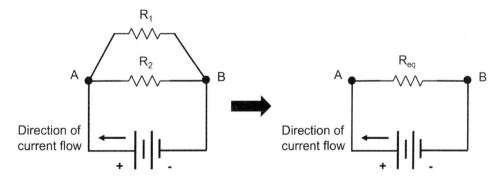

LAB 5

As implied by Equation 5.3, when resistors are arranged in parallel, the equivalent resistance is less than the resistance for the smallest resistor. From a conceptual standpoint, as current flows through the circuit, when each charge comes to a fork in the path (e.g., point A in Figure 5.2), it can only pass through one branch. Thus, each charge only experiences a single resistor and each resistor has charge flowing through it at the same time. Hence, the impediment to the flow of charge is reduced.

Again, and as shown in Equation 5.1, the total current is inversely proportional to the equivalent resistance. When resistors are wired in parallel, the equivalent resistance decreases, allowing for more charges to flow into and out of the voltage source at any given time. Thus, the total current in the system increases when wired in parallel.

In this investigation, students are given four lightbulbs (resistors) and asked to determine how the arrangement of the bulbs affects the current in the circuit. The smallest current and greatest equivalent resistance come from all resistors wired in series. As alternate paths are provided by resistors in parallel, the equivalent resistance is reduced and the total current increases. Thus, the greatest current will occur when all four bulbs are in parallel. Students often find this counterintuitive; an additional resistor in parallel increases the current, despite adding a resistor. Adding traffic lanes to a busy road is a fair analogy—more paths allow for greater current flow.

Timeline

The instructional time needed to complete this lab investigation is 200–280 minutes. Appendix 3 (p. 421) provides options for implementing this lab investigation over several class periods. Option E (280 minutes) should be used if students are unfamiliar with scientific writing, because this option provides extra instructional time for scaffolding the writing process. You can scaffold the writing process by modeling, providing examples, and providing hints as students write each section of the report. Option E can also be used if you are introducing students to the digital interface sensors and/or the data analysis software. Option F (200 minutes) should be used if students are familiar with scientific writing and have developed the skills needed to write an investigation report on their own. In option F, students complete stage 6 (writing the investigation report) and stage 8 (revising the investigation report) as homework.

Materials and Preparation

The materials needed to implement this investigation are listed in Table 5.1. The equipment can be purchased from a science supply company such as Flinn Scientific, PASCO, Vernier, or Ward's Science.

TABLE 5.1

Materials list for Lab 5

Item	Quantity
Consumable	
D battery	1 per group
Equipment and other materials	
Safety glasses with side shields or safety goggles	1 per student
Ammeter	1 per group
Miniature lightbulbs and sockets	4 per group
Copper wire with alligator clips	8 per group
Investigation Proposal C (optional)	1 per group
Whiteboard, 2' × 3'*	1 per group
Lab Handout	1 per student
Peer-review guide and teacher scoring rubric	1 per student
Checkout Questions	1 per student
Equipment for digital interface measurements (optional)	
Digital interface with USB or wireless connections	1 per group
Current measurement sensor	1 per group
Computer or tablet with appropriate data analysis software installed	1 per group

* As an alternative, students can use computer and presentation software such as Microsoft PowerPoint or Apple Keynote to create their arguments.

Be sure to use a set routine for distributing and collecting the materials during the lab investigation. One option is to set up the materials for each group at each group's lab station before class begins. This option works well when there is a dedicated section of the classroom for lab work and the materials are large and difficult to move. A second option is to have all the materials on a table or cart at a central location. You can then assign a member of each group to be the "materials manager." This individual is responsible for collecting all the materials his or her group needs from the table or cart during class and for returning all the materials at the end of the class. This option works well when the materials are small and easy to move (such as magnets, wire, and bulbs). It also makes it easy to inventory the materials at the end of the class before students leave for the day.

LAB 5

Safety Precautions and Laboratory Waste Disposal

Remind students to follow all normal lab safety rules. In addition, tell students to take the following safety precautions:

1. Wear sanitized safety glasses with side shields or goggles during lab setup, hands-on activity, and takedown.
2. Never put consumables in their mouth.
3. Do not connect the terminals of a battery without a lightbulb between them.
4. Wire and other metals with electric current flowing through them may get hot. Use caution when handling components of a closed circuit.
5. Lightbulbs are made of glass. Be careful handling them. If they break, clean them up immediately and place in a broken glass box.
6. Handle electrical wires with caution. They have sharp ends, which can cut or puncture skin.
7. Wash their hands with soap and water when they are done collecting the data.

Batteries, lightbulbs, and wire may be stored for future use. When batteries need replacing, dispose of old batteries according to manufacturer's recommendations.

Topics for the Explicit and Reflective Discussion

Reflecting on the Use of Core Ideas and Crosscutting Concepts During the Investigation

Teachers should begin the explicit and reflective discussion by asking students to discuss what they know about the core ideas they used during the investigation. The following are some important concepts related to the core idea of the conservation of electric charge that students need to use to describe electric current:

- An electric current is a movement of charge through a conductor. Current is the flow of charge per unit time. Resistors use electric current to do work and impede the flow of charge.
- A circuit is a closed loop of electric current.
- The current through a resistor is equal to the potential difference across the resistor divided by its resistance.
- The current in a closed circuit is determined by the properties and arrangement of the individual circuit elements such as voltage sources, resistors, and capacitors.

To help students reflect on what they know about electric current, we recommend showing them two or three images using presentation software that help illustrate these important ideas. You can then ask the students the following questions to encourage them to share how they are thinking about these important concepts:

1. What do we see going on in this image?
2. Does anyone have anything else to add?
3. What might be going on that we can't see?
4. What are some things that we are not sure about here?

You can then encourage students to think about how CCs played a role in their investigation. There are at least two CCs that students need to use to determine how the arrangement of the lightbulbs affects the total current of the system: (a) Patterns and (b) Energy and Matter: Flows, Cycles, and Conservation (see Appendix 2 [p. 417] for a brief description of these CCs). To help students reflect on what they know about these CCs, we recommend asking them the following questions:

1. Did you notice any patterns emerging between variables in your investigation?
2. What predictions can you make using these patterns?
3. How does matter flow in an electric circuit? Where was the matter before the battery was connected?
4. How do you know energy flows in a circuit?

You can then encourage the students to think about how they used all these different concepts to help answer the guiding question and why it is important to use these ideas to help justify their evidence for their final arguments. Be sure to remind your students to explain why they included the evidence in their arguments and make the assumptions underlying their analysis and interpretation of the data explicit in order to provide an adequate justification of their evidence.

Reflecting on Ways to Design Better Investigations

It is important for students to reflect on the strengths and weaknesses of the investigation they designed during the explicit and reflective discussion. Students should therefore be encouraged to discuss ways to eliminate potential flaws, measurement errors, or sources of uncertainty in their investigations. To help students be more reflective about the design of their investigation and what they can do to make their investigations more rigorous in the future, you can ask them the following questions:

1. What were some of the strengths of the way you planned and carried out your investigation? In other words, what made it scientific?

2. What were some of the weaknesses of the way you planned and carried out your investigation? In other words, what made it less scientific?

3. What rules can we make, as a class, to ensure that our next investigation is more scientific?

Reflecting on the Nature of Scientific Knowledge and Scientific Inquiry
This investigation can be used to illustrate two important concepts related to the nature of scientific knowledge and the nature of scientific inquiry: (a) the difference between data and evidence in science and (b) the assumptions made by scientists about order and consistency in nature (see Appendix 2 [p. 417] for a brief description of these concepts). Be sure to review these concepts during and at the end of the explicit and reflective discussion. To help students think about these concepts in relation to what they did during the lab, you can ask them the following questions:

1. You had to talk about data and evidence during your investigation. Can you give me some examples of data and evidence from your investigation?

2. Can you work with your group to come up with a rule that you can use to decide if a piece of information is data or evidence? Be ready to share in a few minutes.

3. Scientists assume that natural laws operate today as they did in the past and that they will continue to do so in the future. Why do you think this assumption is important?

4. Think about what you were trying to do during this investigation. What assumptions did you make regarding how natural laws continue to operate today as they did in the past?

You can also use presentation software or other techniques to encourage your students to think about these concepts. You can show examples of information from the investigation that are either data or evidence and ask students to classify each example and explain their thinking. You can also show images of different scientific laws and ask students if these laws would be the same everywhere if the universe was not a single system, and then ask them to think about what scientists would need to do to be able to study the universe if it was made up of many different systems. Be sure to remind your students that it is important for them to understand what counts as scientific knowledge and how that knowledge develops over time in order to be proficient in science.

Hints for Implementing the Lab

- Allowing students to design their own procedures for collecting data gives students an opportunity to try, to fail, and to learn from their mistakes. However, you can scaffold students as they develop their own procedure by having them fill out an investigation proposal. The proposals provide a way for you to offer students hints and suggestions without telling them how to do it. You can also check the proposals quickly during a class period. For this lab we suggest using Investigation Proposal C.

- Select appropriate batteries and resistors such that the current is large enough for the meters to measure, but not high enough to exceed the power rating of the resistors. Using D batteries with miniature lightbulbs can provide a useful starting point.
- Learn how to use the ammeter before the lab begins. It is important for you to know how to use the equipment so you can help students when technical issues arise.
- Allow the students to become familiar with the ammeter as part of the tool talk before they begin to design their investigation. Give them 5–10 minutes to examine the equipment and materials before they begin designing their investigations. This gives students a chance to see what they can and cannot do with the equipment.
- Each group only needs one battery at a time, but plan to provide fresh batteries for some groups.
- In this investigation, students are looking at the total current drawn from the battery, so the ammeter should be connected in series with the battery and the remainder of the circuit. Lab 6 will expand on this topic by having students explore the individual components of various circuits. If students place the ammeter in parallel with some or all of the other resistors, this can be a good opportunity to help them understand conceptually what is happening when components are arranged in parallel. We suggest asking them to think about how individual electrons move through each component of the system. This also provides a nice opportunity to reinforce the relationship between the question being asked and the arrangement of the equipment.
- Be sure to allow students to go back and re-collect data at the end of the argumentation session. Students often realize that they made numerous mistakes when they were collecting data as a result of their discussions during the argumentation session. The students, as a result, will want a chance to re-collect data, and the re-collection of data should be encouraged when time allows. This also offers an opportunity to discuss what scientists do when they realize a mistake is made inside the lab.

If students use digital interface measurement equipment and analysis

- We suggest allowing students to familiarize themselves with the sensors and the data analysis software before they finalize the procedure for the investigation, especially if they have not used such software previously. This gives students an opportunity to learn how to work with the software and to improve the quality of the data they collect.
- Remind students to begin recording data just before they close the circuits.

- Remind students to follow the user's guide to correctly connect any sensors to avoid damage to lab equipment.

Connections to Standards

Table 5.2 highlights how the investigation can be used to address specific performance expectations from the *NGSS*; learning objectives from AP Physics 1 and 2; learning objectives from AP Physics C: Electricity and Magnetism; *Common Core State Standards for English Language Arts* (*CCSS ELA*); and *Common Core State Standards for Mathematics* (*CCSS Mathematics*).

TABLE 5.2

Lab 5 alignment with standards

NGSS performance expectations	• None
AP Physics 1 and AP Physics 2 learning objectives	• 1.B.1.1 Make claims about natural phenomena based on conservation of electric charge. • 4.E.5.1: Make and justify a quantitative prediction of the effect of a change in values or arrangements of one or two circuit elements on the currents and potential differences in a circuit containing a small number of sources of emf, resistors, capacitors, and switches in series and/or parallel. • 4.E.5.2: Make and justify a qualitative prediction of the effect of a change in values or arrangements of one or two circuit elements on currents and potential differences in a circuit containing a small number of sources of emf, resistors, capacitors, and switches in series and/or parallel. • 4.E.5.3: Plan data collection strategies and perform data analysis to examine the values of currents and potential differences in an electric circuit that is modified by changing or rearranging circuit elements, including sources of emf, resistors, and capacitors. • 5.C.3.2: Design an investigation of an electrical circuit with one or more resistors in which evidence of conservation of electric charge can be collected and analyzed. • 5.C.3.4: Predict or describe current values in series and parallel arrangements of resistors and other branching circuits using Kirchhoff's junction rule, and explain the relationship of the rule to the law of charge conservation.

Continued

Table 5.2 (*continued*)

AP Physics C: Electricity and Magnetism learning objectives	• FIE-3.A.a: Calculate unknown quantities relating to the definition of current. • FIE-3.B.a: Describe the relationship between current, potential difference, and resistance of resistor using Ohm's Law. • FIE-3.B.b: Apply Ohm's Law in an operating circuit with a known resistor or resistances. • CNV-6.A.a: Identify parallel or series arrangement in a circuit containing multiple resistors. • CNV-6.A.b: Describe a series or a parallel arrangement of resistors.
Literacy connections (*CCSS ELA*)	• *Reading*: Key ideas and details, craft and structure, integration of knowledge and ideas • *Writing*: Text types and purposes, production and distribution of writing, research to build and present knowledge, range of writing • *Speaking and listening:* Comprehension and collaboration, presentation of knowledge and ideas
Mathematics connections (*CCSS Mathematics*)	• *Mathematical practices*: Make sense of problems and persevere in solving them, reason abstractly and quantitatively, construct viable arguments and critique the reasoning of others, model with mathematics, use appropriate tools strategically, attend to precision • *Number and quantity*: Reason quantitatively and use units to solve problems, represent and model with vector quantities, perform operations on vectors • *Algebra*: Interpret the structure of expressions, create equations that describe numbers or relationships, understand solving equations as a process of reasoning and explain the reasoning, solve equations and inequalities in one variable, represent and solve equations and inequalities graphically • *Functions*: Understand the concept of a function and use function notation; interpret functions that arise in applications in terms of the context; analyze functions using different representations; build a function that models a relationship between two quantities; construct and compare linear, quadratic, and exponential models and solve problems; interpret expressions for functions in terms of the situation they model • *Statistics and probability*: Summarize, represent, and interpret data on two categorical and quantitative variables; interpret linear models; make inferences and justify conclusions from sample surveys, experiments, and observational studies

LAB 5

Lab Handout

Lab 5. Resistors in Series and Parallel: How Does the Arrangement of Four Lightbulbs in a Circuit Affect the Total Current of the System?

Introduction

Electric circuits are used in countless applications, for different purposes and of different designs. Most circuits are complicated (e.g., a cell phone containing thousands of circuit elements), but they all function using the same basic principles. The circuit is connected to a voltage source that, when the circuit is closed, provides a current. The current flows through the circuit and the electric potential energy contained in the voltage source is converted to other types of energy. In a toaster oven, for example, the electrical energy is converted to heat energy. In power tools, electrical energy is converted to mechanical energy. And in a cell phone, electrical energy is converted to sound and light energy. The voltage source for most electrical devices is either an electrical outlet in a wall or a battery. Figure L5.1 shows the inside of a computer. Notice how many different components are included within the device.

FIGURE L5.1
The inside of a computer

Note: A full-color version of this figure is available on the book's Extras page at *www.nsta.org/adi-physics2*.

Resistors in Series and Parallel
How Does the Arrangement of Four Lightbulbs in a Circuit Affect the Total Current of the System?

In a closed circuit, physicists and engineers are often interested in three quantities: the voltage connected to the circuit, the resistance of the circuit, and the current flowing through the circuit. The *voltage* is the supply of energy to the circuit (note that while voltage and energy are related, they are not the same). The *resistance* is supplied by components, called resistors, that use the energy and impede the flow of current. *Current* is the flow of charge per unit time. The simplest circuits have a single voltage source, such as a battery, and a single resistor, such as a lightbulb. When the circuit is closed, current flows from the battery through the lightbulb. The electrical energy is then converted by the lightbulb (the resistor) to light energy.

When a circuit is made of more than one component, those components are in series, in parallel, or some combination of the two. The arrangement of components can be in series such that current flows through one component and then another. When components are in parallel, the current is shared between components at the same time—some charge flows through one component and some charge flows through the other component.

The total current of a circuit with only one voltage source (which is also the maximum current) is the current flowing in and out of the battery at a given point in time. The total current depends on the arrangement of the resistors in a circuit. It is important to understand how the arrangement of several resistors affects the total current drawn from a voltage source. Too much current can overload electrical devices, causing them to break or to give off sparks, and too little current may be too weak to power electrical devices.

Your Task

Use what you know about energy and energy transfer; the flow of matter and energy into, out of, and within systems; and patterns to design and carry out an investigation to determine the relationship between how resistors are connected in a circuit and the total current flow.

The guiding question of this investigation is, *How does the arrangement of four lightbulbs in a circuit affect the total current of the system?*

Materials

You may use any of the following materials during your investigation:

Consumable
- D battery

Equipment
- Safety glasses with side shields or goggles (required)
- Ammeter
- Miniature lightbulbs and sockets
- Copper wire with alligator clips

LAB 5

If you have access to the following equipment, you may also consider using a digital current sensor with an accompanying interface and a computer or tablet.

Safety Precautions

Follow all normal lab safety rules. In addition, take the following safety precautions:

1. Wear sanitized safety glasses with side shields or goggles during lab setup, hands-on activity, and takedown.
2. Never put consumables in your mouth.
3. Never connect the terminals of a battery without a lightbulb between them.
4. Wire and other metals with electric current flowing through them may get hot. Use caution when handling components of a closed circuit.
5. Lightbulbs are made of glass. Be careful handling them. If they break, clean them up immediately and place in a broken glass box.
6. Handle electrical wires with caution. They have sharp ends, which can cut or puncture skin.
7. Wash your hands with soap and water when you are done collecting the data.

Investigation Proposal Required?   ☐ Yes   ☐ No

Getting Started

To answer the guiding question, you will need to design and carry out an investigation to determine the relationship between how resistors are connected in a circuit and the total current flow. Before you can design your investigation, however, you must determine what type of data you need to collect, how you will collect it, and how you will analyze it.

To determine *what type of data you need to collect*, think about the following questions:

- What are the boundaries and components of the system?
- How do the components of the system interact with each other?
- When is this system stable and under which conditions does it change?
- How does energy flow into, out of, or within this system?
- How could you keep track of changes in this system quantitatively?
- What is going on at the unobservable level that could cause the things that you observe?
- How might changes to the structure of what you are studying change how it functions?
- What could be the underlying cause of this phenomenon?

Resistors in Series and Parallel

How Does the Arrangement of Four Lightbulbs in a Circuit Affect the Total Current of the System?

To determine *how you will collect the data*, think about the following questions:

- What is the independent variable and what is the dependent variable?
- What other factors will you need to control during each experiment?
- What scale or scales should you use when you take your measurements?
- How can you track how energy flows into, out of, or within this system?
- How can you track how matter flows into, out of, or within this system?
- How will you make sure that your data are of high quality (i.e., how will you reduce error)?
- How will you keep track of and organize the data you collect?
- What would make a good measure of lightbulb arrangement?

To determine *how you will analyze the data*, think about the following questions:

- What type of calculations, if any, will you need to make?
- What types of patterns might you look for as you analyze your data?
- What type of table or graph could you create to help make sense of your data?
- Are there any proportional relationships that you can identify?

Connections to the Nature of Scientific Knowledge and Scientific Inquiry

As you work through your investigation, you may want to consider

- the difference between data and evidence in science, and
- the assumptions made by scientists about order and consistency in nature.

Initial Argument

Once your group has finished collecting and analyzing your data, your group will need to develop an initial argument. Your initial argument needs to include a claim, evidence to support your claim, and a justification of the evidence. The *claim* is your group's answer to the guiding question. The *evidence* is an analysis and interpretation of your data. Finally, the *justification* of the evidence is why your group thinks the evidence matters. The justification of the evidence is important because scientists can use different kinds of evidence to support their claims. Your group will create your initial argument on a whiteboard. Your whiteboard should include all the information shown in Figure L5.2.

FIGURE L5.2
Argument presentation on a whiteboard

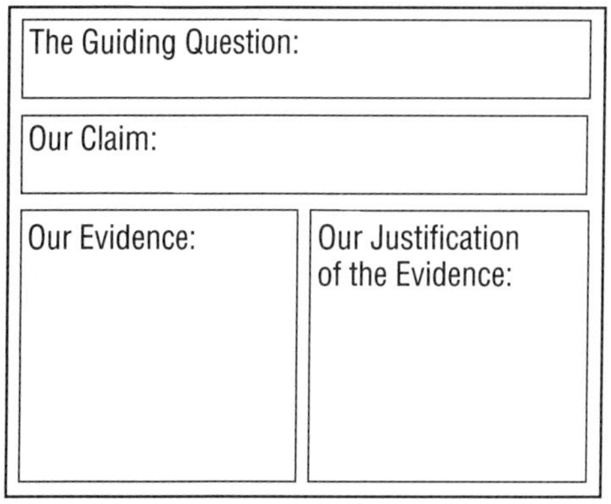

The Guiding Question:	
Our Claim:	
Our Evidence:	Our Justification of the Evidence:

LAB 5

Argumentation Session

The argumentation session allows all of the groups to share their arguments. One or two members of each group will stay at the lab station to share that group's argument, while the other members of the group go to the other lab stations to listen to and critique the other arguments. This is similar to what scientists do when they propose, support, evaluate, and refine new ideas during a poster session at a conference. If you are presenting your group's argument, your goal is to share your ideas and answer questions. You should also keep a record of the critiques and suggestions made by your classmates so you can use this feedback to make your initial argument stronger. You can keep track of specific critiques and suggestions for improvement that your classmates mention in the space below.

Critiques about our initial argument and suggestions for improvement:

If you are critiquing your classmates' arguments, your goal is to look for mistakes in their arguments and offer suggestions for improvement so these mistakes can be fixed. You should look for ways to make your initial argument stronger by looking for things that the other groups did well. You can keep track of interesting ideas that you see and hear during the argumentation in the space below. You can also use this space to keep track of any questions that you will need to discuss with your team.

Interesting ideas from other groups or questions to take back to my group:

Resistors in Series and Parallel
How Does the Arrangement of Four Lightbulbs in a Circuit Affect the Total Current of the System?

Once the argumentation session is complete, you will have a chance to meet with your group and revise your initial argument. Your group might need to gather more data or design a way to test one or more alternative claims as part of this process. Remember, your goal at this stage of the investigation is to develop the best argument possible.

Report

Once you have completed your research, you will need to prepare an *investigation report* that consists of three sections. Each section should provide an answer to the following questions:

1. What question were you trying to answer and why?
2. What did you do to answer your question and why?
3. What is your argument?

Your report should answer these questions in two pages or less. This report must be typed, and any diagrams, figures, or tables should be embedded into the document. Be sure to write in a persuasive style; you are trying to convince others that your claim is acceptable or valid!

LAB 5

Checkout Questions

Lab 5. Resistors in Series and Parallel: How Does the Arrangement of Four Lightbulbs in a Circuit Affect the Total Current of the System?

1. Rank the circuits below in order of greatest to least total current.

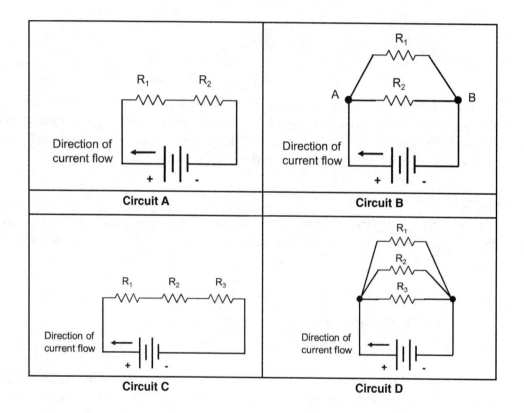

Explain the reasoning behind your ranking.

Resistors in Series and Parallel
How Does the Arrangement of Four Lightbulbs in a Circuit Affect the Total Current of the System?

2. Where could one additional lightbulb be placed to increase the current in the circuit below?

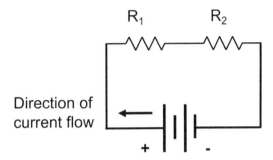

Explain your reasoning.

3. A student said, "Adding a resistor to a circuit makes it harder for the current to flow, so the total current gets smaller."

    a. I agree with this statement.
    b. I disagree with this statement.

Explain your answer, using an example from your investigation about resistors in a circuit.

LAB 5

4. A student said, "The lightbulb lit up, so electrons are flowing in the circuit." Did this student share an observation or an inference? Explain your answer, using an example from your investigation about resistors in a circuit.

5. Many investigations are informed by an assumption that the natural laws operate the same way today as they have in the past, and that they will continue to operate the same way in the future. Why is this assumption important for scientists and engineers? In your answer, be sure to use examples from this investigation and at least one other investigation you have conducted.

6. How do the flow and conservation of matter relate to a circuit? Explain your answer, using an example from your investigation about resistors in a circuit.

7. What patterns are apparent in how the arrangement of lightbulbs in a circuit affect the total current? Explain your answer, using an example from your investigation about resistors in a circuit.

LAB 6

Teacher Notes

Lab 6. Series and Parallel Circuits: How Does the Arrangement of Three Resistors Affect the Voltage Drop Across and the Current Through Each Resistor?

Purpose

The purpose of this lab is for students to *apply* what they know about the disciplinary core idea (DCI) of Conservation of Energy and Energy Transfer (PS3.B) from the *NGSS* to electrical circuits by having them measure voltage and current across three resistors in different arrangements in a circuit. In addition, this lab can be used to help students understand two big ideas from AP Physics: (a) objects and systems have properties such as mass and charge, and systems may have internal structure; and (b) changes that occur as a result of interactions are constrained by conservation laws. This lab also gives students an opportunity to learn about the crosscutting concepts of (a) Energy and Matter: Flows, Cycles, and Conservation and (b) Structure and Function from the *NGSS*. As part of the explicit and reflective discussion, students will also learn about (a) the difference between laws and theories in science and (b) the assumptions made by scientists about order and consistency in nature.

Underlying Physics Concepts

In this investigation, students collect evidence for two relationships regarding circuits developed by 19th-century German physicist Gustav Kirchhoff: the *junction rule* and the *loop rule*. Kirchhoff's first rule, the junction rule, states that electric charge is conserved for any point in a circuit where current flows. In other words, the total current into a point or junction of wires equals the total current leaving the junction (i.e., charge does not accumulate). The junction rule is shown in Equation 6.1, where I_1, I_2, and I_3 are the currents in wires connected to the junction, taking incoming current to be positive and outgoing current to be negative. In SI units, current is measured in amperes (A).

$$\text{(Equation 6.1)} \quad I_1 + I_2 + I_3 + \ldots = 0$$

Because energy is a conserved quantity, the total amount of energy the current gains must be equal to the total energy it loses as it makes its way around a circuit in a closed path. We can't easily measure the energy, but we can measure voltage (i.e., the energy per unit charge), which is also conserved. This is Kirchhoff's second rule, the loop rule, which states that the sum of the changes in voltage equals zero for any closed loop or path in a circuit, as shown in Equation 6.2, where V_1, V_2, and V_3 are the changes in voltage across elements in the loop. In SI units, voltage is measured in volts (V).

Series and Parallel Circuits

How Does the Arrangement of Three Resistors Affect the Voltage Drop Across and the Current Through Each Resistor?

(Equation 6.2) $V_1 + V_2 + V_3 + \ldots = 0$

When the current flows through a voltage source, such as a battery or a power supply, the voltage change is positive (i.e., electrons gain energy). When current flows through a resistor, the voltage change is negative (i.e., electrons lose energy).

Ohm's law can be used to algebraically expand the voltage changes, V, in the loop rule equation to determine the current in individual resistors. Ohm's law is shown in Equation 6.3, where V is the voltage change between the terminals of a resistor, I is the current flowing through it, and R is the resistance. In SI units, voltage is measured in volts (V), current is measured in amperes (A), and resistance is measured in ohms (Ω).

(Equation 6.3) $V = IR$

From a conceptual standpoint, we can also apply Kirchhoff's rules to come up with general heuristics for describing the voltage drop across and current through multiple resistors relative to each other. Figure 6.1 shows a circuit with two resistors (R_1 and R_2) in series. According to Kirchhoff's junction rule, the current entering any point in a circuit must be equal to the current leaving the point. In Figure 6.1, this means that the current entering R_1 must be equal to the current leaving R_1. This is also true for R_2. Because current cannot accumulate, this also means that the current leaving R_1 must be equal to the current entering R_2. This is true for any number of resistors in series. If this were not true, then there would be an accumulation of current in the wire connecting the two resistors.

FIGURE 6.1

A circuit with two resistors in series

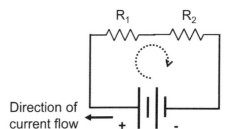

We can then calculate the voltage drop across each resistor using Equation 6.3. Because the current is the same in both resistors, the ratio of the voltage drop across the resistors (V_1/V_2) is the same as the ratio of the resistance of the resistors (R_1/R_2). We can then use Kirchhoff's loop rule to check that we calculated the voltage drop across each resistor correctly. According to Equation 6.2, the voltage of any closed loop must sum to zero. This means that the sum of the voltage drops across the two resistors must be equal in magnitude to the voltage of the battery.

Figure 6.2 shows two resistors in parallel. Notice there are two closed loops of current in this figure—an

FIGURE 6.2

A circuit with two resistors in parallel

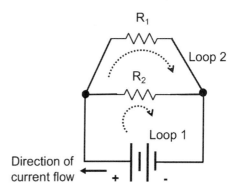

LAB 6

outer loop that contains the voltage source and R_1 and an inner loop that contains the voltage source and R_2. According to Kirchhoff's loop rule, the sum of the voltages in any closed loop must be equal to zero. That means that the voltage drop across the resistors in each loop must be equal in magnitude to the voltage source. Thus, the voltage drop across R_1 is equal to the voltage of the voltage source, and the voltage drop across R_2 is equal to the voltage of the voltage source. V_1 is equal to V_2.

We can then calculate the current through each resistor using Equation 6.3. Because the voltage drop is the same across the two resistors, the ratio of the current through the resistors (I_1/I_2) is equal to the inverse ratio of the resistance of each resistor (R_2/R_1). This makes conceptual sense as well—the larger resistor will have a smaller current running through it. Kirchhoff's junction rule can also be used to understand how the current flows through the two resistors—the total current going into junction A must be equal to the current leaving junction A. Because there are two paths leaving junction A, the current must split.

As a general rule of thumb, when two resistors are in series, they have the same current through them and the voltage drop across them is split. When two resistors are in parallel, they have the same voltage drop and the current is split.

In this investigation, students are given three resistors and asked to see how the arrangement of the three resistors will affect the current and voltage drop through each one. If all three resistors are the same resistance, when the resistors are in series they will all have the same current and the same voltage drop (the voltage drop through each resistor will be one-third the voltage of the battery). When three resistors of equal value are arranged in parallel, all three will have the same voltage drop and current (the current through each will be one-third the current flowing through the battery).

There are two other possible arrangements of the three resistors. Figure 6.3 shows two resistors in parallel with each other with the third resistor arranged in series. Assuming all three resistors have the same resistance, the current through R_1 and R_2 will be the same and equal to half the current through R_3. The current through R_3 will be equal to the total current flowing through the battery. The voltage drop across R_1 and R_2 will be one-third the voltage supplied by the battery, and the voltage drop across R_3 will be equal to two-thirds the total voltage supplied by the battery. If you use resistors of different resistances, the exact values of the current and the voltage drops will change.

Figure 6.4 shows the final possible arrangement of three resistors: a circuit with two resistors in series connected in parallel

FIGURE 6.3
A circuit with two resistors in parallel with each other connected in series to a third resistor

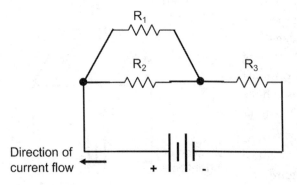

Series and Parallel Circuits

How Does the Arrangement of Three Resistors Affect the Voltage Drop Across and the Current Through Each Resistor?

to a third resistor. Assuming all three resistors have the same resistance, the voltage drop across R_1 and R_2 will be equal and will be half the voltage drop across R_3 (i.e., $V_3 = V_1 + V_2$; $V_1 = V_2$). The voltage drop across R_3 will be equal to the total voltage supplied by the battery. The current flowing through R_1 and R_2 will be the same and will be one-third the total current flowing through the battery. The current flowing through R_3 will be equal to two-thirds the total current flowing through the battery. If you use resistors of different resistances, the exact values of the current through each resistor and the voltage drop across each resistor will change.

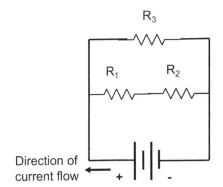

FIGURE 6.4

A circuit with two resistors in series connected in parallel to a third resistor

Timeline

The instructional time needed to complete this lab investigation is 170–230 minutes. Appendix 3 (p. 421) provides options for implementing this lab investigation over several class periods. Option C (230 minutes) should be used if students are unfamiliar with scientific writing, because this option provides extra instructional time for scaffolding the writing process. You can scaffold the writing process by modeling, providing examples, and providing hints as students write each section of the report. Option C can also be used if you are introducing students to the digital interface sensors and/or the data analysis software. Option D (170 minutes) should be used if students are familiar with scientific writing and have developed the skills needed to write an investigation report on their own. In option D, students complete stage 6 (writing the investigation report) and stage 8 (revising the investigation report) as homework.

Materials and Preparation

The materials needed to implement this investigation are listed in Table 6.1 (p. 142). The equipment can be purchased from a science supply company such as Flinn Scientific, PASCO, Vernier, or Ward's Science.

LAB 6

TABLE 6.1
Materials list for Lab 6

Item	Quantity
Consumable	
D battery	1–2 per group
Equipment and other materials	
Safety glasses with side shields or safety goggles	1 per student
Ammeter	1 per group
Voltmeter	1 per group
¼ W identical resistors in the range of 30–100 ohms	3 per group
Copper wire with an alligator clip on each end	7 per group
Investigation Proposal A (optional)	1 per group
Whiteboard, 2'× 3'*	1 per group
Lab Handout	1 per student
Peer-review guide and teacher scoring rubric	1 per student
Checkout Questions	1 per student
Equipment for digital interface measurements (optional)	
Digital interface with USB or wireless connections	1 per group
Current measurement sensor	1 per group
Voltage measurement sensor	1 per group
Computer or tablet with appropriate data analysis software installed	1 per group

* As an alternative, students can use computer and presentation software such as Microsoft PowerPoint or Apple Keynote to create their arguments.

Be sure to use a set routine for distributing and collecting the materials during the lab investigation. One option is to set up the materials for each group at each group's lab station before class begins. This option works well when there is a dedicated section of the classroom for lab work and the materials are large and difficult to move. A second option is to have all the materials on a table or cart at a central location. You can then assign a member of each group to be the "materials manager." This individual is responsible for collecting all the materials his or her group needs from the table or cart during class and for returning all the materials at the end of the class. This option works well when the materials are small and easy to move (such as magnets, wire, and bulbs). It also makes it easy to inventory the materials at the end of the class before students leave for the day.

Series and Parallel Circuits

How Does the Arrangement of Three Resistors Affect the Voltage Drop Across and the Current Through Each Resistor?

Safety Precautions and Laboratory Waste Disposal

Remind students to follow all normal lab safety rules. In addition, tell students to take the following safety precautions:

1. Wear sanitized safety glasses with side shields or goggles during lab setup, hands-on activity, and takedown.
2. Never put consumables in their mouth.
3. Never connect the terminals of a battery without a resistor between them.
4. Wire and other metals with electric current flowing through them may get hot. Use caution when handling components of a closed circuit.
5. Handle electrical wires with caution. They have sharp ends, which can cut or puncture skin.
6. Wash their hands with soap and water when they are done collecting the data.

Batteries and wire may be stored for future use. When batteries need replacing, dispose of old batteries according to manufacturer's recommendations.

Topics for the Explicit and Reflective Discussion

Reflecting on the Use of Core Ideas and Crosscutting Concepts During the Investigation

Teachers should begin the explicit and reflective discussion by asking students to discuss what they know about the core ideas they used during the investigation. The following are some important concepts related to the core ideas of circuits that students need to use to describe electric circuits:

- An electric current is a movement of charge through a conductor. Current is the flow of charge per unit time. Resistors use electric current to do work and impede the flow of charge.
- A circuit is a closed loop of electric current.
- The current through a resistor is equal to the potential difference across the resistor divided by its resistance.
- The current in a closed circuit is determined by the properties and arrangement of the individual circuit elements such as voltage sources, resistors, and capacitors.
- Kirchhoff's junction rule describes the conservation of electric charge in electrical circuits. Since charge is conserved, current must be conserved at each junction in the circuit.
- Kirchhoff's loop rule describes conservation of energy in electrical circuits. Since voltage times charge is energy, and energy is conserved, the sum of the voltages in a closed loop must add to zero. Typically, voltage is considered positive between

the terminals of the voltage source (e.g., a battery). This means that the charge gains energy when it moves inside the voltage source. When charge moves through the resistors in a circuit, the voltage drop across the resistor is considered negative because the charge loses energy in the resistor.

To help students reflect on what they know about electric circuits, we recommend showing them two or three images using presentation software that help illustrate these important ideas. You can then ask the students the following questions to encourage them to share how they are thinking about these important concepts:

1. What do we see going on in this image?
2. Does anyone have anything else to add?
3. What might be going on that we can't see?
4. What are some things that we are not sure about here?

You can then encourage students to think about how CCs played a role in their investigation. There are at least two CCs that students need to use to determine a mathematical relationship between current, voltage, and arrangement of resistors in a circuit: (a) Energy and Matter: Flows, Cycles, and Conservation and (b) Structure and Function (see Appendix 2 [p. 417] for a brief description of these CCs). To help students reflect on what they know about these CCs, we recommend asking them the following questions:

1. What evidence do you have of energy being conserved in an electric circuit?
2. What evidence do you have of electric charge being conserved in an electric circuit?
3. How does the structure of the circuit affect the function of the circuit?
4. What are the advantages of structuring resistors in series or in parallel?

You can then encourage the students to think about how they used all these different concepts to help answer the guiding question and why it is important to use these ideas to help justify their evidence for their final arguments. Be sure to remind your students to explain why they included the evidence in their arguments and make the assumptions underlying their analysis and interpretation of the data explicit in order to provide an adequate justification of their evidence.

Reflecting on Ways to Design Better Investigations

It is important for students to reflect on the strengths and weaknesses of the investigation they designed during the explicit and reflective discussion. Students should therefore be encouraged to discuss ways to eliminate potential flaws, measurement errors, or sources of uncertainty in their investigations. To help students be more reflective about the design of their investigation and what they can do to make their investigations more rigorous in the future, you can ask them the following questions:

1. What were some of the strengths of the way you planned and carried out your investigation? In other words, what made it scientific?

2. What were some of the weaknesses of the way you planned and carried out your investigation? In other words, what made it less scientific?

3. What rules can we make, as a class, to ensure that our next investigation is more scientific?

Reflecting on the Nature of Scientific Knowledge and Scientific Inquiry

This investigation can be used to illustrate two important concepts related to the nature of scientific knowledge and the nature of scientific inquiry: (a) the difference between laws and theories in science and (b) the assumptions made by scientists about order and consistency in nature (see Appendix 2 [p. 417] for a brief description of these concepts). Be sure to review these concepts during and at the end of the explicit and reflective discussion. To help students think about these concepts in relation to what they did during the lab, you can ask them the following questions:

1. Laws and theories are different in science. Why are the junction rule and the loop rule considered laws and not theories?

2. Can you work with your group to come up with a rule that you can use to decide if something is a law or a theory? Be ready to share in a few minutes.

3. Scientists assume that natural laws operate today as they did in the past and that they will continue to do so in the future. Why do you think this assumption is important?

4. Think about what you were trying to do during this investigation. What assumptions did you make regarding how natural laws continue to operate today as they did in the past?

You can also use presentation software or other techniques to encourage your students to think about these concepts. You can show examples of various laws and theories and ask students to classify each one as a law or a theory. You can also show images of different scientific laws and ask students if these laws would be the same everywhere if the universe was not a single system, and then ask them to think about what scientists would need to do to be able to study the universe if it was made up of many different systems. Be sure to remind your students that it is important for them to understand what counts as scientific knowledge and how that knowledge develops over time in order to be proficient in science.

LAB 6

Hints for Implementing the Lab

- Allowing students to design their own procedures for collecting data gives students an opportunity to try, to fail, and to learn from their mistakes. However, you can scaffold students as they develop their own procedure by having them fill out an investigation proposal. The proposals provide a way for you to offer students hints and suggestions without telling them how to do it. You can also check the proposals quickly during a class period. For this lab we suggest using Investigation Proposal A.

- Select appropriate batteries and resistors such that the current is large enough for the meters to measure, but not high enough to exceed the power rating of the resistors. Using D batteries with small ¼ W resistors in the range of 30–100 Ω can provide a useful starting point.

- Learn how to use the ammeter and the voltmeter before the lab begins. It is important for you to know how to use the equipment so you can help students when technical issues arise.

- Allow the students to become familiar with the meters as part of the tool talk before they begin to design their investigation. Give them 5–10 minutes to examine the equipment and materials before they begin designing their investigations. This gives students a chance to see what they can and cannot do with the equipment.

- There are two ways to modify this investigation for more advanced students (such as those in an AP class). First, you can give students resistors of different values. The mathematics of the current and voltage are much more involved when the resistors all have different values. A second approach would be to have students use a second voltage source in their circuits. The second voltage source highlights Kirchhoff's rules, because students must account for the fact that current from each battery may either add or subtract from each other.

- If students have difficulty getting noticeable results, it may be because the equipment is not sensitive enough to pick up small currents. We suggest having students use resistors of smaller resistance values and/or increasing the voltage to 3 V by arranging two D batteries in series.

- Be sure to allow students to go back and re-collect data at the end of the argumentation session. Students often realize that they made numerous mistakes when they were collecting data as a result of their discussions during the argumentation session. The students, as a result, will want a chance to re-collect data, and the re-collection of data should be encouraged when time allows. This also offers an opportunity to discuss what scientists do when they realize a mistake is made inside the lab.

If students use digital interface measurement equipment and analysis

- We suggest allowing students to familiarize themselves with the sensors and the data analysis software before they finalize the procedure for the investigation, especially if they have not used such software previously. This gives students an opportunity to learn how to work with the software and to improve the quality of the data they collect.

- Remind students to begin recording data just before they close the circuits.
- Remind students to follow the user's guide to correctly connect any sensors to avoid damage to lab equipment.

Connections to Standards

Table 6.2 highlights how the investigation can be used to address specific performance expectations from the *NGSS*; learning objectives from AP Physics 1 and 2; learning objectives from AP Physics C: Electricity and Magnetism; *Common Core State Standards for English Language Arts* (*CCSS ELA*); and *Common Core State Standards for Mathematics* (*CCSS Mathematics*).

TABLE 6.2

Lab 6 alignment with standards

NGSS performance expectations	• None
AP Physics 1 and AP Physics 2 learning objectives	• 4.E.5.1: Make and justify a quantitative prediction of the effect of a change in values or arrangements of one or two circuit elements on the currents and potential differences in a circuit containing a small number of sources of emf, resistors, capacitors, and switches in series and/or parallel. • 4.E.5.2. Make and justify a qualitative prediction of the effect of a change in values or arrangements of one or two circuit elements on currents and potential differences in a circuit containing a small number of sources of emf, resistors, capacitors, and switches in series and/or parallel. • 4.E.5.3: Plan data collection strategies and perform data analysis to examine the values of currents and potential differences in an electric circuit that is modified by changing or rearranging circuit elements, including sources of emf, resistors, and capacitors. • 5.B.9.4: Analyze experimental data including an analysis of experimental uncertainty that will demonstrate the validity of Kirchhoff's loop rule: $\sum \Delta V=0$. • 5.B.9.5: Use Describe and make predictions regarding electrical potential difference, charge, and current in steady-state circuits composed of various combinations of resistors and capacitors using conservation of energy principles (Kirchhoff's loop rule). • 5.B.9.6: Mathematically express the changes in electric potential energy of a loop in a multiloop electrical circuit and justify this expression using the principle of the conservation of energy.

Continued

LAB 6

Table 6.2 (*continued*)

AP Physics 1 and AP Physics 2 learning objectives (*continued*)	• 5.C.3.1: Apply conservation of electric charge (Kirchhoff's junction rule) to the comparison of electric current in various segments of an electrical circuit with a single battery and resistors in series and in, at most, one parallel branch and predict how those values would change if configurations of the circuit are changed. • 5.C.3.2: Design an investigation of an electrical circuit with one or more resistors in which evidence of conservation of electric charge can be collected and analyzed. • 5.C.3.4: Predict or describe current values in series and parallel arrangements of resistors and other branching circuits using Kirchhoff's junction rule, and explain the relationship of the rule to the law of charge conservation.
AP Physics C: Electricity and Magnetism learning objectives	• FIE-3.B.a: Describe the relationship between current, potential difference, and resistance of resistor using Ohm's Law. • FIE-3.B.b: Apply Ohm's Law in an operating circuit with a known resistor or resistances. • CNV-6.A.a: Identify parallel or series arrangement in a circuit containing multiple resistors. • CNV-6.A.b: Describe a series or a parallel arrangement of resistors. • CNV-6.B: Calculate equivalent resistances for a network of resistors that can be considered a combination of series and parallel arrangements. • CNV-6.C.a: Calculate voltage, current, and power dissipation for any resistor in a circuit containing a network of known resistors with a single battery or energy source. • CNV-6.C.b: Calculate relationships between the potential difference, current, resistance, and power dissipation for any part of a circuit, given some of the characteristics of the circuit (i.e., battery voltage or current in the battery, or a resistor or branch of resistors).
Literacy connections (*CCSS ELA*)	• *Reading*: Key ideas and details, craft and structure, integration of knowledge and ideas • *Writing*: Text types and purposes, production and distribution of writing, research to build and present knowledge, range of writing • *Speaking and listening*: Comprehension and collaboration, presentation of knowledge and ideas
Mathematics connections (*CCSS Mathematics*)	• *Mathematical practices*: Make sense of problems and persevere in solving them, reason abstractly and quantitatively, construct viable arguments and critique the reasoning of others, model with mathematics, use appropriate tools strategically, attend to precision • *Number and quantity*: Reason quantitatively and use units to solve problems, represent and model with vector quantities, perform operations on vectors

Continued

Table 6.2 (*continued*)

Mathematics connections (CCSS Mathematics) (*continued*)	• *Algebra*: Interpret the structure of expressions, create equations that describe numbers or relationships, understand solving equations as a process of reasoning and explain the reasoning, solve equations and inequalities in one variable, represent and solve equations and inequalities graphically. • *Functions*: Understand the concept of a function and use function notation; interpret functions that arise in applications in terms of the context; analyze functions using different representations; build a function that models a relationship between two quantities; construct and compare linear, quadratic, and exponential models and solve problems; interpret expressions for functions in terms of the situation they model • *Statistics and probability*: Summarize, represent, and interpret data on two categorical and quantitative variables; interpret linear models; make inferences and justify conclusions from sample surveys, experiments, and observational studies

LAB 6

Lab Handout

Lab 6. Series and Parallel Circuits: How Does the Arrangement of Three Resistors Affect the Voltage Drop Across and the Current Through Each Resistor?

Introduction

Electric circuits can be found in many of the tools we use for learning, work, and leisure. Computers, cell phones, cars, and headphones all contain several electric circuits. *Resistors* play an important role in the functioning of electric circuits. They are used to direct the flow of electrons to different parts of the circuit. Resistors are also able to convert electrical energy into light, heat, sound, and mechanical energy, depending on the needs of the device.

The earliest resistors were quite large and provided very little resistance. Figure L6.1 shows a 1-ohm (ohms are the unit of resistance) resistor from around 1900. Modern resistors, by contrast, are small enough to fit in the palm of the hand (see Figure L6.2) and provide thousands of times more resistance. The ability of engineers to build resistors at such a small scale allows for the size of many of our devices to continue to decrease.

The flow of electrons in a circuit is called *electric current*, measured in amperes (A). To measure the amount of current flowing in a wire, the circuit must be temporarily broken and an ammeter placed in series such that the current flows *through* the meter. The electrons gain energy when they pass through a battery and lose energy by passing through a resistor. *Voltage*, measured in volts (V), is the difference in electric potential energy of the electrons, per unit of charge, before and after the current passes through an element in a circuit. Voltage can be measured by touching the leads of a voltmeter to either side of a circuit component, such as the battery or a resistor. Measuring a positive voltage across a circuit element indicates the electrons are gaining energy as they flow through it, such as when they flow through a battery. A negative voltage indicates the electrons are losing energy, such as when they flow through a

FIGURE L6.1
A 1-ohm resistor from around 1900

FIGURE L6.2
A modern resistor

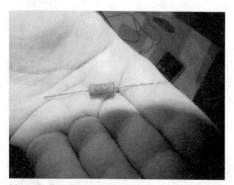

Note: Full-color versions of these figures are available on the book's Extras page at *www.nsta.org/adi-physics2*.

Series and Parallel Circuits

How Does the Arrangement of Three Resistors Affect the Voltage Drop Across and the Current Through Each Resistor?

resistor. Often, a negative voltage indicates that the resistor is using the energy for another purpose.

All circuits obey the law of conservation of energy. As the electrons flow through the circuit, they cannot lose more energy in the resistors than they gain in the battery. In a circuit with one battery and one resistor, the voltage across the battery and across the resistor are the same in magnitude but different in sign (e.g., the battery is positive and the resistor is negative)—the energy gained as the current flows through the battery is lost when the current flows through the resistor. However, the current flowing through both the resistor and the battery is the same in magnitude and in direction.

When building complex circuits containing multiple resistors, it is important to understand if and how the arrangement of the resistors affects the voltage drop across and the current through each resistor. This knowledge allows scientists and engineers to build circuits included in the electric tools we use.

Your Task

Use what you know about conservation of energy, tracking the flow of matter and energy in a system, and the relationship between structure and function to design and carry out an investigation to determine how voltage and current behave in a circuit with multiple resistors.

The guiding question of this investigation is, *How does the arrangement of three resistors affect the voltage drop across and the current through each resistor?*

Materials

You may use any of the following materials during your investigation:

Consumable
- D battery

Equipment
- Safety glasses with side shields or goggles (required)
- Ammeter
- Voltmeter
- Identical resistors
- Copper wire with alligator clips

If you have access to the following equipment, you may also consider using a digital current sensor and digital voltage sensor with an accompanying interface and a computer or tablet.

Safety Precautions

Follow all normal lab safety rules. In addition, take the following safety precautions:

1. Wear sanitized safety glasses with side shields or goggles during lab setup, hands-on activity, and takedown.

LAB 6

2. Never put consumables in your mouth.
3. Never connect the terminals of a battery without a resistor between them.
4. Wire and other metals with electric current flowing through them may get hot. Use caution when handling components of a closed circuit.
5. Handle electrical wires with caution. They have sharp ends, which can cut or puncture skin.
6. Wash your hands with soap and water when you are done collecting the data.

Investigation Proposal Required? ☐ Yes ☐ No

Getting Started

To answer the guiding question, you will need to design and carry out an investigation to determine current and voltage for each resistor in a circuit. Before you can design your investigation, however, you must determine what type of data you need to collect, how you will collect it, and how you will analyze it.

To determine *what type of data you need to collect*, think about the following questions:

- What are the boundaries and components of the system?
- How do the components of the system interact with each other?
- How does energy flow into, out of, or within this system?
- How might the structure of what you are studying relate to its function?
- What is going on at the unobservable level that could cause the things that you observe?

To determine *how you will collect the data*, think about the following questions:

- What scale or scales should you use when you take your measurements?
- How will you make sure that your data are of high quality (i.e., how will you reduce error)?
- How will you keep track of and organize the data you collect?
- How can you track how energy flows into, out of, or within this system?
- Where should you place the voltmeter and ammeter relative to the other components of the circuit?

To determine *how you will analyze the data*, think about the following questions:

- What type of calculations, if any, will you need to make?
- What types of patterns might you look for as you analyze your data?

Series and Parallel Circuits
How Does the Arrangement of Three Resistors Affect the Voltage Drop Across and the Current Through Each Resistor?

- What type of table or graph could you create to help make sense of your data?
- Are there any proportional relationships that you can identify?
- How does the ratio of current and voltage for one resistor compare with the ratio of current and voltage for another resistor?

Connections to the Nature of Scientific Knowledge and Scientific Inquiry

As you work through your investigation, you may want to consider

- the difference between laws and theories in science, and
- the assumptions made by scientists about order and consistency in nature.

Initial Argument

Once your group has finished collecting and analyzing your data, your group will need to develop an initial argument. Your initial argument needs to include a claim, evidence to support your claim, and a justification of the evidence. The *claim* is your group's answer to the guiding question. The *evidence* is an analysis and interpretation of your data. Finally, the *justification* of the evidence is why your group thinks the evidence matters. The justification of the evidence is important because scientists can use different kinds of evidence to support their claims. Your group will create your initial argument on a whiteboard. Your whiteboard should include all the information shown in Figure L6.3.

FIGURE L6.3
Argument presentation on a whiteboard

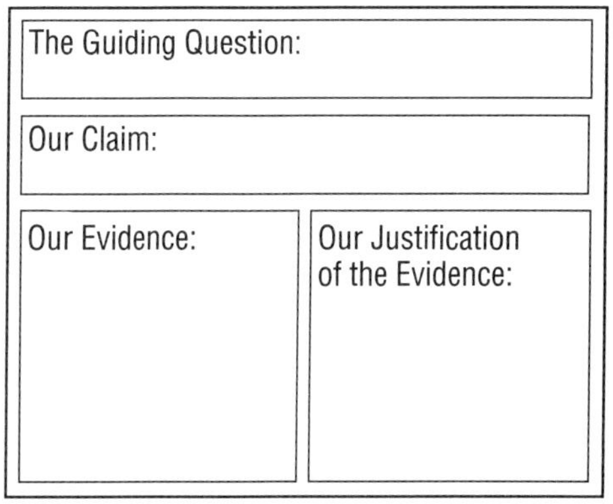

The Guiding Question:	
Our Claim:	
Our Evidence:	Our Justification of the Evidence:

Argumentation Session

The argumentation session allows all of the groups to share their arguments. One or two members of each group will stay at the lab station to share that group's argument, while the other members of the group go to the other lab stations to listen to and critique the other arguments. This is similar to what scientists do when they propose, support, evaluate, and refine new ideas during a poster session at a conference. If you are presenting your group's argument, your goal is to share your ideas and answer questions. You should also keep a record of the critiques and suggestions made by your classmates so you can use this feedback to make your initial argument stronger. You can keep track of specific critiques and suggestions for improvement that your classmates mention in the space below.

LAB 6

Critiques about our initial argument and suggestions for improvement:

If you are critiquing your classmates' arguments, your goal is to look for mistakes in their arguments and offer suggestions for improvement so these mistakes can be fixed. You should look for ways to make your initial argument stronger by looking for things that the other groups did well. You can keep track of interesting ideas that you see and hear during the argumentation in the space below. You can also use this space to keep track of any questions that you will need to discuss with your team.

Interesting ideas from other groups or questions to take back to my group:

Series and Parallel Circuits
How Does the Arrangement of Three Resistors Affect the Voltage Drop Across and the Current Through Each Resistor?

Once the argumentation session is complete, you will have a chance to meet with your group and revise your initial argument. Your group might need to gather more data or design a way to test one or more alternative claims as part of this process. Remember, your goal at this stage of the investigation is to develop the best argument possible.

Report

Once you have completed your research, you will need to prepare an *investigation report* that consists of three sections. Each section should provide an answer to the following questions:

1. What question were you trying to answer and why?
2. What did you do to answer your question and why?
3. What is your argument?

Your report should answer these questions in two pages or less. This report must be typed, and any diagrams, figures, or tables should be embedded into the document. Be sure to write in a persuasive style; you are trying to convince others that your claim is acceptable or valid!

LAB 6

Checkout Questions

Lab 6. Series and Parallel Circuits: How Does the Arrangement of Three Resistors Affect the Voltage Drop Across and the Current Through Each Resistor?

1. The diagram below shows a circuit with three resistors.

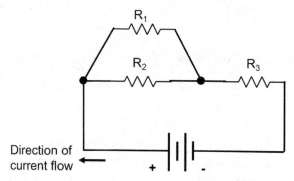

a. Rank the three resistors in order of increasing current.

b. Rank the three resistors in order of increasing voltage.

c. How did you determine your rankings?

Series and Parallel Circuits
How Does the Arrangement of Three Resistors Affect the Voltage Drop Across and the Current Through Each Resistor?

2. The diagram below shows another circuit with three identical resistors. Use the following information to answer questions 2a–d: A voltmeter measures –5 V across R_3, and an ammeter measures 10 A through R_2.

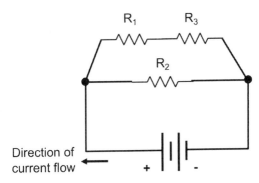

a. How much voltage does the battery provide?

b. How do you know?

c. How much current flows through R_3?

d. How do you know?

LAB 6

3. As students investigate an electric circuit, they report measuring "2 volts across the resistor and 3 amperes through it." Why do you think they used the words *across* and *through* in this context? Are the two words interchangeable? Explain your reasoning.

4. Kirchhoff's loop rule states that the sum of the voltages for a closed loop in a circuit will always equal zero. Would you classify this as a law or a theory? Explain your reasoning.

5. There is an important relationship between structure and function in many natural and engineered systems.

    a. I agree with this statement.
    b. I disagree with this statement.

    Explain your answer, using examples from this investigation and at least one other investigation you have conducted.

6. Why is it useful to understand how energy is conserved during an investigation? In your answer, be sure to use examples from this investigation and at least one other investigation you have conducted.

7. Many investigations are informed by an assumption that the natural laws operate the same way today that they have in the past, and that they will continue to operate the same way in the future. Why is this assumption important for scientists and engineers? In your answer, be sure to use examples from this investigation and at least one other investigation you have conducted.

LAB 7

Teacher Notes

Lab 7. Resistance of a Wire: What Factors Affect the Resistance of a Wire?

Purpose

The purpose of this lab is to *introduce* students to the disciplinary core idea (DCI) of Structure and Properties of Matter (PS1.A) from the *NGSS* and the concepts of resistance and current by having them identify factors that impact the resistance of wire. In addition, this lab can be used to help students understand two big ideas from AP Physics: (a) objects and systems have properties such as mass and charge, and systems may have internal structure; and (b) interactions between systems can result in changes in those systems. This lab also gives students an opportunity to learn about the crosscutting concepts (CCs) of (a) Scale, Proportion, and Quantity and (b) Structure and Function from the *NGSS*. As part of the explicit and reflective discussion, students will also learn about (a) the difference between observations and inferences in science and (b) the difference between laws and theories in science.

Underlying Physics Concepts

When analyzing circuits that contain a resistor, it is often best to start with Ohm's law, which states that the current through any closed circuit is directly proportional to the potential difference (i.e., voltage) applied to the circuit and inversely proportional to the resistance of the circuit. Ohm's law is shown mathematically in Equation 7.1, where V is the potential difference or voltage applied to the circuit, I is the current through the circuit, and R is the resistance of the circuit. In SI units, voltage is measured in volts (V), current is measured in amperes (A), and resistance is measured in ohms (Ω).

$$\text{(Equation 7.1)} \quad V = IR$$

When analyzing circuits, we often assume that the resistor or resistors connected to the wire provide the only resistance impeding the flow of electric charge. In reality, this is not the case, and the wire itself adds resistance to the circuit.

Solid metals, such as copper or silver, exist in a state where the nuclei are arranged in a lattice structure and the electrons are free to move because they are not bound to any specific nucleus. The positively charge nuclei are weakly bonded to each other to maintain the solid structure and are surrounded in a sea of electrons. The electrons move freely, in random directions and with random velocities. When averaged over the entire solid metal, the net velocity of all the electrons is zero (0 m/s). Figure 7.1 shows a simplified version of this phenomenon.

Resistance of a Wire
What Factors Affect the Resistance of a Wire?

When a voltage source is connected to a wire to form a closed circuit, an electric field is induced inside the wire. This electric field is relatively uniform and propagates down the wire at nearly the speed of light. Because electrons have a charge, a force is induced on each electron by the electric field, and the field does work on each electron. Because the electric field is relatively uniform, the force induced on each electron is in the same direction. This leads to each electron accelerating in the same direction, toward the positive terminal of the voltage source. Despite their previous random motion, all the electrons will accelerate in the same direction once the voltage is applied and the electric field propagates down the wire.

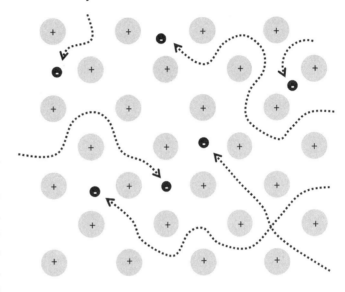

FIGURE 7.1

A solid metal, with the positive nuclei arranged in a lattice surrounded by a sea of electrons

Although the force from the electric field acts on all the electrons in the same direction—toward the positive terminal—each electron had a velocity prior to the application of the electric field to the wire. Therefore, the resulting velocities of the electrons are not all equal, nor are they all in the direction toward the positive terminal of the voltage source. This causes the electrons to collide with each other and, in some cases, the positively charged nuclei. These collisions, combined with the initial velocities in directions other than toward the positive terminal, slow the movement of electrons toward the positive terminal. In other words, the collisions produce resistance to the flow of current through the wire.

After a short period of time, the electrons reach a relatively stable average velocity toward the positive terminal. This average velocity is called the drift velocity, and it is typically much smaller than the random velocities each electron has prior to closing the circuit. Furthermore, the drift velocity and the current are not the same thing; the current is the total charge passing a point per unit time, whereas the drift velocity is the average velocity of the electrons toward the positive terminal.

When a circuit is closed, the current flows almost instantly, despite the drift velocity being minimal. This is because the electric field driving the current propagates down the wire at nearly the speed of light. Thus, work is done on the electrons throughout the wire at almost the same time.

With an idea of how a voltage source causes current to flow, we can then begin to explore the factors that influence the resistance of the wire. The first factor has to do with

TABLE 7.1

Resistivity values for various metals

Metal	Resistivity (Ω·m)
Aluminum	2.65x10⁻⁸
Copper	1.68x10⁻⁸
Gold	2.44x10⁻⁸
Iron	9.71x10⁻⁸
Silver	1.59x10⁻⁸

Source: Values come from Giancoli, D. C. 2005. *Physics: Principles with applications.* 6th ed. Upper Saddle River, NJ: Pearson.

the material composition of the wire. As mentioned above, metals exist in a lattice structure. The exact structure of the lattice influences the flow of electrons, and the lattice structure is a function of the individual atoms. Different metals will have slightly different lattice structures, which influence the resistance of the wire. This property is called the resistivity of the metal and is an intrinsic property of the metal. Thus, the resistance of a wire is a function of the metal the wire is made from. Table 7.1 provides a list of resistivity values for common metals. Silver has a resistivity lower than that of copper, but silver's higher cost offsets its lower resistivity as a useful material for most commercial wires.

Recall that the resistance of the wire arises from the collisions that occur as electrons move down the wire with the drift velocity. As the length of the wire increases, the number of collisions that each electron will experience as it moves down the length of the wire also increases.

Finally, current is carried by electrons. As the cross-sectional area of the wire (for commercial wire, the gauge) increases, the more electrons there are to carry current per unit length of the wire. Thus, as the cross-sectional area of the wire increases, the resistance of the wire decreases.

Taken together, the resistance of a wire is (1) a function of the metal that makes up the wire, (2) directly proportional to the length of the wire, and (3) inversely proportional to the cross-sectional area of the wire. Equation 7.2 shows the mathematical relationship between these four variables, where R is the resistance of the wire, ρ is the resistivity of the metal used in the wire, L is the length of the wire, and A is the cross-sectional area of the wire. In SI units, resistance is measured in ohms (Ω), resistivity in ohm-meters (Ω·m), length in meters (m), and area in meters squared (m²).

(Equation 7.2) $$R = \rho \frac{L}{A}$$

Finally, the resistivity itself is also a function of the temperature of the wire. For most metals, the resistivity increases as the temperature of the metal increases. This is because an increase in the temperature causes the lattice to vibrate, thereby increasing the potential for collisions with the free electrons. The relationship between temperature and resistivity is shown in Equation 7.3, where ρ_T is the resistivity at temperature T, ρ_o is the resistivity at a reference temperature (often 0°C), α is the temperature coefficient of resistivity, T is the temperature of the wire, and T_o is the reference temperature. In SI units, the temperature coefficient of resistivity is measured in units of 1 divided by degrees Celsius (°C^{-1}), and temperature is measured in degrees Celsius (°C).

(Equation 7.3) $\rho_T = \rho_o[1 + \alpha(T - T_o)]$

We mention this relationship between temperature and resistivity because there may be students who hypothesize that temperature does affect the resistance of the wire. However, this is beyond the scope of first-year and AP physics courses.

Timeline

The instructional time needed to complete this lab investigation is 170–230 minutes. Appendix 3 (p. 421) provides options for implementing this lab investigation over several class periods. Option C (230 minutes) should be used if students are unfamiliar with scientific writing, because this option provides extra instructional time for scaffolding the writing process. You can scaffold the writing process by modeling, providing examples, and providing hints as students write each section of the report. Option C can also be used if you are introducing students to the digital interface sensors and/or the data analysis software. Option D (170 minutes) should be used if students are familiar with scientific writing and have developed the skills needed to write an investigation report on their own. In option D, students complete stage 6 (writing the investigation report) and stage 8 (revising the investigation report) as homework.

Materials and Preparation

The materials needed to implement this investigation are listed in Table 7.2. The equipment can be purchased from a science supply company such as Flinn Scientific, PASCO, Vernier, or Ward's Science.

TABLE 7.2

Materials list for Lab 7

Item	Quantity
Consumables	
D batteries	2–3 per group
Masking or cellophane tape	As needed
Equipment and other materials	
Safety glasses with side shields or goggles	1 per student
Voltmeter*	1 per group

Continued

Table 7.2 (*continued*)

Item	Quantity
Ammeter*	1 per group
Copper wire (multiple gauges)	As needed
Aluminum wire (multiple gauges)	As needed
Various resistors	As needed
Ruler	1 per goup
Meterstick	1 per group
Wire cutter	1 per group
Investigation Proposal C (optional)	1 per group
Whiteboard, 2' × 3'†	1 per group
Lab Handout	1 per student
Peer-review guide and teacher scoring rubric	1 per student
Checkout Questions	1 per student
Equipment for digital interface measurements (optional)	
Digital interface with USB or wireless connections	1 per group
Current measurement sensor	1 per group
Voltage measurement sensor	1 per group
Computer or tablet with appropriate data analysis software installed	1 per group

* Instead of giving students a voltmeter and an ammeter, you may choose to give them a multimeter, which is a device that combines the voltmeter and the ammeter into one unit.

† As an alternative, students can use computer and presentation software such as Microsoft PowerPoint or Apple Keynote to create their arguments.

Be sure to use a set routine for distributing and collecting the materials during the lab investigation. One option is to set up the materials for each group at each group's lab station before class begins. This option works well when there is a dedicated section of the classroom for lab work and the materials are large and difficult to move. A second option is to have all the materials on a table or cart at a central location. You can then assign a member of each group to be the "materials manager." This individual is responsible for collecting all the materials his or her group needs from the table or cart during class and for returning all the materials at the end of the class. This option works well when the materials are small and easy to move (such as magnets, wire, and bulbs). It also makes it easy to inventory the materials at the end of the class before students leave for the day.

Resistance of a Wire
What Factors Affect the Resistance of a Wire?

Safety Precautions and Laboratory Waste Disposal

Remind students to follow all normal lab safety rules. In addition, tell students to take the following safety precautions:

1. Wear sanitized safety glasses with side shields or goggles during lab setup, hands-on activity, and takedown.

2. Never put consumables in their mouth.

3. Wire and other metals with electric current flowing through them may get hot. Use caution when handling components of a closed circuit.

4. Wire cutters are very sharp, so make sure to keep fingers out of the way when cutting wire.

5. Use caution when working with sharp objects (e.g., wires) because they can cut or puncture skin.

6. Wash their hands with soap and water when they are done collecting the data.

Batteries and wire may be stored for future use. When batteries need replacing, dispose of old batteries according to manufacturer's recommendations.

Topics for the Explicit and Reflective Discussion
Reflecting on the Use of Core Ideas and Crosscutting Concepts During the Investigation

Teachers should begin the explicit and reflective discussion by asking students to discuss what they know about the core ideas they used during the investigation. The following are some important concepts related to the core ideas of resistance and resistivity of a wire that students need to use to determine the variables that affect the resistance of the wire:

- Electric current is the movement of electrons through the wire. As the electrons move through the wire, they collide with other electrons and, at times, the positively charged nuclei in the wire. These collisions slow the movement of the electrons, giving rise to the resistance in the wire.

- All matter has an intrinsic property called resistivity. The resistivity of a material depends on its molecular and atomic structure and the temperature of the material. The resistivity is related to the frequency of the collisions between electrons.

- The resistance of a length of wire is directly proportional to its length and inversely proportional to its cross-sectional area. The longer the wire, the more collisions each electron will experience as it moves, hence a greater resistance. The larger the cross-sectional area, the more electrons are moving per unit length of the wire, hence a decreased resistance.

LAB 7

- Resistance can be determined experimentally through the application of Ohm's law ($V = IR$) and cannot be measured directly in this investigation.

To help students reflect on what they know about resistance and resistivity, we recommend showing them two or three images using presentation software that help illustrate these important ideas. You can then ask the students the following questions to encourage them to share how they are thinking about these important concepts:

1. What do we see going on in this image?
2. Does anyone have anything else to add?
3. What might be going on that we can't see?
4. What are some things that we are not sure about here?

You can then encourage students to think about how CCs played a role in their investigation. There are at least two CCs that students need to use to determine the factors that affect the resistance of the wire: (a) Scale, Proportion, and Quantity and (b) Structure and Function (see Appendix 2 [p. 417] for a brief description of these CCs). To help students reflect on what they know about these CCs, we recommend asking them the following questions:

1. What proportional relationships did you uncover in this investigation? Were they all direct proportions?
2. What units do you need to use for the constant of proportionality in order to combine your various proportional relationships into a single equation? (Note that in this lab, the constant of proportionality is the resistivity of the metal.)
3. How does the atomic structure of metals affect how a wire functions when it carries an electric current?
4. How does the structure of a wire affect how the wire functions when it carries an electric current?

You can then encourage the students to think about how they used all these different concepts to help answer the guiding question and why it is important to use these ideas to help justify their evidence for their final arguments. Be sure to remind your students to explain why they included the evidence in their arguments and make the assumptions underlying their analysis and interpretation of the data explicit in order to provide an adequate justification of their evidence.

Reflecting on Ways to Design Better Investigations

It is important for students to reflect on the strengths and weaknesses of the investigation they designed during the explicit and reflective discussion. Students should therefore be encouraged to discuss ways to eliminate potential flaws, measurement errors, or sources of uncertainty in their investigations. To help students be more reflective about the design of their investigation and what they can do to make their investigations more rigorous in the future, you can ask them the following questions:

1. What were some of the strengths of the way you planned and carried out your investigation? In other words, what made it scientific?

2. What were some of the weaknesses of the way you planned and carried out your investigation? In other words, what made it less scientific?

3. What rules can we make, as a class, to ensure that our next investigation is more scientific?

Reflecting on the Nature of Scientific Knowledge and Scientific Inquiry

This investigation can be used to illustrate two important concepts related to the nature of scientific knowledge and the nature of scientific inquiry: (a) the difference between observations and inferences in science and (b) the difference between laws and theories in science (see Appendix 2 [p. 417] for a brief description of these concepts). Be sure to review these concepts during and at the end of the explicit and reflective discussion. To help students think about these concepts in relation to what they did during the lab, you can ask them the following questions:

1. You made observations and inferences during your investigation. Can you give me some examples of these observations and inferences?

2. Can you work with your group to come up with a rule that you can use to tell the difference between an observation and an inference? Be ready to share in a few minutes.

3. Laws and theories are different in science. What are some examples of laws that you used in your investigation, and what are some examples of theories that you used?

4. Can you work with your group to come up with a rule that you can use to decide if something is a law or a theory? Be ready to share in a few minutes.

You can show examples of information from the investigation that are either observations or inferences and ask students to classify each example and explain their thinking. You can also encourage the students to think about these concepts by showing examples of laws (such as Ohm's law or Newton's law of universal gravitation) and theories (such as electromagnetic theory, atomic theory, or Einstein's theory of gravity) and ask students

LAB 7

to indicate if they think each example is a law or a theory and explain their thinking. Be sure to remind your students that it is important for them to understand what counts as scientific knowledge and how that knowledge develops over time in order to be proficient in science.

Hints for Implementing the Lab

- Allowing students to design their own procedures for collecting data gives students an opportunity to try, to fail, and to learn from their mistakes. However, you can scaffold students as they develop their own procedure by having them fill out an investigation proposal. The proposals provide a way for you to offer students hints and suggestions without telling them how to do it. You can also check the proposals quickly during a class period. Using an investigation proposal is also highly recommended for this lab, which has a potential for waste (using up batteries), because it will allow you to review each group's procedure and provide suggestions for how they can improve their procedure to reduce waste. For this lab we suggest using Investigation Proposal C.

- Learn how to use the ammeter, voltmeter, and/or multimeter before the lab begins. It is important for you to know how to use the equipment so you can help students when technical issues arise.

- Allow the students to become familiar with the meters as part of the tool talk before they begin to design their investigation. Give them 5–10 minutes to examine the equipment and materials before they begin designing their investigations. This gives students a chance to see what they can and cannot do with the equipment.

- Because the resistance of the wire is minimal, the batteries will lose charge quickly. We recommend using fresh batteries each class period. To mitigate the loss of charge, you can also have students insert a resistor into the circuit. This will moderate the current flow and keep the batteries fresh. However, the larger the resistance value of the resistor, the less sensitive the circuit will be to changes in the wire, such as increasing the length. If the resistance of a resistor added to the circuit is too large, the effect of changing the wire will be harder to measure.

- For students who want to collect temperature data, the best way to do this is to allow the wire to heat up naturally when a circuit is closed and current flows through the wire. Thus, students could look at how the resistance of the wire changes over time. However, the battery will quickly lose charge and the observed decrease in current may be partly due to the decrease in voltage of the battery and not due to the increased resistance of the wire. Students can control for the voltage of the battery by periodically inserting a new battery into the circuit.

- To see how length affects the resistance of the wire, we suggest using extended lengths of up to (or even greater than) 5 m. To reduce the cost, you can give each group a segment of wire that is 5 m long and tell them that they should cut smaller pieces from that wire, and work backward, testing the longest segment first. Alternatively, you can pre-cut various lengths of wire and allow students to use those segments by checking the segments out from you.

- Be sure to allow students to go back and re-collect data at the end of the argumentation session. Students often realize that they made numerous mistakes when they were collecting data as a result of their discussions during the argumentation session. The students, as a result, will want a chance to re-collect data, and the re-collection of data should be encouraged when time allows. This also offers an opportunity to discuss what scientists do when they realize a mistake is made inside the lab.

If students use digital interface measurement equipment and analysis

- We suggest allowing students to familiarize themselves with the sensors and the data analysis software before they finalize the procedure for the investigation, especially if they have not used such software previously. This gives students an opportunity to learn how to work with the software and to improve the quality of the data they collect.
- Remind students to begin recording data just before they close the circuit.
- Remind students to follow the user's guide to correctly connect any sensors to avoid damage to lab equipment.

Connections to Standards

Table 7.3 (p. 170) highlights how the investigation can be used to address specific performance expectations from the *NGSS*; learning objectives from AP Physics 1 and 2; learning objectives from AP Physics C: Electricity and Magnetism; *Common Core State Standards for English Language Arts* (*CCSS ELA*); and *Common Core State Standards for Mathematics* (*CCSS Mathematics*).

LAB 7

TABLE 7.3
Lab 7 alignment with standards

NGSS performance expectation	• HS-PS2-6: Communicate scientific and technical information about why the molecular-level structure is important in the functioning of designed materials.
AP Physics 1 and AP Physics 2 learning objectives	• 1.E.2.1: Choose and justify the selection of data needed to determine resistivity for a given material. • 4.E.4.1: Make predictions about the properties of resistors and/or capacitors when placed in a simple circuit based on the geometry of the circuit element and supported by scientific theories and mathematical relationships. • 4.E.4.2: Design a plan for the collection of data to determine the effect of changing the geometry and/or materials on the resistance or capacitance of a circuit element and relate results to the basic properties of resistors and capacitors. • 4.E.4.3: Analyze data to determine the effect of changing the geometry and/or materials on the resistance or capacitance of a circuit element and relate results to the basic properties of resistors and capacitors.
AP Physics C: Electricity and Magnetism learning objectives	• FIE-3.C.a: Explain how the properties of a conductor affect resistance. • FIE-3.C.b: Compare resistances of conductors with different geometries or material. • FIE-3.C.c: Calculate the resistance of a conductor of known resistivity and geometry.
Literacy connections (*CCSS ELA*)	• *Reading*: Key ideas and details, craft and structure, integration of knowledge and ideas • *Writing:* Text types and purposes, production and distribution of writing, research to build and present knowledge, range of writing • *Speaking and listening:* Comprehension and collaboration, presentation of knowledge and ideas

Continued

Table 7.3 (*continued*)

Mathematics connections (CCSS Mathematics)	• *Mathematical practices:* Make sense of problems and persevere in solving them, reason abstractly and quantitatively, construct viable arguments and critique the reasoning of others, model with mathematics, use appropriate tools strategically, attend to precision • *Number and quantity:* Reason quantitatively and use units to solve problems, represent and model with vector quantities, perform operations on vectors • *Algebra:* Interpret the structure of expressions, create equations that describe numbers or relationships, understand solving equations as a process of reasoning and explain the reasoning, solve equations and inequalities in one variable, represent and solve equations and inequalities graphically • *Functions:* Understand the concept of a function and use function notation; interpret functions that arise in applications in terms of the context; analyze functions using different representations; build a function that models a relationship between two quantities; construct and compare linear, quadratic, and exponential models and solve problems; interpret expressions for functions in terms of the situation they model • *Statistics and probability:* Summarize, represent, and interpret data on two categorical and quantitative variables; interpret linear models; make inferences and justify conclusions from sample surveys, experiments, and observational studies

LAB 7

Lab Handout

Lab 7. Resistance of a Wire: What Factors Affect the Resistance of a Wire?

Introduction

One of the great feats of human engineering was the development of the transatlantic telegraph cable. The first successful transatlantic telegraph cable was made up of seven copper wires of different cross-sectional areas covered in a latex produced from the sap of the gutta-percha plant and then wrapped in tarred hemp and iron cable. When completed in 1858, the cable had a mass density of approximately 550 kg/km, equivalent to 1,951 lb/mile (Dibner 1959). Figure L7.1 shows a cross section of the copper wires covered in the latex. The transatlantic telegraph cable was a tremendous improvement in communication between Europe and North America because it allowed a message to be transmitted across the Atlantic Ocean in only 17 hours, compared with 10 days by boat.

FIGURE L7.1
A cross section of the first transatlantic telegraph cable

To send a message, a telegraph operator would use Morse code (or another code) to send each letter of the message. When sending a letter, the operator would briefly close a circuit, thereby sending an electric current down the wire. When we say it took 17 hours to transmit a message via the first transatlantic telegraph cable, we mean that the electric current took 17 hours to travel the length of the wire.

Electric current is the flow of electric charge per second. The current in a circuit is dependent on two factors: the voltage sources connected to the circuit and the resistance of the circuit. The greater the voltage, the more current will flow; the greater the resistance of the circuit, the less current will flow. Mathematically, these two relationships are described by Ohm's law, where V is the potential difference or voltage applied to the circuit, I is the current flowing through the circuit, and R is the total resistance of the circuit. In SI units, voltage is measured in volts (V), current is measured in amperes or amps (A), and resistance is measured in ohms (Ω).

Ohm's law $V = IR$

Resistance of a Wire
What Factors Affect the Resistance of a Wire?

In metals, charge flows because electrons are free to move between atoms. Most solid metals exist in a lattice structure, where the electrons are not bound to any one atom or compound (unlike in ionic and covalent bonds). When a circuit consisting of a voltage source (e.g., a battery) is closed, the voltage source does work on the electrons, causing them to move. The electrons have a negative charge, and they move toward the positive terminal of the voltage source. As they move toward the positive terminal, the electrons collide with each other and the positively charged nucleus and slowly move down the length of the wire. This causes a slight resistance to the flow of electric charge.

In many cases, the effects of the electron collisions can be ignored when analyzing circuits, and the flow of electrons is assumed to be instantaneous. In other cases, such as the transatlantic telegraph cable, this added resistance of the wire cannot be ignored. This added resistance is also important in the design of electrical infrastructure in cities and when wiring electrical devices together.

Your Task

Use what you know about electric current, proportional relationships, and structure and function to design and carry out an investigation to determine the variables that affect the resistance of the wire.

The guiding question of this investigation is, **What factors affect the resistance of a wire?**

Materials

You may use any of the following materials during your investigation:

Consumables
- D batteries
- Masking or cellophane tape

Equipment
- Safety glasses with side shields or goggles (required)
- Voltmeter and ammeter (or multimeter)
- Copper wire (multiple gauges)
- Aluminum wire (multiple gauges)
- Various resistors
- Ruler
- Meterstick
- Wire cutter

If you have access to the following equipment, you may also consider using a digital current sensor and digital voltage sensor with an accompanying interface and a computer or tablet.

Safety Precautions

Follow all normal lab safety rules. In addition, take the following safety precautions:

1. Wear sanitized safety glasses with side shields or goggles during lab setup, hands-on activity, and takedown.

LAB 7

2. Never put consumables in your mouth.

3. Wire and other metals with electric current flowing through them may get hot. Use caution when handling components of a closed circuit.

4. Wire cutters are very sharp. Make sure your fingers are out of the way when cutting wire.

5. Use caution when working with sharp objects (e.g., wires) because they can cut or puncture skin.

6. Wash your hands with soap and water when you are done collecting the data.

Investigation Proposal Required?   ☐ Yes   ☐ No

Getting Started

To answer the guiding question, you will need to design and carry out an investigation to determine what variables affect the resistance of the wire. Before you can design your investigation, however, you must determine what type of data you need to collect, how you will collect it, and how you will analyze it.

To determine *what type of data you need to collect,* think about the following questions:

- What variables might affect the resistance of the wire?
- Can you directly measure the resistance of the wire, or will you need to calculate it based on other measurements?
- What are the boundaries and components of the system?
- How do the components of the system interact with each other?
- How could you keep track of changes in this system quantitatively?
- How might the structure of what you are studying relate to its function?

To determine *how you will collect the data,* think about the following questions:

- What is the independent variable and what is the dependent variable?
- What other factors will you need to control during each experiment?
- What scale or scales should you use when you take your measurements?
- How will you make sure that your data are of high quality (i.e., how will you reduce error)?
- How will you keep track of and organize the data you collect?

To determine *how you will analyze the data,* think about the following questions:

- What type of calculations, if any, will you need to make?

- What types of patterns might you look for as you analyze your data?
- What type of table or graph could you create to help make sense of your data?
- What types of proportional relationships might you look for as you analyze your data?

Connections to the Nature of Scientific Knowledge and Scientific Inquiry

As you work through your investigation, you may want to consider

- the difference between observations and inferences in science, and
- the difference between laws and theories in science.

Initial Argument

Once your group has finished collecting and analyzing your data, your group will need to develop an initial argument. Your initial argument needs to include a claim, evidence to support your claim, and a justification of the evidence. The *claim* is your group's answer to the guiding question. The *evidence* is an analysis and interpretation of your data. Finally, the *justification* of the evidence is why your group thinks the evidence matters. The justification of the evidence is important because scientists can use different kinds of evidence to support their claims. Your group will create your initial argument on a whiteboard. Your whiteboard should include all the information shown in Figure L7.2.

FIGURE L7.2
Argument presentation on a whiteboard

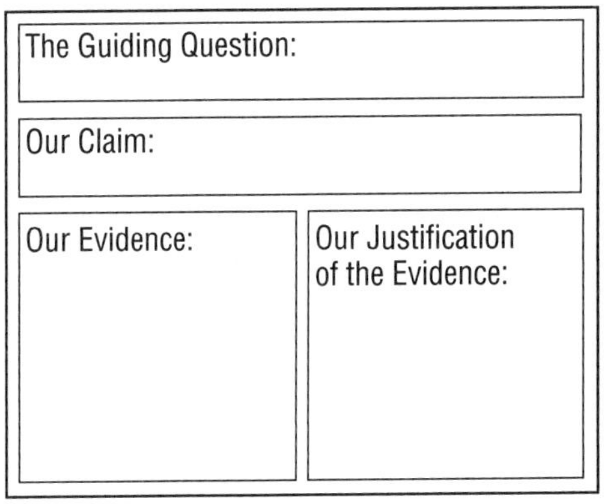

The Guiding Question:	
Our Claim:	
Our Evidence:	Our Justification of the Evidence:

Argumentation Session

The argumentation session allows all of the groups to share their arguments. One or two members of each group will stay at the lab station to share that group's argument, while the other members of the group go to the other lab stations to listen to and critique the other arguments. This is similar to what scientists do when they propose, support, evaluate, and refine new ideas during a poster session at a conference. If you are presenting your group's argument, your goal is to share your ideas and answer questions. You should also keep a record of the critiques and suggestions made by your classmates so you can use this feedback to make your initial argument stronger. You can keep track of specific critiques and suggestions for improvement that your classmates mention in the space below.

LAB 7

Critiques about our initial argument and suggestions for improvement:

If you are critiquing your classmates' arguments, your goal is to look for mistakes in their arguments and offer suggestions for improvement so these mistakes can be fixed. You should look for ways to make your initial argument stronger by looking for things that the other groups did well. You can keep track of interesting ideas that you see and hear during the argumentation in the space below. You can also use this space to keep track of any questions that you will need to discuss with your team.

Interesting ideas from other groups or questions to take back to my group:

Once the argumentation session is complete, you will have a chance to meet with your group and revise your initial argument. Your group might need to gather more data or design a way to test one or more alternative claims as part of this process. Remember, your goal at this stage of the investigation is to develop the best argument possible.

Report

Once you have completed your research, you will need to prepare an *investigation report* that consists of three sections. Each section should provide an answer to the following questions:

1. What question were you trying to answer and why?
2. What did you do to answer your question and why?
3. What is your argument?

Your report should answer these questions in two pages or less. This report must be typed, and any diagrams, figures, or tables should be embedded into the document. Be sure to write in a persuasive style; you are trying to convince others that your claim is acceptable or valid!

Reference

Dibner, B. 1959. The Atlantic cable. Norwalk, CT: Burndy Library. *www.sil.si.edu/DigitalCollections/hst/atlantic-cable/ac-index.htm.*

LAB 7

Checkout Questions

Lab 7. Resistance of a Wire: What Factors Affect the Resistance of a Wire?

1. The internal resistance of a wire with length L is 4 Ω. What will the resistance of the wire be if the length is doubled?

2. The internal resistance of a wire of length L is 4 Ω when the radius of the wire is r. If the same material is used to make a wire of length L with a radius of 2 r, what will the resistance of the wire be?

3. A theory turns into a law once it has been proven to be true.

   a. I agree with this statement.
   b. I disagree with this statement.

   Explain your answer, using examples from this investigation and at least one other investigation you have conducted.

4. *Observation* and *inference* are terms that have the same meaning in science.

   a. I agree with this statement.
   b. I disagree with this statement.

   Explain your answer, using examples from this investigation and at least one other investigation you have conducted.

5. Why is it important to identify constants of proportionality and the units associated with the constants in science? In your answer, be sure to use examples from this investigation and at least one other investigation you have conducted.

6. In nature, the structure of materials often affects how they function in designed systems. How did the structure of the metal wire affect how it functioned when it carried current? In your answer, be sure to use examples from this investigation and at least one other investigation you have conducted.

Application Labs

LAB 8

Teacher Notes

Lab 8. Power, Voltage, and Resistance in a Circuit: What Is the Mathematical Relationship Between the Voltage of a Battery, the Total Resistance of a Circuit, and the Power Output of a Motor?

Purpose

The purpose of this lab is for students to *apply* what they know about power, voltage, and resistance and the disciplinary core idea (DCI) of Conservation of Energy and Energy Transfer (PS3.B) from the *NGSS* to determine the relationship between voltage, total resistance of a circuit, and the power output of the circuit, which in this lab is a small electric motor. In addition, this lab can be used to help students understand two big ideas from AP Physics: (a) objects and systems have properties such as mass and charge, and systems may have internal structure; and (b) changes that occur as a result of interactions are constrained by conservation laws. This lab also gives students an opportunity to learn about the crosscutting concepts (CCs) of (a) Scale, Proportion, and Quantity and (b) Energy and Matter: Flows, Cycles, and Conservation from the *NGSS*. As part of the explicit and reflective discussion, students will also learn about (a) the difference between observations and inferences in science and (b) how the culture of science, societal needs, and current events influence the work of scientists.

Underlying Physics Concepts

In this lab, students apply the concepts of voltage, resistance, current, and power that were developed prior to this lab. Students are likely familiar with mechanical power from units on work and energy. In this lab, students will use that knowledge to determine the relationship between voltage, resistance, and electric power.

Underlying much of this lab is Ohm's law, which states that voltage and current are directly proportional in a circuit and resistance and current are inversely proportional in a circuit. As electric power is dependent on voltage, current, and resistance, it is important that the relationship between these three properties of a circuit is well understood. Current and resistance in a circuit are related to the voltage of a circuit using Equation 8.1, where V is the voltage drop (or potential difference) across the circuit, I is the current through the circuit, and R is the resistance of the circuit. In SI units, voltage is measured in volts (V), current is measured in amperes (A), and resistance is measured in ohms (Ω).

(Equation 8.1) $V = IR$

Power, Voltage, and Resistance in a Circuit
What Is the Mathematical Relationship Between the Voltage of a Battery, the Total Resistance of a Circuit, and the Power Output of a Motor?

Power, in the general sense, is defined as the rate of doing work or the amount of energy transferred per unit time. This is shown in Equation 8.2, where P is power, E is energy, and t is time. In SI units, power is measured in watts (W), energy is measured in joules (J), and time is measured in seconds (s).

(Equation 8.2) $P = E/t$

The energy carried by a moving electric charge is the charge multiplied by the potential difference (voltage) as the charge moves. Thus, we can rewrite Equation 8.2 as Equation 8.3, where Q is the electric charge. In SI units, charge is measured in coulombs (C).

(Equation 8.3) $P = QV/t$

Finally, charge per unit time is current. Thus, the power of the circuit as it relates to the total current flowing through the circuit is shown in Equation 8.4.

(Equation 8.4) $P = IV$

If we rearrange Equation 8.1 such that we get current in terms of the voltage of the circuit and the resistance of the circuit (i.e., $I = V/R$), and then replace the current with voltage over resistance in Equation 8.4, we get Equation 8.5.

(Equation 8.5) $P = V^2/R$

Thus, the total power of the circuit is directly proportional to the square of the voltage. As the voltage source increases, the power of the components in the circuit will increase. At the same time, the power of the components in the circuit is inversely proportional to the total resistance of the circuit.

It is also possible to use Ohm's law to express the power of the circuit relative to the current and resistance of the circuit. In this case, the power of the circuit is $P = I^2R$. We suggest that students use Equation 8.5 instead, because the current is an emergent property of the circuit—you cannot directly manipulate the current. Students can, however, change the resistance of the circuit or the voltage source that the circuit is connected to.

Finally, the power of the circuit can be related to the mechanical energy of a motor (or the light energy produced by a lightbulb, the sound energy produced by a speaker, and so forth). In the case of a motor, the electric power produced by the circuit equals the mechanical power of the motor (assuming a perfectly efficient circuit and motor where no energy is lost to heat). The power output related to mechanical energy is shown in Equation 8.6, where **F** is the force applied to an object and **d** is the displacement of the object due to the force (the quantity **Fd** is the amount of work done on an object). In SI units, force is measured in newtons (N) and displacement is measured in meters (m).

LAB 8

(Equation 8.6) $P = Fd/t$

If we attach a string and a mass to the motor and allow the string to hang from the motor, when the motor is connected to the circuit, it will wind the string and lift the mass. We can set Equation 8.6 equal to Equation 8.5, as shown in Equation 8.7, where *m* is the mass hanging from the motor, **g** is the acceleration due to gravity, and **h** is the vertical displacement that the mass moves in time *t*. In SI units, mass is measured in kilograms (kg) and acceleration due to gravity near Earth's surface has a value of 9.8 m/s².

(Equation 8.7) $mgh/t = V^2/R$

Note that the left side of Equation 8.7 is equal to the potential energy of an object in Earth's gravitational field due to its height **h** above the ground. This is explained by the work-energy theorem, which states that work done on an object or system results in a change to the energy of the object or system. In this case, the motor does work on the hanging mass *m*, lifting it to a height **h** above the ground. From the law of conservation of energy, we can say that the energy needed to lift the mass must be equal to the energy output by the motor.

Students must understand how each load will dissipate energy. An electric motor will rotate more rapidly when consuming more power and will rotate more slowly when consuming less power. This will result in the motor lifting the hanging mass to the same height in different times—the greater the power output of the motor, the less time it will take to lift the mass. Figure 8.1 shows a possible circuit setup; it is simple and prevents other circuit components from interfering with the test. Both the voltage source and the resistor should be easy to modify.

FIGURE 8.1
Simple circuit for testing

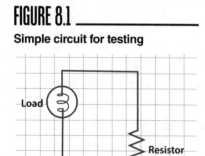

When increasing the voltage, students should see a significant positive effect on the load. As shown in Equation 8.5, power is proportional to the square of voltage. Thus, if the voltage is doubled, students should observe a fourfold increase in power output in the load. If using D batteries, students can start with one battery and double from one to two and then from two to four batteries.

When increasing resistance, students should see the power consumption of the load decrease. As shown in Equation 8.5, power is inversely proportional to resistance. Thus, if the resistance of the resistor is doubled, students should observe a twofold decrease in power output in the load.

We are making several assumptions in this investigation:

1. The internal resistance of the load is constant and does not change with current or voltage.

2. The load's visible power output responds directly and regularly to the amount of power that the load is consuming.

3. Energy is not lost to heat in the circuit, and the resistance of each component in the circuit is not a function of their temperature.

You may want to mention these assumptions during the explicit and reflective discussion as well.

Timeline

The instructional time needed to complete this lab investigation is 200–280 minutes. Appendix 3 (p. 421) provides options for implementing this lab investigation over several class periods. Option E (280 minutes) should be used if students are unfamiliar with scientific writing, because this option provides extra instructional time for scaffolding the writing process. You can scaffold the writing process by modeling, providing examples, and providing hints as students write each section of the report. Option E can also be used if you are introducing students to the video analysis programs. Option F (200 minutes) should be used if students are familiar with scientific writing and have developed the skills needed to write an investigation report on their own. In option F, students complete stage 6 (writing the investigation report) and stage 8 (revising the investigation report) as homework.

Materials and Preparation

The materials needed to implement this investigation are listed in Table 8.1 (p. 186). The equipment can be purchased from a science supply company such as Flinn Scientific, PASCO, Vernier, or Ward's Science. Video analysis software can be purchased from Vernier (Logger *Pro*) or PASCO (SPARKvue or Capstone). These companies also have apps that can be used on Apple- or Android-based tablets and cell phones. We recommend consulting with your school's information technology coordinator to determine the best option for your students.

LAB 8

TABLE 8.1

Materials list for Lab 8

Item	Quantity
Consumables	
D batteries	4–6 per group
9V battery	1 per group
String	50–100 cm per group
Tape	As needed per group
Equipment and other materials	
Safety glasses with side shields or safety goggles	1 per student
Battery holders	4–6 per group
Resistors (various resistances)	4–6 per group
Copper wires with an alligator clip on each end	4–6 per group
Small motor	1 per group
Washers or other masses	1 per group
Stopwatch	1 per group
Investigation Proposal C (optional)	1 per group
Whiteboard, 2' × 3'*	1 per group
Lab Handout	1 per student
Peer-review guide and teacher scoring rubric	1 per student
Checkout Questions	1 per student
Equipment for video analysis (optional)	
Video camera	1 per group
Computer or tablet with video analysis software	1 per group

* As an alternative, students can use computer and presentation software such as Microsoft PowerPoint or Apple Keynote to create their arguments.

Use of video analysis software is optional, but using this software will allow students to relate the output of the motor in the form of mechanical energy to the input to the motor in the form of electrical energy.

Be sure to use a set routine for distributing and collecting the materials during the lab investigation. One option is to set up the materials for each group at each group's lab

Power, Voltage, and Resistance in a Circuit

What Is the Mathematical Relationship Between the Voltage of a Battery, the Total Resistance of a Circuit, and the Power Output of a Motor?

station before class begins. This option works well when there is a dedicated section of the classroom for lab work and the materials are large and difficult to move. A second option is to have all the materials on a table or cart at a central location. You can then assign a member of each group to be the "materials manager." This individual is responsible for collecting all the materials his or her group needs from the table or cart during class and for returning all the materials at the end of the class. This option works well when the materials are small and easy to move (such as magnets, wire, and bulbs). It also makes it easy to inventory the materials at the end of the class before students leave for the day.

Safety Precautions and Laboratory Waste Disposal

Remind students to follow all normal lab safety rules. In addition, tell students to take the following safety precautions:

1. Wear sanitized safety glasses with side shields or goggles during lab setup, hands-on activity, and takedown.

2. Never put consumables in their mouth.

3. Keep their fingers and toes out of the way of the moving objects.

4. Wire and other metals with electric current flowing through them may get hot. Use caution when handling components of a closed circuit.

5. Handle electrical wires with caution. They have sharp ends, which can cut or puncture skin.

6. Wash their hands with soap and water when they are done collecting the data.

Batteries and wire may be stored for future use. When batteries need replacing, dispose of old batteries according to manufacturer's recommendations.

Topics for the Explicit and Reflective Discussion

Reflecting on the Use of Core Ideas and Crosscutting Concepts During the Investigation

Teachers should begin the explicit and reflective discussion by asking students to discuss what they know about the core ideas they used during the investigation. The following are some important concepts related to the core idea of energy transfer in circuits that students need to use to develop the relationship between power, voltage, and resistance:

- The values of currents and electric potential differences in an electric circuit are determined by the properties and arrangement of the individual circuit elements such as sources of electromotive force (emf) and resistors.

- The rate at which energy is transferred in a circuit is directly proportional to the square of the voltage source and inversely proportional to the resistance of the circuit.

LAB 8

- Energy is conserved in all closed systems.
- Energy transfer in mechanical or electrical systems may occur at different rates. Power is defined as the rate of energy transfer into, out of, or within a system.

To help students reflect on what they know about electric power, we recommend showing them two or three images using presentation software that help illustrate these important ideas. You can then ask the students the following questions to encourage them to share how they are thinking about these important concepts:

1. What do we see going on in this image?
2. Does anyone have anything else to add?
3. What might be going on that we can't see?
4. What are some things that we are not sure about here?

You can then encourage students to think about how CCs played a role in their investigation. There are at least two CCs that students need to use to determine a mathematical relationship between power, voltage, and resistance: (a) Scale, Proportion, and Quantity and (b) Energy and Matter: Flows, Cycles, and Conservation (see Appendix 2 [p. 417] for a brief description of these CCs). To help students reflect on what they know about these CCs, we recommend asking them the following questions:

1. What types of proportional relationships did you uncover in your investigation? Are all proportional relationships linear?
2. Why is it important to identify proportional relationships in science?
3. What assumptions did you use to help you determine the relationship between power, voltage, and resistance in your investigation?
4. How does the law of conservation of energy relate to power?

You can then encourage the students to think about how they used all these different concepts to help answer the guiding question and why it is important to use these ideas to help justify their evidence for their final arguments. Be sure to remind your students to explain why they included the evidence in their arguments and make the assumptions underlying their analysis and interpretation of the data explicit in order to provide an adequate justification of their evidence.

Reflecting on Ways to Design Better Investigations

It is important for students to reflect on the strengths and weaknesses of the investigation they designed during the explicit and reflective discussion. Students should therefore be encouraged to discuss ways to eliminate potential flaws, measurement errors, or sources of uncertainty in their investigations. To help students be more reflective about the design of their investigation and what they can do to make their investigations more rigorous in the future, you can ask them the following questions:

1. What were some of the strengths of the way you planned and carried out your investigation? In other words, what made it scientific?

2. What were some of the weaknesses of the way you planned and carried out your investigation? In other words, what made it less scientific?

3. What rules can we make, as a class, to ensure that our next investigation is more scientific?

Reflecting on the Nature of Scientific Knowledge and Scientific Inquiry

This investigation can be used to illustrate two important concepts related to the nature of scientific knowledge and the nature of scientific inquiry: (a) the difference between observations and inferences in science and (b) how the culture of science, societal needs, and current events influence the work of scientists (see Appendix 2 [p. 417] for a brief description of these concepts). Be sure to review these concepts during and at the end of the explicit and reflective discussion. To help students think about these concepts in relation to what they did during the lab, you can ask them the following questions:

1. You made observations and inferences during your investigation. Can you give me some examples of these observations and inferences?

2. Can you work with your group to come up with a rule that you can use to tell the difference between an observation and an inference? Be ready to share in a few minutes.

3. People view some types of research as being more important than other types of research because of cultural values and current events. Can you come up with some examples of how cultural values and current events have influenced the work of scientists?

4. Scientists share a set of values, norms, and commitments that shape what counts as knowing, how to represent or communicate information, and how to interact with other scientists. Can you work with your group to come up with a rule that you can use to decide if something is science or not science? Be ready to share in a few minutes.

You can also encourage the students to think about these concepts by showing examples of observations and inferences using presentation software and then asking students to classify each one. You can also show images of different types of science and engineering projects and ask them to indicate why the scientists or engineers decided to pursue that project—what cultural values and current events influenced the choices of the scientists?

LAB 8

Be sure to remind your students that it is important for them to understand what counts as scientific knowledge and how that knowledge develops over time in order to be proficient in science.

Hints for Implementing the Lab

- Allowing students to design their own procedures for collecting data gives students an opportunity to try, to fail, and to learn from their mistakes. However, you can scaffold students as they develop their own procedure by having them fill out an investigation proposal. The proposals provide a way for you to offer students hints and suggestions without telling them how to do it. You can also check the proposals quickly during a class period. For this lab we suggest using Investigation Proposal C.

- Allow the students to become familiar with all of the materials as part of the tool talk before they begin to design their investigation. Give them 5–10 minutes to examine the equipment and materials before they begin designing their investigations. This gives students a chance to see what they can and cannot do with the equipment.

- Make sure the motor is appropriately rated for the mass that it will lift. If you use too heavy of a mass, the motor may burn out. Similarly, make sure that the motor can handle the current supplied by the voltage source. Too large of a current will overload the circuit.

- If multiples of the same types of resistors are available, doubling the amount each time would allow for the best collection of data. Students can insert the second resistor in series with the other resistor and the motor.

- The equations used to describe the relationships in the "Underlying Physics Concepts" section represent idealized versions of the relationships governed by some important assumptions. Depending on the data collection procedures students use, their results may deviate from the idealized condition. We suggest discussing the assumptions underlying their investigations during the explicit and reflective discussion and asking students to mention this in their reports.

- Be sure to allow students to go back and re-collect data at the end of the argumentation session. Students often realize that they made numerous mistakes when they were collecting data as a result of their discussions during the argumentation session. The students, as a result, will want a chance to re-collect data, and the re-collection of data should be encouraged when time allows. This also offers an opportunity to discuss what scientists do when they realize a mistake is made inside the lab.

If students use video analysis

- We suggest allowing students to familiarize themselves with the video analysis software before they finalize the procedure for the investigation, especially if they have not used such software previously. This gives students an opportunity to learn how to work with the software and to improve the quality of the video they take.

- Remind students to hold the video camera as still as possible. Any movement of the camera will introduce error into their analysis. If using actual camcorders, we recommend using a tripod to hold the camera steady. If students are using a camera on a cell phone or tablet, we recommend using a table to help steady the camera.

- Remind students to place a meterstick in the same field of view as the motion they are capturing. Also, the meterstick should be approximately the same distance from the camera as the motion. Most video analysis software requires the user define a scale in the video (this allows the software to establish distances and, subsequently, other variables dependent on distance and displacement).

Connections to Standards

Table 8.2 highlights how the investigation can be used to address specific performance expectations from the *NGSS;* learning objectives from AP Physics 1 and 2; learning objectives from AP Physics C: Electricity and Magnetism; *Common Core State Standards for English Language Arts* (*CCSS ELA*); and *Common Core State Standards for Mathematics* (*CCSS Mathematics*).

TABLE 8.2

Lab 8 alignment with standards

NGSS performance expectation	- HS-PS3-1: Create a computational model to calculate the change in the energy of one component in a system when the change in energy of the other component(s) and energy flows in and out of the system are known.
AP Physics 1 and AP Physics 2 learning objectives	- 4.E.5.1: Make and justify a quantitative prediction of the effect of a change in values or arrangements of one or two circuit elements on the currents and potential differences in a circuit containing a small number of sources of emf, resistors, capacitors, and switches in series and/or parallel. - 4.E.5.2: Make and justify a qualitative prediction of the effect of a change in values or arrangements of one or two circuit elements on currents and potential differences in a circuit containing a small number of sources of emf, resistors, capacitors, and switches in series and/or parallel. - 4.E.5.3: Plan data collection strategies and perform data analysis to examine the values of currents and potential differences in an electric circuit that is modified by changing or rearranging circuit elements, including sources of emf, resistors, and capacitors. - 5.B.9.8: Translate between graphical and symbolic representations of experimental data describing relationships among power, current, and potential difference across a resistor.

Continued

Table 8.2 (continued)

AP Physics C: Electricity and Magnetism learning objectives	• CNV-6.C.a: Calculate voltage, current, and power dissipation for any resistors in a circuit containing a network of known resistors with a single battery or energy source. • CNV-6.C.b: Calculate relationships between the potential difference, current, resistance, and power dissipation for any part of a circuit, given some of the characteristics of the circuit (i.e., battery voltage or current in the battery, or a resistor or branch of resistors).
Literacy connections (*CCSS ELA*)	• *Reading:* Key ideas and details, craft and structure, integration of knowledge and ideas • *Writing:* Text types and purposes, production and distribution of writing, research to build and present knowledge, range of writing • *Speaking and listening:* Comprehension and collaboration, presentation of knowledge and ideas
Mathematics connections (*CCSS Mathematics*)	• *Mathematical practices:* Make sense of problems and persevere in solving them, reason abstractly and quantitatively, construct viable arguments and critique the reasoning of others, model with mathematics, use appropriate tools strategically, attend to precision • *Number and Quantity:* Reason quantitatively and use units to solve problems, represent and model with vector quantities, perform operations on vectors • *Algebra:* Interpret the structure of expressions, create equations that describe numbers or relationships, understand solving equations as a process of reasoning and explain the reasoning, solve equations and inequalities in one variable, represent and solve equations and inequalities graphically • *Functions:* Understand the concept of a function and use function notation; interpret functions that arise in applications in terms of the context; analyze functions using different representations; build a function that models a relationship between two quantities; construct and compare linear, quadratic, and exponential models and solve problems; interpret expressions for functions in terms of the situation they model, • *Statistics and Probability:* Summarize, represent, and interpret data on two categorical and quantitative variables; interpret linear models; make inferences and justify conclusions from sample surveys, experiments, and observational studies

Lab Handout

Lab 8. Power, Voltage, and Resistance in a Circuit: What Is the Mathematical Relationship Between the Voltage of a Battery, the Total Resistance of a Circuit, and the Power Output of a Motor?

Introduction

Previously, you have learned how changing the voltage or resistance of an electric circuit affects the current through the circuit. Mathematically, these three quantities are related through *Ohm's law*, $V = IR$, where V is voltage (measured in volts), I is current (measured in amperes, or amps), and R is resistance (measured in ohms). From a practical context, it is often easier to directly change the voltage of a circuit (through the addition or subtraction of batteries) or the resistance of a circuit (through the addition, subtraction, and position of resistors) than it is to change the current of a circuit. While voltage, resistance, and current are all very useful pieces of information to know from a scientific perspective, they do not directly tell us what a particular circuit can do for us. That is because electrical energy produced by a circuit must first be transformed into another source of energy—light, sound, heat, or mechanical energy.

To understand how useful a circuit component is, engineers are often interested in the rate with which the energy can be transformed by circuit components, such as resistors and motors. The rate of energy transfer is energy per unit time, or the *power* of the circuit (measured in watts). In our daily lives, we also talk about electrical devices in terms of their power consumption—how much electrical energy is transferred by a device per unit time. As an example, Figure L8.1 shows a lightbulb package that says the bulbs are "60 watt," which means that they transform 60 joules of electrical energy to light energy every second. Many other electronic devices are rated by their wattage. Hair dryers, portable heaters, and microwave ovens also include their wattage in the packaging.

FIGURE L8.1

A 60-watt lightbulb. This lightbulb transforms 60 joules of electrical energy to light energy every second.

Knowing how to increase, decrease, or more generally manage the power of a circuit is incredibly important when designing properly functioning electronics. One way to explore the relationship between voltage, resistance, current, and power is to change one factor and

LAB 8

then measure the effect on a load. An electric load is something that consumes electric power. Common loads include lightbulbs, motors, buzzers, and the common resistor. By determining how each load expresses its power consumption and then quantifying those results, it is possible to create mathematical relationships between (1) voltage and power and (2) resistance and power.

Your Task

Use what you know about circuits, energy transfer, and proportional relationships to design and carry out an investigation to determine the relationship between voltage, resistance, and power.

The guiding question of this investigation is, *What is the mathematical relationship between the voltage of a battery, the total resistance of a circuit, and the power output of a motor?*

Materials

You may use any of the following materials during your investigation:

Consumables
- D batteries
- 9V battery
- String
- Tape

Equipment
- Safety glasses with side shields or goggles (required)
- Battery holders
- Resistors (various resistances)
- Wires with alligator clips
- Small motor
- Washers or other masses
- Stopwatch

If you have access to the following equipment, you may also consider using a video camera and a computer or tablet.

Safety Precautions

Follow all normal lab safety rules. In addition, take the following safety precautions:

1. Wear sanitized safety glasses with side shields or goggles during lab setup, hands-on activity, and takedown.

2. Never put consumables in your mouth.

3. Keep fingers and toes out of the way of the moving objects.

4. Wire and other metals with electric current flowing through them may get hot. Use caution when handling components of a closed circuit.

5. Handle electrical wires with caution. They have sharp ends, which can cut or puncture skin.

6. Wash your hands with soap and water when you are done collecting the data.

Power, Voltage, and Resistance in a Circuit

What Is the Mathematical Relationship Between the Voltage of a Battery, the Total Resistance of a Circuit, and the Power Output of a Motor?

Investigation Proposal Required? ☐ Yes ☐ No

Getting Started

To answer the guiding question, you will need to design and carry out two separate experiments to determine how the power consumed by the load is affected by changes in voltage and resistance. Figure L8.2 illustrates how you can use the available equipment to study this relationship. Before you can design your investigation, however, you must determine what type of data you need to collect, how you will collect it, and how you will analyze it.

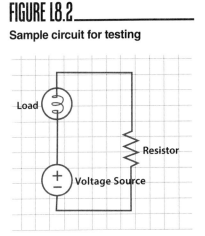

FIGURE L8.2

Sample circuit for testing

To determine *what type of data you need to collect*, think about the following questions:

- How could you keep track of changes in this system quantitatively?
- How useful is it to track how energy flows into, out of, or within this system?
- How might your load respond to increases in power? Decreases in power?
- How might changes to the structure of the circuit change how it functions?
- What quantities are vectors and what quantities are scalars?

To determine *how you will collect the data*, think about the following questions:

- What is the independent variable and what is the dependent variable?
- What other factors will you need to control during each experiment?
- What scale or scales should you use when you take your measurements?
- How can you track how energy flows into, out of, or within this system?
- How will you make sure that your data are of high quality (i.e., how will you reduce error)?
- How will you keep track of and organize the data you collect?

To determine *how you will analyze the data*, think about the following questions:

- What type of calculations, if any, will you need to make?
- What types of patterns might you look for as you analyze your data?
- What type of table or graph could you create to help make sense of your data?
- What types of proportional relationships might you look for as you analyze your data?
- How will you model the system to indicate under what parameters the power production is stable?

LAB 8

Connections to the Nature of Scientific Knowledge and Scientific Inquiry

As you work through your investigation, you may want to consider

- the difference between observations and inferences in science, and
- how the culture of science, societal needs, and current events influence the work of scientists.

Initial Argument

Once your group has finished collecting and analyzing your data, your group will need to develop an initial argument. Your initial argument needs to include a claim, evidence to support your claim, and a justification of the evidence. The *claim* is your group's answer to the guiding question. The *evidence* is an analysis and interpretation of your data. Finally, the *justification* of the evidence is why your group thinks the evidence matters. The justification of the evidence is important because scientists can use different kinds of evidence to support their claims. Your group will create your initial argument on a whiteboard. Your whiteboard should include all the information shown in Figure L8.3.

FIGURE L8.3
Argument presentation on a whiteboard

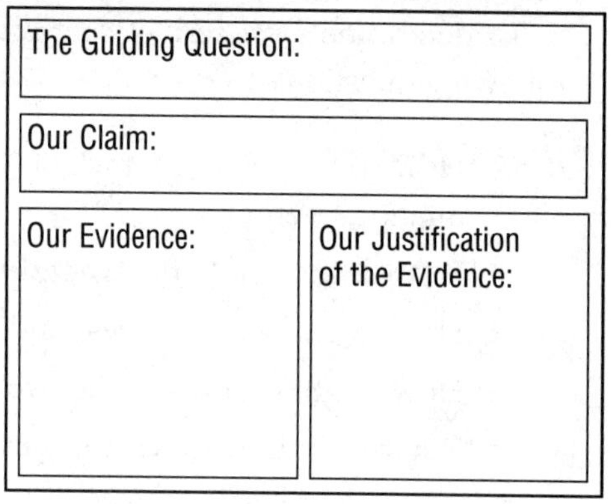

The Guiding Question:	
Our Claim:	
Our Evidence:	Our Justification of the Evidence:

Argumentation Session

The argumentation session allows all of the groups to share their arguments. One or two members of each group will stay at the lab station to share that group's argument, while the other members of the group go to the other lab stations to listen to and critique the other arguments. This is similar to what scientists do when they propose, support, evaluate, and refine new ideas during a poster session at a conference. If you are presenting your group's argument, your goal is to share your ideas and answer questions. You should also keep a record of the critiques and suggestions made by your classmates so you can use this feedback to make your initial argument stronger. You can keep track of specific critiques and suggestions for improvement that your classmates mention in the space below.

Power, Voltage, and Resistance in a Circuit
What Is the Mathematical Relationship Between the Voltage of a Battery, the Total Resistance of a Circuit, and the Power Output of a Motor?

Critiques about our initial argument and suggestions for improvement:

If you are critiquing your classmates' arguments, your goal is to look for mistakes in their arguments and offer suggestions for improvement so these mistakes can be fixed. You should look for ways to make your initial argument stronger by looking for things that the other groups did well. You can keep track of interesting ideas that you see and hear during the argumentation in the space below. You can also use this space to keep track of any questions that you will need to discuss with your team.

Interesting ideas from other groups or questions to take back to my group:

LAB 8

Once the argumentation session is complete, you will have a chance to meet with your group and revise your initial argument. Your group might need to gather more data or design a way to test one or more alternative claims as part of this process. Remember, your goal at this stage of the investigation is to develop the best argument possible.

Report

Once you have completed your research, you will need to prepare an *investigation report* that consists of three sections. Each section should provide an answer to the following questions:

1. What question were you trying to answer and why?
2. What did you do to answer your question and why?
3. What is your argument?

Your report should answer these questions in two pages or less. This report must be typed, and any diagrams, figures, or tables should be embedded into the document. Be sure to write in a persuasive style; you are trying to convince others that your claim is acceptable or valid!

Checkout Questions

Lab 8. Power, Voltage, and Resistance in a Circuit: What Is the Mathematical Relationship Between the Voltage of a Battery, the Total Resistance of a Circuit, and the Power Output of a Motor?

1. The figures below show two circuits. Each circuit is comprised of two resistors of resistance R wired with a battery of voltage V.

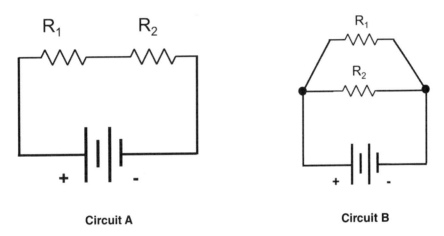

Circuit A   Circuit B

   a. In which circuit will the total power output from the battery be the greatest?

   b. How do you know?

LAB 8

2. The figures below show two circuits. Each circuit is composed of three resistors of resistance R wired with a battery of voltage V.

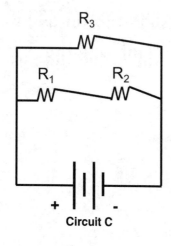
Circuit C

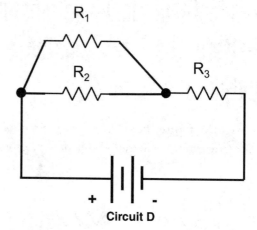
Circuit D

   a. In which circuit will the total power output from the battery be the greatest?

   b. How do you know?

3. A motor is attached to a crane so that it can lift a mass m to a height h in time t. The motor is wired to a circuit with a total resistance of R and powered by a voltage source V. At what velocity does the mass rise as a result of being lifted by the crane, in terms of **m, h, t, R,** and **V**?

4. The difference between observations and inferences is that observations are based on data and inferences are one's opinion.

   a. I agree with this statement.
   b. I disagree with this statement.

   Explain your answer, using examples from this investigation and at least one other investigation you have conducted.

5. Scientists are often interested in uncovering proportional relationships in science. What types of proportional relationships exist in science, and why are they important to discover? Explain your answer, using examples from this investigation and at least one other investigation you have conducted.

6. The law of conservation of energy cuts across many different areas of physics and all scientific fields. Explain how the law of conservation of energy allows scientists to make claims about one type of system (e.g., electrical, mechanical, chemical, biological) that interacts with a second system of a different kind. In your answer, be sure to use examples from this investigation and at least one other investigation you have conducted.

7. The research done by a scientist is often influenced by what is important in society.

   a. I agree with this statement.
   b. I disagree with this statement.

   Explain your answer, using examples from this investigation and at least one other investigation you have conducted.

LAB 9

Teacher Notes

Lab 9. Unknown Resistors in a Circuit: Given One Known Resistor and a Voltmeter, How Can You Determine the Resistance of the Unknown Resistors in a Circuit?

Purpose

The purpose of this lab is for students to *apply* what they know about the disciplinary core idea (DCI) of Conservation of Energy and Energy Transfer (PS3.B) from the *NGSS* and about resistors in series and in parallel by having them find the resistance of several resistors in a complex circuit. In addition, this lab can be used to help students understand two big ideas from AP Physics: (a) interactions between systems can result in changes in those systems and (b) changes that occur as a result of interactions are constrained by conservation laws. This lab also gives students an opportunity to learn about the crosscutting concepts (CCs) of (a) Structure and Function and (b) Energy and Matter: Flows, Cycles, and Conservation from the *NGSS*. As part of the explicit and reflective discussion, students will also learn about (a) how scientific knowledge changes over time and (b) how the culture of science, societal needs, and current events influence the work of scientists.

Underlying Physics Concepts

Ohm's law provides the starting point for this lab. Ohm's law states that the current in a closed circuit is directly proportional to the voltage source for the circuit and inversely proportional to the resistance of the circuit. Mathematically, Ohm's law is given in Equation 9.1, where V is the voltage of the voltage source, I is the current in the circuit, and R is the resistance of a resistor. In SI units, voltage is measured in volts (V), current is measured in amperes (A), and resistance is measured in ohms (Ω).

$$\text{(Equation 9.1) } V = IR$$

To analyze circuits, it helps to define current, resistance, and voltage. *Current* is the flow of charge per unit time, and 1 A of current corresponds to 1 coulomb (the unit of charge, abbreviated as "C"; there are approximately 6.24×10^{18} electrons in 1 C) passing a given point per second. *Resistance* can be defined as the tendency of materials to impede the movement of electrons. The greater the resistance, the more difficult it is for electrons to pass by a given point (for more on this, see the Teacher Notes for Lab 7). *Voltage* is the electric potential difference between two points in an electric field and is analogous to the gravitational potential due to an object's position. One volt is equal to 1 joule (a unit of energy, abbreviated as "J") per coulomb.

To understand what happens when we have multiple resistors in a single circuit, we can use circuit diagrams to analyze the situation. Also, it is important to point out that

Unknown Resistors in a Circuit
Given One Known Resistor and a Voltmeter, How Can You Determine the Resistance of the Unknown Resistors in a Circuit?

traditionally we assume that current flows from positive charge to negative charge, even though we also know that current is carried by electrons.

We start by analyzing the situation where two resistors are in series, shown in Figure 9.1. A series circuit as shown in Figure 9.1 requires each electron to pass through both resistors. As such, the total, or equivalent, resistance of the circuit adds together, because each electron first experiences the resistance in resistor 1 and then the resistance in resistor 2. Mathematically, this is shown in Equation 9.2, where R_s is the equivalent resistance of the resistors in series and R_n is the resistance of the nth resistor.

FIGURE 9.1
A circuit diagram showing two resistors in series

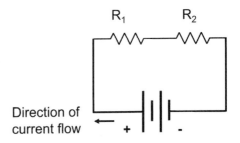

Direction of current flow

(Equation 9.2) $R_s = R_1 + R_2 + \ldots R_n$

Using Equation 9.2 to find the equivalent resistance of resistors in series, we can then find the total current flowing through the circuit (I_T) if we know the voltage of the battery. To do this, we use Equation 9.1 and set the resistance equal to the equivalent resistance of the entire circuit.

When two resistors are in series, the current passing through each resistor is the same. Using Figure 9.1, this means that I_{R1} (the current through resistor 1) is equal to I_{R2}. Because this circuit only contains two resistors in series, the current through each resistor is also equal to the total current through the circuit (I_T).

Finally, we can find the voltage drop across each resistor (i.e., the energy used per coulomb of charge flowing through a resistor) by again using Equation 9.1. The voltage drop across R_1 is the current through the resistor multiplied by the resistance of the resistor. Note that for resistors in series, the ratio of the voltage drop across each resistor is equal to the ratio of the resistances.

Assume that $R_1 = 10\ \Omega$, $R_2 = 20\ \Omega$, and the voltage of the battery is 10 V. We can find the equivalent resistance, the total current, and the voltage drop across each resistor. The solutions are shown below:

$R_s = 10\ \Omega + 20\ \Omega = 30\ \Omega$

$I_T = V/R_s = 10\ \text{V}/30\ \Omega = 0.333\ \text{A}$

$V_{R1} = I_T R_1 = 0.333\ \text{A} \cdot 10\ \Omega = 3.33\ \text{V}$

$V_{R2} = I_T R_2 = 0.333\ \text{A} \cdot 20\ \Omega = 6.667\ \text{V}$

LAB 9

Again, note that the ratio R_1/R_2 is equal to the ratio V_1/V_2 for resistors in series. Also note that the total voltage V_T is equal to the sum of the voltage drop across each resistor in series. This is because the law of conservation of energy states that energy must be conserved in a closed system, and when applied to electric circuits, the voltage drop (or potential difference) through the circuit must be equal to the voltage added by the voltage source.

When analyzing parallel circuits, the situation is more complex. Figure 9.2 shows two resistors in parallel. When two resistors are in parallel, each electron will only pass through one of the two resistors. Thus, there are multiple paths that the electrons can take, and the equivalent resistance of the circuit will be less than the resistance of the smallest individual resistor. Equation 9.3 shows the formula for finding the equivalent resistance for resistors in parallel, where R_p is the equivalent resistance of resistors in parallel and R_n is the resistance of the nth resistor.

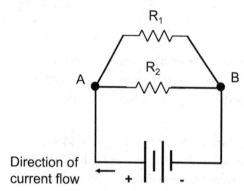

FIGURE 9.2
A circuit diagram showing two resistors in parallel

(Equation 9.3) $1/R_p = 1/R_1 + 1/R_2 + \ldots 1/R_n$

Using Equation 9.3 to find the equivalent resistance of resistors in parallel, we can then find the total current flowing through the circuit (I_T) if we know the voltage of the battery. To do this, we use Equation 9.1 and set the resistance equal to the equivalent resistance of the entire circuit.

When two resistors are in parallel, the voltage drop across each resistor is the same. In Figure 9.2, this means that V_{R1} (the voltage drop across resistor 1) is equal to V_{R2}. Because this circuit only contains two resistors in parallel, the voltage drop across each resistor is also equal to the total voltage of the battery (V_T). If there were three resistors all in parallel to each other instead of only two, the voltage drop would be the same for all three resistors. This makes sense, because the voltage is the potential difference between two points on the wire. Thus, the potential difference between point A and point B in our circuit must be the same, regardless of the path an electron takes. And, this again follows from the law of conservation of energy, because the voltage drop through the circuit must be equal to the voltage supplied by the voltage source.

Finally, we can find the current through each resistor by again using Equation 9.1. The current through R_1 is the voltage drop across the resistor divided by the resistance of the resistor. Note that for resistors in parallel, the ratio of the current through each resistor is the inverse of the ratio of the resistance of each resistor; that is, $I_1/I_2 = R_2/R_1$. This also makes sense because the lower the resistance, the greater the current flowing through a resistor. So we would expect more current to flow through the resistor with the smaller value of resistance.

Unknown Resistors in a Circuit
Given One Known Resistor and a Voltmeter, How Can You Determine the Resistance of the Unknown Resistors in a Circuit?

Let's again assume that $R_1 = 10\ \Omega$, $R_2 = 20\ \Omega$, and the voltage of the battery is 10 V. We can find the equivalent resistance, the total current, and the current through each resistor using Equations 9.1 and 9.3. The solutions are shown below:

$1/R_p = 1/10\ \Omega + 1/20\ \Omega = 3/20\ \Omega$. The equivalent resistance is then equal to $20\ \Omega/3 = 6.67\ \Omega$.

$I_T = V_T/R_p = 10\ \text{V}/6.67\ \Omega = 1.5\ \text{A}$

$I_{R1} = V_T/R_1 = 10\ \text{V}/10\ \Omega = 1\ \text{A}$

$I_{R2} = V_T/R_2 = 10\ \text{V}/20\ \Omega = 0.5\ \text{A}$

There are two additional things to note in this setup: (1) the total current I_T is equal to the sum of the current through each resistor in parallel, and (2) the total resistance of the circuit is lower than the resistance of each individual resistor.

In summary, for resistors in series, the voltage drop across each resistor must sum to the total voltage drop of the circuit, and the current is the same through each resistor. For resistors in parallel, the current through each resistor must sum to the total current through the circuit, and the voltage drop is the same across each resistor.

In this lab, students must apply their knowledge of resistors in parallel and series to determine the resistance of each resistor in the circuit. Students should be given known resistors that they can add into the circuit and a voltmeter to measure the voltage drop across resistors. The key to analyzing the complex circuit is to recognize that when placing a resistor in series with any number of other resistors, the total current in each resistor will be the same. In a series circuit with any number of unknown resistors, by adding a resistor of known value and measuring the voltage drop across the known resistor, we can calculate the current flowing through the known resistor by using Equation 9.1. This current will be the same in all resistors in series. We can then find the resistance of each unknown resistor by measuring the voltage drop across each individual resistor (the voltage drop will not be the same for all resistors in series if the resistance of each resistor is different) and using Equation 9.1 to find the resistance of each unknown resistor.

Timeline

The instructional time needed to complete this lab investigation is 170–230 minutes. Appendix 3 (p. 421) provides options for implementing this lab investigation over several class periods. Option C (230 minutes) should be used if students are unfamiliar with scientific writing, because this option provides extra instructional time for scaffolding the writing process. You can scaffold the writing process by modeling, providing examples, and providing hints as students write each section of the report. Option C can also be used

LAB 9

if you are introducing students to the digital interface sensors and/or the data analysis software. Option D (170 minutes) should be used if students are familiar with scientific writing and have developed the skills needed to write an investigation report on their own. In option D, students complete stage 6 (writing the investigation report) and stage 8 (revising the investigation report) as homework.

Materials and Preparation

The materials needed to implement this investigation are listed in Table 9.1. The equipment can be purchased from a science supply company such as Flinn Scientific, PASCO, Vernier, or Ward's Science.

TABLE 9.1

Materials list for Lab 9

Item	Quantity
Consumable	
D battery	1 per group
Equipment and other materials	
Safety glasses with side shields or goggles	1 per student
Battery holder	1 per group
Copper wire	1 m per group
Wire connectors	As needed
Resistors (various resistances)	2–5 per group
Miniature lightbulb and socket	1 per group
Voltmeter or multimeter	1 per group
Test circuit	1 per group
Investigation Proposal C (optional)	1 per group
Whiteboard, 2' × 3'*	1 per group
Lab Handout	1 per student
Peer-review guide and teacher scoring rubric	1 per student
Checkout Questions	1 per student
Equipment for digital interface measurements (optional)	
Digital interface with USB or wireless connections	1 per group

Continued

Table 9.1 (*continued*)

Item	Quantity
Voltage measurement sensor	1 per group
Computer or tablet with appropriate data analysis software installed	1 per group

* As an alternative, students can use computer and presentation software such as Microsoft PowerPoint or Apple Keynote to create their arguments.

Be sure to use a set routine for distributing and collecting the materials during the lab investigation. One option is to set up the materials for each group at each group's lab station before class begins. This option works well when there is a dedicated section of the classroom for lab work and the materials are large and difficult to move. A second option is to have all the materials on a table or cart at a central location. You can then assign a member of each group to be the "materials manager." This individual is responsible for collecting all the materials his or her group needs from the table or cart during class and for returning all the materials at the end of the class. This option works well when the materials are small and easy to move (such as magnets, wire, and bulbs). It also makes it easy to inventory the materials at the end of the class before students leave for the day.

Safety Precautions and Laboratory Waste Disposal

Remind students to follow all normal lab safety rules. In addition, tell students to take the following safety precautions:

1. Wear sanitized safety glasses with side shields or goggles during lab setup, hands-on activity, and takedown.

2. Never put consumables in their mouth.

3. Wire and other metals with electric current flowing through them may get hot. Use caution when handling components of a closed circuit.

4. Lightbulbs are made of glass. Be careful handling them. If they break, clean them up immediately and place in a broken glass box.

5. Handle electrical wires with caution. They have sharp ends, which can cut or puncture skin.

6. Wash their hands with soap and water when they are done collecting the data.

Batteries, lightbulbs, and wire may be stored for future use. When batteries need replacing, dispose of old batteries according to manufacturer's recommendations.

LAB 9

Topics for the Explicit and Reflective Discussion

Reflecting on the Use of Core Ideas and Crosscutting Concepts During the Investigation

Teachers should begin the explicit and reflective discussion by asking students to discuss what they know about the core ideas they used during the investigation. The following are some important concepts related to the core ideas of (a) conservation of energy and (b) resistors in series and parallel that students need to use to determine the value of the unknown resistors:

- An electric current is the movement of charge through a circuit. Current is equal to the amount of charge per unit time and measured in amperes.
- The current through a resistor is equal to the voltage, or potential difference, across the resistor divided by its resistance. Voltage is measured in units of volts, and resistance is measured in units of ohms.
- The current and voltage in an electric circuit are determined by the properties and arrangement of the individual components of the circuit, such as the voltage source and the resistors.
- The law of conservation of energy applies to circuits. The sum of the voltage drop across an entire circuit must be equal to the total voltage supplied by the voltage source. This also means that the voltage drop between any two points on a circuit must be the same, regardless of the path an electron takes between those two points.
- The voltage drop across a resistor is the product of the current through the resistor and its resistance.
- Resistors in series have the same current. Resistors in parallel have the same voltage drop across them.

To help students reflect on what they know about resistors in series and parallel, we recommend showing them two or three images using presentation software that help illustrate these important ideas. You can then ask the students the following questions to encourage them to share how they are thinking about these important concepts:

1. What do we see going on in this image?
2. Does anyone have anything else to add?
3. What might be going on that we can't see?
4. What are some things that we are not sure about here?

You can then encourage students to think about how CCs played a role in their investigation. There are at least two CCs that students need to use to develop a method for determining the value of the unknown resistors: (a) Energy and Matter: Flows, Cycles, and Conservation and (b) Structure and Function (see Appendix 2 [p. 417] for a brief description

Unknown Resistors in a Circuit
Given One Known Resistor and a Voltmeter, How Can You Determine the Resistance of the Unknown Resistors in a Circuit?

of these CCs). To help students reflect on what they know about these CCs, we recommend asking them the following questions:

1. How does the law of conservation of energy help us to analyze resistors in series?
2. How does the law of conservation of energy help us to analyze resistors in parallel?
3. How does the structure of a complex circuit containing resistors in both series and parallel affect how the circuit functions?
4. The function of a voltmeter is to measure the voltage drop across two points in a circuit. How should the resistance of the voltmeter be structured to achieve its function?

You can then encourage the students to think about how they used all these different concepts to help answer the guiding question and why it is important to use these ideas to help justify their evidence for their final arguments. Be sure to remind your students to explain why they included the evidence in their arguments and make the assumptions underlying their analysis and interpretation of the data explicit in order to provide an adequate justification of their evidence.

Reflecting on Ways to Design Better Investigations

It is important for students to reflect on the strengths and weaknesses of the investigation they designed during the explicit and reflective discussion. Students should therefore be encouraged to discuss ways to eliminate potential flaws, measurement errors, or sources of uncertainty in their investigations. To help students be more reflective about the design of their investigation and what they can do to make their investigations more rigorous in the future, you can ask them the following questions:

1. What were some of the strengths of the way you planned and carried out your investigation? In other words, what made it scientific?
2. What were some of the weaknesses of the way you planned and carried out your investigation? In other words, what made it less scientific?
3. What rules can we make, as a class, to ensure that our next investigation is more scientific?

Reflecting on the Nature of Scientific Knowledge and Scientific Inquiry

This investigation can be used to illustrate two important concepts related to the nature of scientific knowledge and the nature of scientific inquiry: (a) how scientific knowledge changes over time and (b) how the culture of science, societal needs, and current events influence the work of scientists (see Appendix 2 [p. 417] for a brief description of these concepts). Be sure to review these concepts during and at the end of the explicit and reflective discussion. To help students think about these concepts in relation to what they did during the lab, you can ask them the following questions:

1. Scientific knowledge can and does change over time. Can you tell me why it changes?

2. Can you work with your group to come up with some examples of how scientific knowledge related to electricity and electric circuits has changed over time? Be ready to share in a few minutes.

3. People view some types of research as being more important than other types of research because of cultural values and current events. Can you come up with some examples of how cultural values and current events have influenced the work of scientists?

4. Scientists share a set of values, norms, and commitments that shape what counts as knowing, how to represent or communicate information, and how to interact with other scientists. Can you work with your group to come up with a rule that you can use to decide if something is science or not science? Be ready to share in a few minutes.

You can also use presentation software or other techniques to encourage your students to think about these concepts. You can show examples of how our thinking about electricity and electric circuits has changed over time and ask students to discuss what they think led to those changes. You can also show examples of research projects that were influenced by cultural values and current events and ask students to think about what was going on in society when the research was conducted and why that research was viewed as being important for the greater good. Be sure to remind your students that it is important for them to understand what counts as scientific knowledge and how that knowledge develops over time in order to be proficient in science.

Hints for Implementing the Lab

- Allowing students to design their own procedures for collecting data gives students an opportunity to try, to fail, and to learn from their mistakes. However, you can scaffold students as they develop their own procedure by having them fill out an investigation proposal. The proposals provide a way for you to offer students hints and suggestions without telling them how to do it. You can also check the proposals quickly during a class period. For this lab we suggest using Investigation Proposal C.

- Learn how to use the voltmeter (or multimeter, if used) before the lab begins. It is important for you to know how to use the equipment so you can help students when technical issues arise.

Unknown Resistors in a Circuit
Given One Known Resistor and a Voltmeter, How Can You Determine the Resistance of the Unknown Resistors in a Circuit?

- We suggest providing students with a stand-alone voltmeter instead of a multimeter, if this equipment is available. This is because students can use the multimeter to directly measure resistance of unknown resistors. However, if only multimeters are available, we suggest making it clear that students must only use the voltmeter setting when testing their test circuit.

- Allow the students to become familiar with the voltmeter (or multimeter, if used) as part of the tool talk before they begin to design their investigation. Also, allow them to measure current and voltage through circuits using known values for the battery and the resistor before designing their method. Give them 5–10 minutes to examine the equipment and materials before they begin designing their investigations. This gives students a chance to see what they can and cannot do with the equipment.

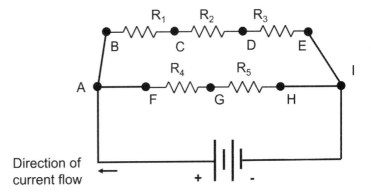

FIGURE 9.3

A suggested setup for the test circuit

- In the lab, students must first devise a method that will allow them to find the resistance of each resistor in the test circuit you give them. Initially, you want to provide students with resistors of known resistance that they can use to check their method. Then, they can use one known resistor to check their method on the test circuit.

- We suggest giving students a test circuit that is made up of two parallel branches of resistors in series. This is the circuit they will use to determine if their method is correct for determining the resistance of the unknown resistors. Figure 9.3 shows a schematic of the setup we suggest. To make the lab more challenging, you can add an increasing number of resistors to each branch or an additional branch.

- Another approach to making the lab more challenging is to provide different test circuits to different groups. In this case, we suggest using the same arrangement of resistors, but with different resistances. This will help to ensure that students focus on the methods for determining the resistance and, consequently, focus on the core ideas during the argumentation session.

- Students should be given an additional resistor that they can add to the circuit. If you give students the value of a resistor already in the circuit, the investigation becomes a series of math problems.

- Make sure that the components of the test circuit are rated appropriately for the battery they will be connected to. Many resistors have a limit to the current that they can handle. Thus, a battery of too high a voltage may lead the circuit components to overheat. We suggest using a D battery.

- We also suggest limiting the length of wire students use to connect the resistors. Over a small length, the internal resistance of the wire will be small enough to be ignored. However, if the students use too much wire to connect their resistors, they may inadvertently affect their results.
- Be sure to allow students to go back and re-collect data at the end of the argumentation session. Students often realize that they made numerous mistakes when they were collecting data as a result of their discussions during the argumentation session. The students, as a result, will want a chance to re-collect data, and the re-collection of data should be encouraged when time allows. This also offers an opportunity to discuss what scientists do when they realize a mistake is made inside the lab.

If students use digital interface measurement equipment and analysis

We suggest allowing students to familiarize themselves with the sensors and the data analysis software before they finalize the procedure for the investigation, especially if they have not used such software previously. This gives students an opportunity to learn how to work with the software and to improve the quality of the data they collect.

- Remind students to begin recording data just before they close the circuits.
- Remind students to follow the user's guide to correctly connect any sensors to avoid damage to lab equipment.

Connections to Standards

Table 9.2 highlights how the investigation can be used to address specific performance expectations from the *NGSS*; learning objectives from AP Physics 1 and 2; learning objectives from AP Physics C: Electricity and Magnetism; *Common Core State Standards for English Language Arts* (*CCSS ELA*); and *Common Core State Standards for Mathematics* (*CCSS Mathematics*).

Unknown Resistors in a Circuit
Given One Known Resistor and a Voltmeter, How Can You Determine the Resistance of the Unknown Resistors in a Circuit?

TABLE 9.2

Lab 9 alignment with standards

NGSS performance expectations	• None
AP Physics 1 and AP Physics 2 learning objectives	• 4.E.4.1: Make predictions about the properties of resistors and/or capacitors when placed in a simple circuit based on the geometry of the circuit element and supported by scientific theories and mathematical relationships. • 4.E.4.2: Plan for the collection of data to determine the effect of changing the geometry and/or materials on the resistance or capacitance of a circuit element and relate results to the basic properties of resistors and capacitors. • 4.E.4.3: Analyze data to determine the effect of changing the geometry and/or materials on the resistance or capacitance of a circuit element and relate results to the basic properties of resistors and capacitors. • 4.E.5.1: Make and justify a quantitative prediction of the effect of a change in values or arrangements of one or two circuit elements on the currents and potential differences in a circuit containing a small number of sources of emf, resistors, capacitors, and switches in series and/or parallel. • 4.E.5.2: Make and justify a qualitative prediction of the effect of a change in values or arrangements of one or two circuit elements on currents and potential differences in a circuit containing a small number of sources of emf, resistors, capacitors, and switches in series and/or parallel. • 4.E.5.3: Plan data collection strategies and perform data analysis to examine the values of currents and potential differences in an electric circuit that is modified by changing or rearranging circuit elements, including sources of emf, resistors, and capacitors. • 5.B.9.6: Mathematically express the changes in electric potential energy of a loop in a multiloop electrical circuit and justify this expression using the principle of the conservation of energy • 5.B.9.8: Translate between graphical and symbolic representations of experimental data describing relationships among power, current, and potential difference across a resistor.
AP Physics C: Electricity and Magnetism learning objectives	• CNV-6.A.a: Identify parallel or series arrangement in a circuit containing multiple resistors. • CNV-6.A.b: Describe a series or a parallel arrangement of resistors. • CNV-6.B: Calculate equivalent resistances for a network of resistors that can be considered a combination of series and parallel arrangements. • CNV-6.C.a: Calculate voltage, current, and power dissipation for any resistors in a circuit containing a network of known resistors with a single battery or energy source. • CNV-6.C.b: Calculate relationships between the potential difference, current, resistance, and power dissipation for any part of a circuit, given some of the characteristics of the circuit (i.e., battery voltage or current in the battery, or a resistor or branch of resistors).

Continued

Table 9.2 (*continued*)

Literacy connections (*CCSS ELA*)	• *Reading:* Key ideas and details, craft and structure, integration of knowledge and ideas • *Writing:* Text types and purposes, production and distribution of writing, research to build and present knowledge, range of writing • *Speaking and listening:* Comprehension and collaboration, presentation of knowledge and ideas
Mathematics connections (*CCSS Mathematics*)	• *Mathematical practices:* Make sense of problems and persevere in solving them, reason abstractly and quantitatively, construct viable arguments and critique the reasoning of others, model with mathematics, use appropriate tools strategically, attend to precision • *Number and quantity:* Reason quantitatively and use units to solve problems, represent and model with vector quantities, perform operations on vectors • *Algebra:* Interpret the structure of expressions, create equations that describe numbers or relationships, understand solving equations as a process of reasoning and explain the reasoning, solve equations and inequalities in one variable, represent and solve equations and inequalities graphically • *Functions:* Understand the concept of a function and use function notation; interpret functions that arise in applications in terms of the context; analyze functions using different representations; build a function that models a relationship between two quantities; construct and compare linear, quadratic, and exponential models and solve problems; interpret expressions for functions in terms of the situation they model • *Statistics and probability:* Summarize, represent, and interpret data on two categorical and quantitative variables; interpret linear models; make inferences and justify conclusions from sample surveys, experiments, and observational studies

Lab Handout

Lab 9. Unknown Resistors in a Circuit: Given One Known Resistor and a Voltmeter, How Can You Determine the Resistance of the Unknown Resistors in a Circuit?

Introduction

Historians of science and technology have identified the 50-year period from 1880 to 1930 as the Second Industrial Revolution (Hughes 1993). Like the first Industrial Revolution (circa 1760–1820), the Second Industrial Revolution brought about massive technological change and greatly improved the quality of life for people in the United States, Western Europe, and Japan. One of the defining characteristics of the Second Industrial Revolution (and a feature that distinguishes it from the first one) is the prevalence of electrification. The term *electrification* is used to describe both the proliferation of products that required electricity and the development of infrastructure in cities to provide power to homes and business for the use of such products requiring electric power. One of the most important products of the Second Industrial Revolution was the lightbulb. Figure L9.1 shows a picture of Thomas Edison's first working lightbulb.

FIGURE L9.1
Edison's first working lightbulb

Note: A full-color version of this figure is available on the book's Extras page at *www.nsta.org/adi-physics2*.

Several important scientific discoveries led to development of the new technologies emerging during the Second Industrial Revolution. First, Alessandro Volta (1745–1827) conducted a number of important experiments with electric charge, hence the unit of volts being named in his honor. Volta's most important contribution was the invention of the electric battery and, as a result, the production of the first continuous flow of electric charge—an electric current. Georg Simon Ohm (1789–1854) then conducted a series of experiments using Volta's discovery. Ohm's research showed that the current in a closed circuit is directly proportional to the voltage of the battery and inversely proportional to the resistance of a device. This has become known as Ohm's law and is expressed mathematically as $V = IR$, where V is the voltage, measured in volts; I is the current, measured in amperes; and R is the resistance, measured in ohms.

It is also important to precisely define *resistance*. We define resistance as the tendency of the individual electrons in electrical current to have their flow restricted by interaction with atoms in the materials through which current flows. We use the term *resistor* to denote any device that impedes the flow of electricity through a circuit. Often, a resistor is a device that uses electrical energy, such as a

LAB 9

microwave or a television. However, some electronic equipment has resistors to help direct the flow of current. And it is important to note that the wire itself that carries the current has a resistance, although for most calculations, we can ignore the resistance of the wire.

Subsequent work by Ohm and other scientists further explored the relationship between current, voltage, and resistance of circuits. When two resistors are arranged in series, the equivalent resistance of the circuit is the sum of the resistance of each resistor. When resistors are connected in parallel, the equivalent resistance of the circuit is a bit more complex. In this case, we can use the following formula to find the equivalent resistance: $1/R_p = 1/R_1 + 1/R_2 + \ldots 1/R_n$, where R_n is the nth resistor in parallel.

The proliferation of modern electronics has caused an increased demand for electrical energy. Not only that, we often place several electronic devices near each other in homes. For example, in your kitchen, you may have a toaster, a blender, and a microwave all placed near each other on the kitchen counter. In your living room, you may have a TV, a DVR, and a video game console all on one TV stand. Because we want all our devices to work at the same time, we often use a power strip to allow us to plug all the devices into the same electrical outlet. Power strips are also wired in parallel, because we want one device to work even if another device plugged into the power strip is off. This, however, increases the current, because resistors wired in parallel lead to a decrease in the equivalent resistance of the circuit. And drawing too much current through an electrical outlet can cause the devices to burn out or, even worse, start a fire.

When designing homes and offices or wiring in new devices, engineers and electricians may need to determine the resistance of a device. For example, when wiring in speakers to a home, engineers and electricians need to know the resistance of the speakers to make sure they are not drawing too much current.

Your Task

Use what you know about circuits with resistors in series and parallel, how energy and matter flow in an electric circuit, and the relationship between structure and function to design a method for finding the resistance of each resistor in a complex circuit. Your method must allow you to then find the value of all the resistors given to you by your teacher. You cannot remove any of the resistors from the complex circuit your teacher gives you, but you can add one resistor with a known resistance either in parallel or in series at any point in the circuit. When you test your final circuit, you will be able to add one resistor of known resistance to the circuit as well. Furthermore, you may only use a voltmeter to take measurements from the circuits.

The guiding question of this investigation is, *Given one known resistor and a voltmeter, how can you determine the resistance of the unknown resistors in a circuit?*

Unknown Resistors in a Circuit
Given One Known Resistor and a Voltmeter, How Can You Determine the Resistance of the Unknown Resistors in a Circuit?

Materials
You may use any of the following materials during your investigation:

Consumable
- Battery

Equipment
- Safety glasses with side shields or goggles (required)
- Battery holder
- Copper wire
- Wire connectors
- Resistors (various resistances)
- Miniature lightbulb and socket
- Voltmeter or multimeter
- Test circuit

If you have access to the following equipment, you may also consider using a digital voltage sensor with an accompanying interface and a computer or tablet.

Safety Precautions
Follow all normal lab safety rules. In addition, take the following safety precautions:

1. Wear sanitized safety glasses with side shields or goggles during lab setup, hands-on activity, and takedown.
2. Never put consumables in your mouth.
3. Wire and other metals with electric current flowing through them may get hot. Use caution when handling components of a closed circuit.
4. Lightbulbs are made of glass. Be careful handling them. If they break, clean them up immediately and place in a broken glass box.
5. Handle electrical wires with caution. They have sharp ends, which can cut or puncture skin.
6. Wash your hands with soap and water when you are done collecting the data.

Investigation Proposal Required?   ☐ Yes   ☐ No

Getting Started
To answer the guiding question, you will need to design a method to determine the resistance of a resistor using a voltmeter and a second resistor of known resistance. Once you have developed a method for determining the resistance of unknown resistors, you can use your method to determine the resistance of each resistor in the complex circuit your teacher gives you. Before you can design your investigation, however, you must determine what type of data you need to collect, how you will collect it, and how you will analyze it.

To determine *what type of data you need to collect*, think about the following questions:

LAB 9

- Can you directly measure the resistance of the resistors, or will you need to calculate it based on other measurements?
- How could you keep track of changes in this system quantitatively?
- How useful is it to track how energy flows into, out of, or within this system?
- How might changes to the structure of a circuit change how it functions?
- What is the relationship between resistance, voltage, and current in a circuit?

To determine *how you will collect the data*, think about the following questions:

- What is the independent variable and what is the dependent variable?
- What scale or scales should you use when you take your measurements?
- How will you make sure that your data are of high quality (i.e., how will you reduce error)?
- How will you keep track of and organize the data you collect?

To determine *how you will analyze the data*, think about the following questions:

- What types of patterns might you look for as you analyze your data?
- What type of table or graph could you create to help make sense of your data?

Connections to the Nature of Scientific Knowledge and Scientific Inquiry

As you work through your investigation, you may want to consider

- how scientific knowledge changes over time, and
- how the culture of science, societal needs, and current events influence the work of scientists.

Initial Argument

Once your group has finished collecting and analyzing your data, your group will need to develop an initial argument. Your initial argument needs to include a claim, evidence to support your claim, and a justification of the evidence. The *claim* is your group's answer to the guiding question. The *evidence* is an analysis and interpretation of your data. Finally, the *justification* of the evidence is why your group thinks the evidence matters. The justification of the evidence is important because scientists can use different kinds of evidence to support their claims. Your group will create your initial argument on a whiteboard. Your whiteboard should include all the information shown in Figure L9.2.

Argumentation Session

The argumentation session allows all of the groups to share their arguments. One or two members of each group will stay at the lab station to share that group's argument, while

Unknown Resistors in a Circuit
Given One Known Resistor and a Voltmeter, How Can You Determine the Resistance of the Unknown Resistors in a Circuit?

the other members of the group go to the other lab stations to listen to and critique the other arguments. This is similar to what scientists do when they propose, support, evaluate, and refine new ideas during a poster session at a conference. If you are presenting your group's argument, your goal is to share your ideas and answer questions. You should also keep a record of the critiques and suggestions made by your classmates so you can use this feedback to make your initial argument stronger. You can keep track of specific critiques and suggestions for improvement that your classmates mention in the space below.

FIGURE L9.2
Argument presentation on a whiteboard

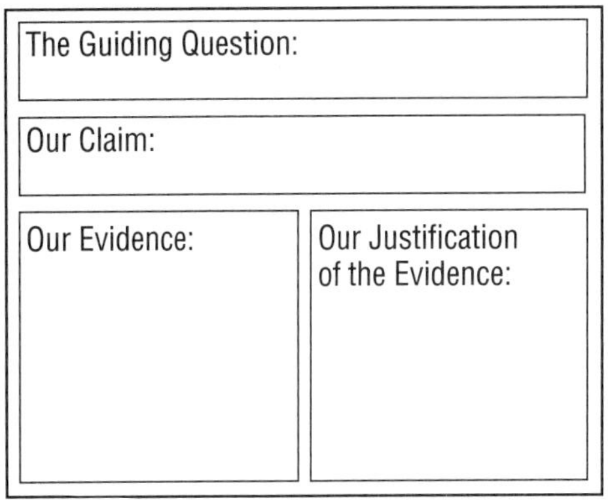

The Guiding Question:	
Our Claim:	
Our Evidence:	Our Justification of the Evidence:

Critiques about our initial argument and suggestions for improvement:

If you are critiquing your classmates' arguments, your goal is to look for mistakes in their arguments and offer suggestions for improvement so these mistakes can be fixed. You should look for ways to make your initial argument stronger by looking for things that the other groups did well. You can keep track of interesting ideas that you see and hear during the argumentation in the space below. You can also use this space to keep track of any questions that you will need to discuss with your team.

Interesting ideas from other groups or questions to take back to my group:

Once the argumentation session is complete, you will have a chance to meet with your group and revise your initial argument. Your group might need to gather more data or design a way to test one or more alternative claims as part of this process. Remember, your goal at this stage of the investigation is to develop the best argument possible.

Report

Once you have completed your research, you will need to prepare an *investigation report* that consists of three sections. Each section should provide an answer to the following questions:

1. What question were you trying to answer and why?
2. What did you do to answer your question and why?
3. What is your argument?

Your report should answer these questions in two pages or less. This report must be typed, and any diagrams, figures, or tables should be embedded into the document. Be sure to write in a persuasive style; you are trying to convince others that your claim is acceptable or valid!

Reference

Hughes, T. P. 1993. *Networks of power: Electrification in western society, 1880–1930*. Baltimore, MD: Johns Hopkins University Press.

Checkout Questions

Lab 9. Unknown Resistors in a Circuit: Given One Known Resistor and a Voltmeter, How Can You Determine the Resistance of the Unknown Resistors in a Circuit?

Use the diagram below to answer questions 1–4.

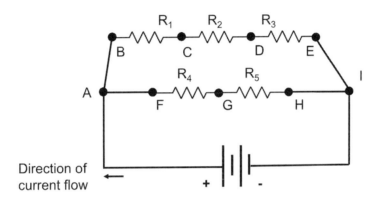

1. At which two points should a person attach the leads of a voltmeter in order to measure the voltage drop through R_1?

    How do you know?

2. At which two points should a person attach the leads of a voltmeter in order to measure the voltage drop through R_2?

    How do you know?

3. At which two points should a person attach the leads of a voltmeter in order to measure the voltage drop through R_5?

   How do you know?

4. At which two points should a person attach the leads of a voltmeter in order to measure the voltage drop across the entire circuit?

   How do you know?

5. Once scientists agree on an idea, the idea is proven true for all time.

   a. I agree with this statement.
   b. I disagree with this statement.

   Explain your answer, using examples from this investigation and at least one other investigation you have conducted.

Unknown Resistors in a Circuit
Given One Known Resistor and a Voltmeter, How Can You Determine the Resistance of the Unknown Resistors in a Circuit?

6. Societal values and needs sometimes influence the questions scientists ask.

    a. I agree with this statement.

    b. I disagree with this statement.

    Explain your answer, using examples from this investigation and at least one other investigation you have conducted.

7. How does the law of conservation of energy assist scientists and engineers in analyzing complex circuits? In your answer, include examples from your investigation on unknown resistors.

8. The function of a voltmeter is to measure the voltage drop across two points in a circuit. The function of an ammeter, on the other hand, is to measure the current through a circuit. How should the resistance of the voltmeter and ammeter be structured to achieve their respective functions? In your answer, include examples from your investigation on unknown resistors.

SECTION 4
Forces and Interactions

Magnetic Fields and Electromagnetism

Introduction Labs

LAB 10

Teacher Notes

Lab 10. Magnetic Field Around a Permanent Magnet: How Does the Strength and Direction of a Magnetic Field Change as One Moves Around a Permanent Magnet?

Purpose

The purpose of this lab is to *introduce* students to the disciplinary core idea (DCI) of Motion and Stability: Forces and Interactions (PS2) from the *NGSS* by having them determine the strength and direction of the magnetic field around a permanent magnet. In addition, this lab can be used to help students understand two big ideas from AP Physics: (a) fields existing in space can be used to explain interactions and (b) the interactions of an object with other objects can be described by forces. This lab also gives students an opportunity to learn about the crosscutting concepts (CCs) of (a) Patterns and (b) Stability and Change from the *NGSS*. As part of the explicit and reflective discussion, students will also learn about (a) the difference between data and evidence in science and (b) the assumptions made by scientists about order and consistency in nature.

Underlying Physics Concepts

The magnetic field due to a permanent magnet is similar to the electric field and the gravitational field in that it permeates space due to the presence of a magnetic source. This allows us to use the concept of the field to analyze magnetic interactions (see Lab 11). The magnetic source arises due to atomic properties of the magnetic material. When drawing magnetic field lines (which are a way to model the magnetic field), the convention is to draw the field lines pointing out from the north pole of a magnet and in toward the south pole of a magnet.

The magnetic field is different from the electric and gravitational field in some very important ways. The first difference surrounds the creation of the field. Electric fields can be created by a single charge: a negative charge, for example, can exist independently from a positive charge. Thus, an electric field can arise due to a single positive or negative charge. The magnetic field, however, cannot be created by a north pole independently of a south pole. A north pole is always accompanied by a south pole, and vice versa. In more technical parlance, a magnetic monopole does not exist, and all magnetic fields are established by a magnetic dipole.

The second difference between the magnetic field and other fields is how their fields are distributed in space. Electric and gravitational fields have an origin—the source of the charge or mass. Because the magnetic field is created by a dipole, the field does not have an origin, nor does it have an end. Instead, it forms closed loops. Figure 10.1 shows the magnetic field due to a permanent bar magnet, including the closed loops of the magnetic field. Also notice how the field lines pass through the permanent magnet. This means

Magnetic Field Around a Permanent Magnet

How Does the Strength and Direction of a Magnetic Field Change as One Moves Around a Permanent Magnet?

that the magnetic field exists inside the permanent magnet. Finally, the strength of the magnetic field is represented by the density of the field lines. The greater the density of field lines, the stronger the magnetic field.

The final difference between the magnetic field and the electric and gravitational fields is how the field varies with the distance from the magnet. The electric and gravitational fields follow an inverse square law (i.e., $E \propto 1/r^2$; $G \propto 1/r^2$). Because the magnetic field is created by a dipole, it does not follow an inverse square law. Instead, the strength of the magnetic field obeys an inverse cube law and can be calculated by Equation 10.1, where B is the magnetic field strength; μ_o is the permeability constant; m is the magnetic moment, a measure of the strength of the permanent magnet (this is analogous to the magnitude of the charge creating an electric field); and r is the distance from the magnet. In SI units, magnetic field is measured in teslas (T); the permeability constant has a value of $4\pi \times 10^{-7}$ tesla-meters per ampere (T·m/A); the magnetic moment is measured in tesla-meters squared (T·m^2); and the distance is measured in meters (m). Note that in this equation, the factor of 2 holds for the magnetic field along the dipole axis (i.e., the axis that runs through the north and south poles). For the magnetic field at the equator of a permanent bar magnet (i.e., a line perpendicular to the dipole axis that passes through the point midway between the two dipoles), the factor of 2 shown in Equation 10.1 is instead a factor of 1.

FIGURE 10.1

Magnetic field due to a permanent bar magnet (S = south pole; N = north pole)

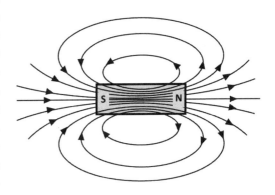

$$\text{(Equation 10.1)} \quad B = \frac{2\mu_o m}{4\pi r^3}$$

As an interesting aside, notice how the units of the permeability constant are given in terms of amperes, the unit of current, hinting at the relationship between magnetic fields and electric current.

A magnetic field of 1 T is a particularly strong magnetic field. The Earth's magnetic field has a strength of 1×10^{-4} T, and the strongest commercially available magnets are usually around 0.5 T. For this reason, many people instead use the gauss (G) as a unit for the magnetic field, where $1 \text{ G} = 1 \times 10^{-4}$ T. The gauss is not an SI unit, so calculations involving the magnetic field strength should be done using the tesla.

Timeline

The instructional time needed to complete this lab investigation is 170–230 minutes. Appendix 3 (p. 421) provides options for implementing this lab investigation over several class periods. Option C (230 minutes) should be used if students are unfamiliar with

LAB 10

scientific writing, because this option provides extra instructional time for scaffolding the writing process. You can scaffold the writing process by modeling, providing examples, and providing hints as students write each section of the report. Option C should also be used if you are introducing students to the magnetic field sensor and/or data analysis software. Option D (170 minutes) should be used if students are familiar with scientific writing and have developed the skills needed to write an investigation report on their own. In option D, students complete stage 6 (writing the investigation report) and stage 8 (revising the investigation report) as homework.

Materials and Preparation

The materials needed to implement this investigation are listed in Table 10.1. The equipment can be purchased from a science supply company such as Flinn Scientific, PASCO, Vernier, or Ward's Science.

TABLE 10.1

Materials list for Lab 10

Item	Quantity
Consumables	
Tape	As needed
String	As needed
Equipment and other materials	
Safety glasses with side shields or safety goggles	1 per student
Bar magnet	1 per group
Large cow magnet	1 per group
Compass	1 per group
Ruler	1 per group
Magnetic field sensor*	1 per group
Digital interface with USB or wireless connections*	1 per group
Computer or tablet with appropriate data analysis software installed*	1 per group
Spring scale	1 per group
Investigation Proposal C (optional)	1 per group
Whiteboard, 2' × 3'†	1 per group
Lab Handout	1 per student
Peer-review guide and teacher scoring rubric	1 per student

Continued

Table 10.1 (*continued*)

Item	Quantity
Checkout Questions	1 per student

* Use of this sensor and the data analysis software is highly recommended, but see the "Hints" section if you do not have access to this equipment.

† As an alternative, students can use computer and presentation software such as Microsoft PowerPoint or Apple Keynote to create their arguments.

Be sure to use a set routine for distributing and collecting the materials during the lab investigation. One option is to set up the materials for each group at each group's lab station before class begins. This option works well when there is a dedicated section of the classroom for lab work and the materials are large and difficult to move. A second option is to have all the materials on a table or cart at a central location. You can then assign a member of each group to be the "materials manager." This individual is responsible for collecting all the materials his or her group needs from the table or cart during class and for returning all the materials at the end of the class. This option works well when the materials are small and easy to move (such as magnets, wire, and bulbs). It also makes it easy to inventory the materials at the end of the class before students leave for the day.

Safety Precautions and Laboratory Waste Disposal

Remind students to follow all normal lab safety rules. In addition, tell students to take the following safety precautions:

1. Wear sanitized safety glasses with side shields or goggles during lab setup, hands-on activity, and takedown.

2. Wash their hands with soap and water when they are done collecting the data.

There is no laboratory waste associated with this activity.

Topics for the Explicit and Reflective Discussion

Reflecting on the Use of Core Ideas and Crosscutting Concepts During the Investigation

Teachers should begin the explicit and reflective discussion by asking students to discuss what they know about the core ideas they used during the investigation. The following are some important concepts related to the core idea that permanent magnetism is a property resulting from the alignment of magnetic dipole moments within the system that students need to use to find the strength and direction of the magnetic field around a permanent magnet:

- A field associates a value of some physical quantity with every point in space. Fields are a model that physicists use to describe interactions that occur over a distance.

LAB 10

Fields permeate space, and objects experience forces due to their interaction with a field.

- Fields are represented by field lines indicating direction and magnitude.
- The density of the field lines is used to model the strength of the magnetic field. For a given area, the stronger the magnetic field, the higher the number of field lines passing through the area.
- All magnets produce a magnetic field.
- Magnetic field lines point out from a north pole and in toward a south pole.
- Magnetic poles always come in pairs—there are no known magnetic monopoles in the universe.

To help students reflect on what they know about magnetic fields, we recommend showing them two or three images using presentation software that help illustrate these important ideas. You can then ask the students the following questions to encourage them to share how they are thinking about these important concepts:

1. What do we see going on in this image?
2. Does anyone have anything else to add?
3. What might be going on that we can't see?
4. What are some things that we are not sure about here?

You can then encourage students to think about how CCs played a role in their investigation. There are at least two CCs that students need to use to determine how the magnetic field strength changes as one moves around a permanent magnet: (a) Patterns and (b) Stability and Change (see Appendix 2 [p. 417] for a brief description of these CCs). To help students reflect on what they know about these CCs, we recommend asking them the following questions:

1. What patterns do magnetic field lines represent?
2. How are the magnetic field lines similar to or different from electric or gravitational field lines?
3. How does the strength of a magnetic field due to a permanent magnet change with respect to space?

You can then encourage the students to think about how they used all these different concepts to help answer the guiding question and why it is important to use these ideas to help justify their evidence for their final arguments. Be sure to remind your students to explain why they included the evidence in their arguments and make the assumptions underlying their analysis and interpretation of the data explicit in order to provide an adequate justification of their evidence.

Reflecting on Ways to Design Better Investigations

It is important for students to reflect on the strengths and weaknesses of the investigation they designed during the explicit and reflective discussion. Students should therefore be encouraged to discuss ways to eliminate potential flaws, measurement errors, or sources of uncertainty in their investigations. To help students be more reflective about the design of their investigation and what they can do to make their investigations more rigorous in the future, you can ask them the following questions:

1. What were some of the strengths of the way you planned and carried out your investigation? In other words, what made it scientific?

2. What were some of the weaknesses of the way you planned and carried out your investigation? In other words, what made it less scientific?

3. What rules can we make, as a class, to ensure that our next investigation is more scientific?

Reflecting on the Nature of Scientific Knowledge and Scientific Inquiry

This investigation can be used to illustrate two important concepts related to the nature of scientific knowledge and the nature of scientific inquiry (a) the difference between data and evidence and (b) the assumptions made by scientists about order and consistency in nature (see Appendix 2 [p. 417] for a brief description of these concepts). Be sure to review these concepts during and at the end of the explicit and reflective discussion. To help students think about these concepts in relation to what they did during the lab, you can ask them the following questions:

1. You had to talk about data and evidence during your investigation. Can you give me some examples of data and evidence from your investigation?

2. Can you work with your group to come up with a rule that you can use to decide if a piece of information is data or evidence? Be ready to share in a few minutes.

3. Scientists assume that natural laws operate today as they did in the past and that they will continue to do so in the future. Why do you think this assumption is important?

4. Think about what you were trying to do during this investigation. What would you have had to do differently if you could not assume natural laws operate today as they have in the past?

LAB 10

You can also use presentation software or other techniques to encourage your students to think about these concepts. You can show examples of information from the investigation that are either data or evidence and ask students to classify each example and explain their thinking. You can also show images of different scientific laws (such as the law of universal gravitation or Coulomb's law) and ask students if they think these laws have been the same throughout Earth's history. Then ask them to think about what scientists would need to do to be able to study the past if laws are not consistent throughout time and space. Be sure to remind your students that it is important for them to understand what counts as scientific knowledge and how that knowledge develops over time in order to be proficient in science.

Hints for Implementing the Lab

- Allowing students to design their own procedures for collecting data gives students an opportunity to try, to fail, and to learn from their mistakes. However, you can scaffold students as they develop their own procedure by having them fill out an investigation proposal. The proposals provide a way for you to offer students hints and suggestions without telling them how to do it. You can also check the proposals quickly during a class period. For this lab we suggest using Investigation Proposal C.

- Learn how to use the magnetic field sensor before the lab begins. It is important for you to know how to use the equipment so you can help students when technical issues arise.

- Allow the students to become familiar with the magnetic field sensors and compasses as part of the tool talk before they begin to design their investigation. Give them 5–10 minutes to examine the equipment and materials before they begin designing their investigations. This gives students a chance to see what they can and cannot do with the equipment.

- Allow the students to become familiar with the equipment and materials as part of the tool talk, to give them a chance to see what they can and cannot do with the equipment. In particular, we suggest allowing students to familiarize themselves with the magnetic field sensor and the data analysis software before they finalize the procedure for the investigation, especially if they have not used such software previously. This gives students an opportunity to learn how to work with the software and to improve the quality of the data they collect.

- Remind students to follow the user's guide to correctly connect any sensors to avoid damage to lab equipment.

- For this lab, we suggest using strong magnets. The stronger the magnet, the more pronounced student results will be—particularly for measuring changes in the strength of the magnetic field.

Magnetic Field Around a Permanent Magnet
How Does the Strength and Direction of a Magnetic Field Change as One Moves Around a Permanent Magnet?

- If you do not have access to a magnetic field sensor, students can use a compass to obtain the direction of the magnetic field and the spring scale with a magnet attached to it to get the strength of the magnetic field. In this case, students will actually be obtaining the force acting on one permanent magnet due to the magnetic field created by the second permanent magnet. Thus, their measurements will be in newtons, not in teslas. This is a good opportunity to reinforce the importance of units and the relationship between forces and fields during the explicit and reflective discussion.

- Be sure to allow students to go back and re-collect data at the end of the argumentation session. Students often realize that they made numerous mistakes when they were collecting data as a result of their discussions during the argumentation session. The students, as a result, will want a chance to re-collect data, and the re-collection of data should be encouraged when time allows. This also offers an opportunity to discuss what scientists do when they realize a mistake is made inside the lab.

Connections to Standards

Table 10.2 highlights how the investigation can be used to address specific performance expectations from the *NGSS*; learning objectives from AP Physics 1 and 2; learning objectives from AP Physics C: Electricity and Magnetism; *Common Core State Standards for English Language Arts* (*CCSS ELA*); and *Common Core State Standards for Mathematics* (*CCSS Mathematics*).

TABLE 10.2

Lab 10 alignment with standards

NGSS performance expectations	• None
AP Physics 1 and AP Physics 2 learning objectives	• 2.C.4.1: Distinguish the characteristics that differ between monopole fields (gravitational field of spherical mass and electrical field due to single-point charge) and dipole fields (electric dipole field and magnetic field) and make claims about the spatial behavior of the fields using qualitative or semiquantitative arguments based on vector addition of fields due to each point source, including identifying the locations and signs of sources from a vector diagram of the field. • 2.D.3.1: Describe the orientation of a magnetic dipole placed in a magnetic field in general and the particular cases of a compass in the magnetic field of Earth and iron filings surrounding a bar magnet. • 2.D.4.1: Qualitatively analyze the magnetic behavior of a bar magnet composed of ferromagnetic material. • 3.A.3.3: Describe a force as an interaction between two objects and identify both objects for any force.

Continued

Table 10.2 (continued)

AP Physics C: Electricity and Magnetism learning objectives	• None
Literacy connections (CCSS ELA)	• *Reading*: Key ideas and details, craft and structure, integration of knowledge and ideas • *Writing*: Text types and purposes, production and distribution of writing, research to build and present knowledge, range of writing • *Speaking and listening:* Comprehension and collaboration, presentation of knowledge and ideas
Mathematics connections (CCSS Mathematics)	• *Mathematical practices*: Make sense of problems and persevere in solving them, reason abstractly and quantitatively, construct viable arguments and critique the reasoning of others, model with mathematics, use appropriate tools strategically, attend to precision • *Number and quantity*: Reason quantitatively and use units to solve problems, represent and model with vector quantities, perform operations on vectors • *Algebra*: Interpret the structure of expressions, understand solving equations as a process of reasoning and explain the reasoning, solve equations and inequalities in one variable, represent and solve equations and inequalities graphically • *Functions*: Understand the concept of a function and use function notation; interpret functions that arise in applications in terms of the context; analyze functions using different representations; construct and compare linear, quadratic, and exponential models and solve problems; interpret expressions for functions in terms of the situation they model • *Statistics and probability*: Summarize, represent, and interpret data on two categorical and quantitative variables; interpret linear models; make inferences and justify conclusions from sample surveys, experiments, and observational studies

Lab Handout

Lab 10. Magnetic Field Around a Permanent Magnet: How Does the Strength and Direction of a Magnetic Field Change as One Moves Around a Permanent Magnet?

Introduction

The earliest known compass was invented in China during the Han dynasty, which ruled for over 400 years between 206 BCE and 220 CE (Bodde 1986). Figure L10.1 shows a picture of a recreated Han-era Si Nan, or south-facing ladle, which was designed to take advantage of the magnetic properties of lodestone (an ore containing iron). When the lodestone spoon was placed in the center of a bronze disc, it would spin around; when it stopped, the handle part of the spoon pointed south and the bowl part of the spoon pointed north.

It was not until approximately a thousand years later that Chinese explorers began to use the compass as a navigational tool (Needham 1954). Navigation by compass is made possible because of the Earth's magnetic field, which acts as if there is a bar magnet underneath the Earth's surface (see Figure L10.2). There's not an actual bar magnet inside of the Earth, but rather the magnetic field is caused by the motion of molten iron in the outer core. There are a few things to note about the Earth's magnetic pole as shown in Figure L.10.2. First, the geographic and magnetic poles do not exactly line up—the magnetic pole in the Northern Hemisphere is approximately 500 km from the

FIGURE L10.1

A recreated model of a Han-era Si Nan

Note: A full-color version of this figure is available on the book's Extras page at *www.nsta.org/adi-physics2*.

FIGURE L10.2

A model of the Earth's magnetic field. Notice how the northern magnetic pole (i.e., the magnetic pole nearest to geographic North) is a south pole for the magnetic field. Similarly, the southern magnetic pole is, in fact, a north pole for the magnetic field.

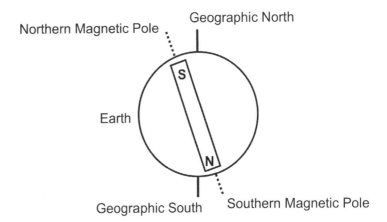

LAB 10

geographic North Pole. Second, the magnetic polarity of the Northern Hemisphere's magnetic pole is, in actuality, a south pole. When a compass needle points toward the North Pole, it is because the north pole of the compass is attracted to the south pole of the Earth's magnetic field, which happens to be closer to geographic North.

Magnetic fields are similar to electric and gravitational fields in that a magnetic field permeates space due to the presence of a magnetic source. The magnetic field is different from the electric and gravitational field in that a magnetic field always contains a pair of magnetic sources—a north pole and south pole. That is, a north pole cannot exist without a corresponding south pole (contrast this with the electric field, where a positive charge can create an electric field without the existence of a corresponding negative charge). In fact, if you were to cut a bar magnet in half, you would create two new bar magnets, each with a north and south pole. When mapping the magnetic field, scientists use field lines with arrows pointing away from the north pole and toward the south pole. Finally, opposite poles are attracted to each other, whereas like poles repel. This is how a compass works—the compass needle interacts with the Earth's magnetic field, and the north pole of the compass is attracted to the magnetic south pole of the Earth's magnetic field (which happens to be near geographic North).

Understanding the magnetic field created by a permanent magnet is important for many applications of modern life. Airplane guidance partially relies on the Earth's magnetic field, and many modern electronic components utilize magnets as part of their design. Understanding how magnetic fields change around permanent magnets includes understanding the relationship of the field strength and direction to the location in space from the source of the field.

Your Task

Use what you know about magnetic fields, patterns, and stability and change in systems to design and carry out an investigation to discover how the strength and direction of the magnetic field changes around a permanent magnet.

The guiding question of this investigation is, *How does the strength and direction of a magnetic field change as one moves around a permanent magnet?*

Materials

You may use any of the following materials during your investigation:

Consumables
- Tape
- String

Equipment
- Safety glasses with side shields or safety goggles (required)
- Bar magnet
- Large cow magnet
- Compass
- Ruler
- Magnetic field sensor
- Spring scale

Magnetic Field Around a Permanent Magnet

How Does the Strength and Direction of a Magnetic Field Change as One Moves Around a Permanent Magnet?

Safety Precautions

Follow all normal lab safety rules. In addition, take the following safety precautions:

1. Wear sanitized safety glasses with side shields or goggles during lab setup, hands-on activity, and takedown.

2. Wash your hands with soap and water when you are done collecting the data.

Investigation Proposal Required? ☐ Yes ☐ No

Getting Started

To answer the guiding question, you will need to design and carry out an investigation to understand the magnetic field around the permanent magnet. Before you can design your investigation, however, you must determine what type of data you need to collect, how you will collect it, and how you will analyze it.

To determine *what type of data you need to collect*, think about the following questions:

- What are the boundaries and components of the system?
- When is this system stable and under which conditions does it change?
- What is going on at the unobservable level that could cause the things that you observe?
- What quantities are vectors and what quantities are scalars?

To determine *how you will collect the data*, think about the following questions:

- What factors will you need to control during the investigation?
- What scale or scales should you use when you take your measurements?
- How will you make sure that your data are of high quality (i.e., how will you reduce error)?
- How will you keep track of and organize the data you collect?
- What are the components of this phenomenon or system and how do they interact?

To determine *how you will analyze the data*, think about the following questions:

- What types of patterns might you look for as you analyze your data?
- What type of table or graph could you create to help make sense of your data?
- What potential proportional relationships can you find in the data?
- How could you use mathematics to describe a relationship between variables?

LAB 10

Connections to the Nature of Scientific Knowledge and Scientific Inquiry

As you work through your investigation, you may want to consider

- the difference between data and evidence in science, and
- the assumptions made by scientists about order and consistency in nature.

Initial Argument

Once your group has finished collecting and analyzing your data, your group will need to develop an initial argument. Your initial argument needs to include a claim, evidence to support your claim, and a justification of the evidence. The *claim* is your group's answer to the guiding question. The *evidence* is an analysis and interpretation of your data. Finally, the justification of the evidence is why your group thinks the evidence matters. The *justification* of the evidence is important because scientists can use different kinds of evidence to support their claims. Your group will create your initial argument on a whiteboard. Your whiteboard should include all the information shown in Figure L10.3.

FIGURE L10.3
Argument presentation on a whiteboard

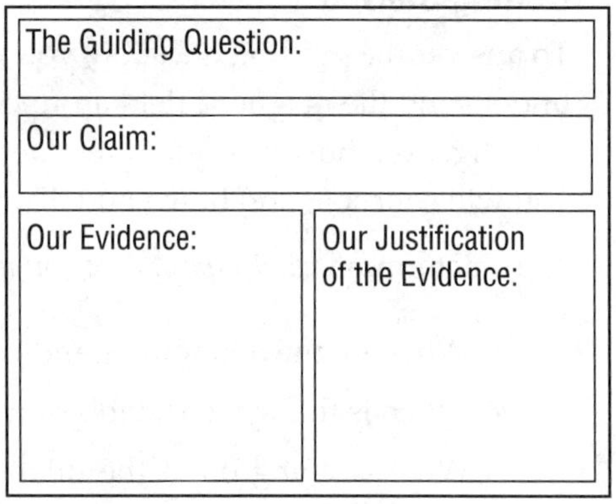

The Guiding Question:	
Our Claim:	
Our Evidence:	Our Justification of the Evidence:

Argumentation Session

The argumentation session allows all of the groups to share their arguments. One or two members of each group will stay at the lab station to share that group's argument, while the other members of the group go to the other lab stations to listen to and critique the other arguments. This is similar to what scientists do when they propose, support, evaluate, and refine new ideas during a poster session at a conference. If you are presenting your group's argument, your goal is to share your ideas and answer questions. You should also keep a record of the critiques and suggestions made by your classmates so you can use this feedback to make your initial argument stronger. You can keep track of specific critiques and suggestions for improvement that your classmates mention in the space below.

Critiques about our initial argument and suggestions for improvement:

Magnetic Field Around a Permanent Magnet
How Does the Strength and Direction of a Magnetic Field Change as One Moves Around a Permanent Magnet?

If you are critiquing your classmates' arguments, your goal is to look for mistakes in their arguments and offer suggestions for improvement so these mistakes can be fixed. You should look for ways to make your initial argument stronger by looking for things that the other groups did well. You can keep track of interesting ideas that you see and hear during the argumentation in the space below. You can also use this space to keep track of any questions that you will need to discuss with your team.

Interesting ideas from other groups or questions to take back to my group:

Once the argumentation session is complete, you will have a chance to meet with your group and revise your initial argument. Your group might need to gather more data or design a way to test one or more alternative claims as part of this process. Remember, your goal at this stage of the investigation is to develop the best argument possible.

Report

Once you have completed your research, you will need to prepare an *investigation report* that consists of three sections. Each section should provide an answer to the following questions:

1. What question were you trying to answer and why?
2. What did you do to answer your question and why?
3. What is your argument?

Your report should answer these questions in two pages or less. This report must be typed, and any diagrams, figures, or tables should be embedded into the document. Be sure to write in a persuasive style; you are trying to convince others that your claim is acceptable or valid!

References

Bodde, D. 1986. The state and empire of Ch'in. In *The Cambridge history of China*, ed. D. Twitchett and M. Loewe, 20–102. Cambridge, UK: Cambridge University Press.

Needham, J. 1954. *Science and civilization in China*. Cambridge, UK: Cambridge University Press.

LAB 10

Checkout Questions

Lab 10. Magnetic Field Around a Permanent Magnet: How Does the Strength and Direction of a Magnetic Field Change as One Moves Around a Permanent Magnet?

Use the image below to answer questions 1 and 2.

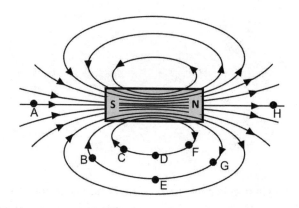

1. How strong is the magnetic field at point A in comparison with the strength of the magnetic field at point H?

    a. A is greater than H.
    b. A is less than H.
    c. A is equal to H.

    How do you know?

2. What point on the picture above has the weakest magnetic field?

   How do you know?

3. Scientific investigations are designed based on the assumption that natural laws operate today as they did in the past and they will continue to do so in the future.

   a. I agree with this statement.
   b. I disagree with this statement.

   Explain your answer, using examples from this investigation and at least one other investigation you have conducted.

LAB 10

4. The difference between data and evidence is that all data are numerical, whereas evidence is words or pictures.

   a. I agree with this statement.
   b. I disagree with this statement.

   Explain your answer, using examples from this investigation and at least one other investigation you have conducted.

5. Why is it useful to understand how fields change over space? In your answer, be sure to include examples from this investigation and at least one other investigation you have conducted.

6. What kinds of patterns do scientists identify during their investigations? In your answer, be sure to use examples from this investigation and at least one other investigation you have conducted.

LAB 11

Teacher Notes

Lab 11. Magnetic Forces: What Is the Mathematical Relationship Between the Distance Between Two Magnets and the Strength of the Force Acting on Them?

Purpose

The purpose of this lab is to introduce students to the disciplinary core idea (DCI) of Motion and Stability: Forces and Interactions (PS2) from the *NGSS* by having them determine the relationship between the force of the magnetic interaction between two magnets and the distance between the two magnets. In addition, this lab can be used to help students understand two big ideas from AP Physics: (a) fields existing in space can be used to explain interactions and (b) the interactions of an object with other objects can be described by forces. This lab also gives students an opportunity to learn about the crosscutting concepts (CCs) of (a) Patterns and (b) Cause and Effect: Mechanism and Explanation from the *NGSS*. As part of the explicit and reflective discussion, students will also learn about (a) the difference between observations and inferences in science and (b) the nature and role of experiments in science.

Underlying Physics Concepts

The starting point for this lab is the recognition that permanent magnets establish a magnetic field. The field lines point out from the north pole of a magnet and into the south pole of a magnet. Figure 11.1 shows the field established by a permanent bar magnet.

FIGURE 11.1

Magnetic field due to a permanent bar magnet (S = south pole; N = north pole)

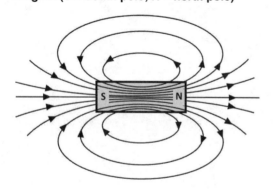

When two magnets are brought near each other, their fields will interact, producing a force on each magnet. When two opposite poles are brought near each other, an attractive field is produced, as shown in Figure 11.2. When two of the same poles are brought near each other—either north-north or south-south—a repulsive field is produced, as shown in Figure 11.3.

In this lab, students will investigate how the force between the two magnets changes as the distance between them changes. The lab setup has students place one magnet on top of an electronic balance. The balance will initially read the true mass of the magnet placed on the balance. From this, they can extrapolate the force of gravity and the normal force acting on the magnet using Newton's second law, shown in Equation 11.1, where $\sum F$ is the sum

Magnetic Forces

What Is the Mathematical Relationship Between the Distance Between Two Magnets and the Strength of the Force Acting on Them?

FIGURE 11.2
An attractive field

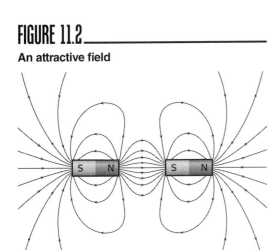

FIGURE 11.3
A repulsive field

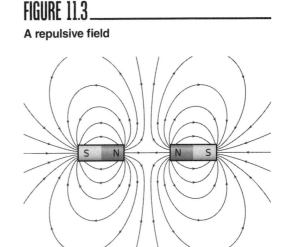

of the forces acting on the magnet, m is the mass of the magnet, and **a** is the acceleration of the magnet. In SI units, force is measured in newtons (N), mass is measured in kilograms (kg), and acceleration is measured in meters per second squared (m/s²).

$$\text{(Equation 11.1)} \quad \sum F = ma$$

Students need to find the force of gravity and the normal force acting on the magnet when it is not interacting with a second magnet; this is important to do because the reading on the electronic balance reflects the normal force acting on the object (in this case, the object is the first magnet) resting on the balance. That is, the mass reported on the balance is the normal force that the balance exerts on the magnet divided by the acceleration due to gravity (9.8 m/s²).

The students then set up the equipment as shown in the Lab Handout. They vary the distance between the magnets on the dowel rod and the magnet on the balance, and they will notice that the reading on the balance will change. If the poles facing each other are opposite, they will find that the mass shown on the balance will decrease as the distance between the magnets decreases. If the poles facing each other are the same, they will find that the mass shown on the balance will increase as the distance between them increases.

Figure 11.4 shows a free-body diagram for the case where a repulsive field is established between the two magnets because they have the same poles facing each other. In this figure, F_N is

FIGURE 11.4
A free-body diagram for the magnet on the balance

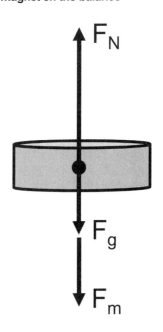

the normal force, F_g is the force due to gravity, and F_m is the force due to the interaction with the magnet on the dowel. Because the magnet on the balance is not accelerating (the magnet is at rest), the sum of the forces must be equal to zero. To find the force of the interaction between the two magnets, we can use Equation 11.2; we subtract F_g and F_m from F_N because they are pointed in an opposite direction from F_N and in the down direction.

$$\text{(Equation 11.2)} \quad \sum F = F_N - F_g - F_m = 0 \text{ N}$$

Because the mass of the magnet on the balance remains constant, we can solve for the force of the interaction between the two magnets by rearranging Equation 11.2 into Equation 11.3. Note that we pull out the common factor g from both the normal force and the force of gravity. In this case, m_a is the apparent mass shown by the balance and m_t is the true mass; g is the acceleration due to gravity.

$$\text{(Equation 11.3)} \quad F_m = (m_a - m_t)g$$

FIGURE 11.5

A free-body diagram for an attractive field

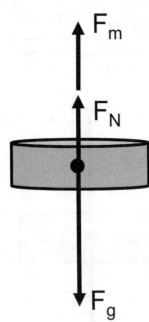

Figure 11.5 shows the free-body diagram for the case where an attractive field is established between the two magnets because opposite poles are facing each other. Because the field is attractive, the normal force decreases, as Newton's second law and, by extension, Equation 11.2 still hold true. The force of the interactions between the two magnets is shown in Equation 11.4. Because the force of the magnetic interaction will be attractive, the apparent mass read by the balance will decrease as the distance between the two magnets decreases. This is indicative of the force between the two magnets increasing. This is also why Equation 11.4 shows the apparent mass being subtracted from the true mass.

$$\text{(Equation 11.4)} \quad F_m = (m_t - m_a)g$$

After collecting data on the masses, computing the change in mass, and the distances from the magnets at these changes, students can plot force as a function of distance. Via regression, students will produce a mathematical model where force (F_m) is inversely proportional to the square of the distance between the magnets, r, and k is the constant of proportionality as shown in Equation 11.5. Different groups will produce different values for the constant of proportionality based on the type and number of magnets used in the exploration.

$$\text{(Equation 11.5)} \quad F_m \propto \frac{k}{r^2}$$

Magnetic Forces
What Is the Mathematical Relationship Between the Distance Between Two Magnets and the Strength of the Force Acting on Them?

Timeline

The instructional time needed to complete this lab investigation is 170–230 minutes. Appendix 3 (p. 421) provides options for implementing this lab investigation over several class periods. Option C (230 minutes) should be used if students are unfamiliar with scientific writing, because this option provides extra instructional time for scaffolding the writing process. You can scaffold the writing process by modeling, providing examples, and providing hints as students write each section of the report. Option C can also be used if you are introducing students to the video analysis programs. Option D (170 minutes) should be used if students are familiar with scientific writing and have developed the skills needed to write an investigation report on their own. In option D, students complete stage 6 (writing the investigation report) and stage 8 (revising the investigation report) as homework.

Materials and Preparation

The materials needed to implement this investigation are listed in Table 11.1. The equipment can be purchased from a science supply company such as Flinn Scientific, PASCO, Vernier, or Ward's Science. Video analysis software can be purchased from Vernier (*Logger Pro*) or PASCO (SPARKvue or Capstone). These companies also have apps that can be used on Apple- or Android-based tablets and cell phones. We recommend consulting with your school's information technology coordinator to determine the best option for your students.

TABLE 11.1
Materials list for Lab 11

Item	Quantity
Consumable	
Tape	As needed
Equipment and other materials	
Safety glasses with side shields or safety goggles	1 per student
Magnets (cylindrical with hole for dowel rod)	5 per group
Ring stand	1 per group
Clamps	2 per group
Dowel rod	1 per group
Electronic balance	1 per group
Magnetic field sensor	1 per group

Continued

LAB 11

Table 11.1 (*continued*)

Item	Quantity
Ruler	1 per group
Investigation Proposal B (optional)	1 per group
Whiteboard, 2' × 3'*	1 per group
Lab Handout	1 per student
Peer-review guide and teacher scoring rubric	1 per student
Checkout Questions	1 per student
Equipment for video analysis (optional)	
Video camera	1 per group
Computer or tablet with video analysis software	1 per group

* As an alternative, students can use computer and presentation software such as Microsoft PowerPoint or Apple Keynote to create their arguments.

Use of video analysis software is optional, but using this software will allow students to more precisely measure any movement of the magnets.

Be sure to use a set routine for distributing and collecting the materials during the lab investigation. One option is to set up the materials for each group at each group's lab station before class begins. This option works well when there is a dedicated section of the classroom for lab work and the materials are large and difficult to move. A second option is to have all the materials on a table or cart at a central location. You can then assign a member of each group to be the "materials manager." This individual is responsible for collecting all the materials his or her group needs from the table or cart during class and for returning all the materials at the end of the class. This option works well when the materials are small and easy to move (such as magnets, wire, and bulbs). It also makes it easy to inventory the materials at the end of the class before students leave for the day.

Safety Precautions and Laboratory Waste Disposal

Remind students to follow all normal lab safety rules. In addition, tell students to take the following safety precautions:

1. Wear sanitized safety glasses with side shields or goggles during lab setup, hands-on activity, and takedown.

2. Wash their hands with soap and water when they are done collecting the data.

There is no laboratory waste associated with this activity.

Magnetic Forces
What Is the Mathematical Relationship Between the Distance Between Two Magnets and the Strength of the Force Acting on Them?

Topics for the Explicit and Reflective Discussion
Reflecting on the Use of Core Ideas and Crosscutting Concepts During the Investigation

Teachers should begin the explicit and reflective discussion by asking students to discuss what they know about the core ideas they used during the investigation. The following are some important concepts related to the core idea of forces and interactions that students need to use to determine the relationship between the distance and the strength of repulsive force:

- A force exerted on an object is always due to the interaction of that object with another object.
- If an object of interest interacts with multiple other objects, the net force is the vector sum of the individual forces acting on the object of interest. If the object of interest is at rest, it may be because no forces are acting on it. It may also be because forces exerted on that object by other objects sum to zero.
- Newton's second law is expressed mathematically as $\sum \mathbf{F} = m\mathbf{a}$.
- The reading on an electronic balance displays the normal force—and not the force of gravity—with which the balance is pushing back up on the object resting on the balance. The reading on the balance may change depending on the interaction of the object on the balance with other objects. However, the gravitational interaction between the object and Earth remains unchanged.
- Free-body diagrams help identify the forces acting on an object of interest.
- Forces arise from fields. The stronger the field, the larger the force the field exerts on an object placed in the field.

To help students reflect on what they know about magnetic forces and interactions, we recommend showing them two or three images using presentation software that help illustrate these important ideas. You can then ask the students the following questions to encourage them to share how they are thinking about these important concepts:

1. What do we see going on in this image?
2. Does anyone have anything else to add?
3. What might be going on that we can't see?
4. What are some things that we are not sure about here?

You can then encourage students to think about how CCs played a role in their investigation. There are at least two CCs that students need to use to determine a mathematical relationship between the magnetic force and the distance between the two magnets: (a Patterns and (b) Cause and Effect: Mechanism and Explanation (see Appendix 2 [p. 417] for a brief description of these CCs). To help students reflect on what they know about these CCs, we recommend asking them the following questions:

1. Are the patterns you identified similar to other patterns you are familiar with in physics?

2. Why do scientists want to identify patterns in natural phenomena?

3. Natural phenomena have causes, where a change in one variable produces a change in a second variable. Did you identify any cause-and-effect relationships in this investigation?

4. Why is it important for scientists to identify cause-and-effect relationships in the natural world?

You can then encourage the students to think about how they used all these different concepts to help answer the guiding question and why it is important to use these ideas to help justify their evidence for their final arguments. Be sure to remind your students to explain why they included the evidence in their arguments and make the assumptions underlying their analysis and interpretation of the data explicit in order to provide an adequate justification of their evidence.

Reflecting on Ways to Design Better Investigations

It is important for students to reflect on the strengths and weaknesses of the investigation they designed during the explicit and reflective discussion. Students should therefore be encouraged to discuss ways to eliminate potential flaws, measurement errors, or sources of uncertainty in their investigations. To help students be more reflective about the design of their investigation and what they can do to make their investigations more rigorous in the future, you can ask them the following questions:

1. What were some of the strengths of the way you planned and carried out your investigation? In other words, what made it scientific?

2. What were some of the weaknesses of the way you planned and carried out your investigation? In other words, what made it less scientific?

3. What rules can we make, as a class, to ensure that our next investigation is more scientific?

Reflecting on the Nature of Scientific Knowledge and Scientific Inquiry

This investigation can be used to illustrate two important concepts related to the nature of scientific knowledge and the nature of scientific inquiry: (a) the difference between observations and inferences in science and (b) the nature and role of experiments (see Appendix 2 [p. 417] for a brief description of these concepts). Be sure to review these concepts during and at the end of the explicit and reflective discussion. To help students think about these concepts in relation to what they did during the lab, you can ask them the following questions:

1. You made observations and inferences during your investigation. Can you give me some examples of these observations and inferences?

2. Can you work with your group to come up with a rule that you can use to tell the difference between an observation and an inference? Be ready to share in a few minutes.

3. I asked you to design an experiment as part of your investigation. Why do scientists conduct experiments?

4. Can you work with your group to come up with a rule that you can use to decide if an investigation is an experiment or not? Be ready to share in a few minutes.

You can also encourage the students to think about these concepts by showing examples of observations and inferences using presentation software and then asking students to classify each one. You can also show images of different experimental designs and ask students if the design meets the requirements for a good experiment. Be sure to remind your students that it is important for them to understand what counts as scientific knowledge and how that knowledge develops over time in order to be proficient in science.

Hints for Implementing the Lab

- Allowing students to design their own procedures for collecting data gives students an opportunity to try, to fail, and to learn from their mistakes. However, you can scaffold students as they develop their own procedure by having them fill out an investigation proposal. The proposals provide a way for you to offer students hints and suggestions without telling them how to do it. You can also check the proposals quickly during a class period. For this lab we suggest using Investigation Proposal B.

- Learn how to use the magnetic field sensor before the lab begins. It is important for you to know how to use the equipment so you can help students when technical issues arise.

- Allow the students to become familiar with the equipment as part of the tool talk before they begin to design their investigation. Give them 5–10 minutes to examine the equipment and materials before they begin designing their investigations. This gives students a chance to see what they can and cannot do with the equipment.

- We suggest having students create a repulsive field between the magnets. This will provide more pronounced results than an attractive field.

- The ranges for the distance between the two magnets do not need to be more than 0.5 m. Any greater distances will result in a weak interaction of the magnetic fields.
- Students using a magnetic field sensor may choose to investigate how the magnetic field changes with distance. This is an opportunity to reinforce the connection between fields and forces. In this case, students will also want to add a statement to their justification about the relationship between forces and fields. We suggest also mentioning this during the explicit and reflective discussion.
- Be sure to allow students to go back and re-collect data at the end of the argumentation session. Students often realize that they made numerous mistakes when they were collecting data as a result of their discussions during the argumentation session. The students, as a result, will want a chance to re-collect data, and the re-collection of data should be encouraged when time allows. This also offers an opportunity to discuss what scientists do when they realize a mistake is made inside the lab.

If students use video analysis

- We suggest allowing students to familiarize themselves with the video analysis software before they finalize the procedure for the investigation, especially if they have not used such software previously. This gives students an opportunity to learn how to work with the software and to improve the quality of the video they take.
- Remind students to hold the video camera as still as possible. Any movement of the camera will introduce error into their analysis. If using actual camcorders, we recommend using a tripod to hold the camera steady. If students are using a camera on a cell phone or tablet, we recommend using a table to help steady the camera.
- Remind students to place a meterstick in the same field of view as the motion they are capturing with the video camera. Also, the meterstick should be approximately the same distance from the camera as the motion. Most video analysis software requires the user to define a scale in the video (this allows the software to establish distances and, subsequently, other variables dependent on distance and displacement).

Connections to Standards

Table 11.2 highlights how the investigation can be used to address specific performance expectations from the *NGSS*; learning objectives from AP Physics 1 and 2; learning objectives from AP Physics C: Electricity and Magnetism; *Common Core State Standards for English Language Arts* (*CCSS ELA*); and *Common Core State Standards for Mathematics* (*CCSS Mathematics*).

Magnetic Forces

What Is the Mathematical Relationship Between the Distance Between Two Magnets and the Strength of the Force Acting on Them?

TABLE 11.2

Lab 11 alignment with standards

NGSS performance expectation	• HS-PS2-1: Analyze data to support the claim that Newton's second law of motion describes the mathematical relationship among the net force on a macroscopic object, its mass, and its acceleration.
AP Physics 1 and AP Physics 2 learning objectives	• 3.A.2.1: Represent forces in diagrams or mathematically using appropriately labeled vectors with magnitude, direction, and units during the analysis of a situation. • 3.A.3.1: Analyze a scenario and make claims (develop arguments, justify assertions) about the forces exerted on an object by other objects for different types of forces or components of forces. • 3.A.3.3: Describe a force as an interaction between two objects, and identify both objects for any force.
AP Physics C: Electricity and Magnetism learning objectives	• None
Literacy connections (*CCSS ELA*)	• *Reading*: Key ideas and details, craft and structure, integration of knowledge and ideas • *Writing*: Text types and purposes, production and distribution of writing, research to build and present knowledge, range of writing • *Speaking and listening:* Comprehension and collaboration, presentation of knowledge and ideas
Mathematics connections (*CCSS Mathematics*)	• *Mathematical practices*: Make sense of problems and persevere in solving them, reason abstractly and quantitatively, construct viable arguments and critique the reasoning of others, model with mathematics, use appropriate tools strategically, attend to precision • *Number and quantity*: Reason quantitatively and use units to solve problems, represent and model with vector quantities, perform operations on vectors • *Algebra*: Interpret the structure of expressions, create equations that describe numbers or relationships, understand solving equations as a process of reasoning and explain the reasoning, solve equations and inequalities in one variable, represent and solve equations and inequalities graphically • *Functions*: Understand the concept of a function and use function notation; interpret functions that arise in applications in terms of the context; analyze functions using different representations; build a function that models a relationship between two quantities; construct and compare linear, quadratic, and exponential models and solve problems; interpret expressions for functions in terms of the situation they model • *Statistics and probability*: Summarize, represent, and interpret data on two categorical and quantitative variables; interpret linear models; make inferences and justify conclusions from sample surveys, experiments, and observational studies

LAB 11

Lab Handout

Lab 11. Magnetic Forces: What Is the Mathematical Relationship Between the Distance Between Two Magnets and the Strength of the Force Acting on Them?

Introduction

Many people enjoy visiting amusement parks and riding roller coasters. Many roller coasters use a chain attached to a motor to lift the roller coaster cars up the first hill. Recently, amusement parks have begun using a launched start for roller coasters, where the cars are propelled forward. There are many types of propelled launch systems for roller coasters, including linear induction motor (LIM) systems. In a LIM system, a series of magnets are attached to the undercarriage of the roller coaster cars. There are also electromagnets embedded in the track that can be controlled by the ride operator. When the ride is started, the electromagnets produce a strong magnetic field. The magnets on the bottom of the car interact with the magnetic field produced by the track, and a repulsive force is exerted on the car, accelerating it to velocities of over 60 mph in under two seconds.

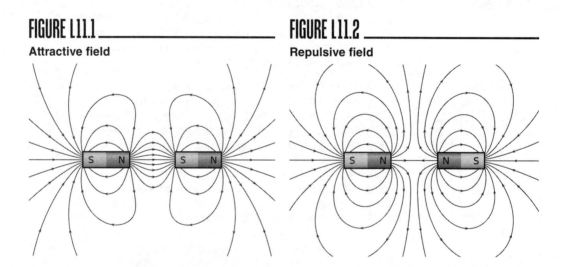

FIGURE L11.1
Attractive field

FIGURE L11.2
Repulsive field

As anyone who has ever played with magnets can attest, magnets exert forces that either attract or repel, based on the direction of the magnets. When the north pole of a cylindrical bar magnet comes near the south pole of another bar magnet, an attractive field is formed (see Figure L11.1). Conversely, when the north pole of one magnet comes near the north pole of another magnet, a repulsive field is formed (see Figure L11.2). Through experimentation, one notices that as the north pole of one magnet gets closer and closer to the south pole of another, the magnetic force becomes stronger and stronger until the

Magnetic Forces

What Is the Mathematical Relationship Between the Distance Between Two Magnets and the Strength of the Force Acting on Them?

magnets snap together. The opposite is true when one pushes two north poles toward each other (or two south poles). As the two north poles get closer, it becomes harder to continue to push them closer. This phenomenon is seen and used in a vast number of operations in our everyday world, from those as simple as ensuring our cabinets stay shut when we close them, to sorting recyclables on a conveyor line, to the dampening of suspension for better performance and safety in the vehicles we travel in daily.

It is important for scientists and engineers to understand how two magnets interact when designing systems that use magnets. For example, the suspension mechanisms in vehicles need to use magnets of precise strength, because magnets that are too strong or too weak will result in an uncomfortable ride. And, roller coaster designers need to understand how the magnets in the car and the magnets in the track will interact to produce a force on the car. Thus, a mathematical relationship relating the force acting on two magnets as they approach each other is important to find.

Your Task

Use what you know about forces and interactions, cause and effect, and patterns to design and carry out an experiment to determine the relationship between the distance two magnets are from one another and the force acting on the magnets. Your mathematical model should be designed for both interpolation and extrapolation of forces given known distances between two uniform magnets.

The guiding question of this investigation is, **What is the mathematical relationship between the distance between two magnets and the strength of the magnetic force acting on them?**

Materials

You may use any of the following materials during your investigation:

Consumable
- Tape

Equipment
- Safety glasses with side shields or safety goggles (required)
- Magnets
- Ring stand
- Clamps
- Dowel rod
- Electronic balance
- Magnetic field sensor
- Ruler

Safety Precautions

Follow all normal lab safety rules. In addition, take the following safety precautions:

- Wear sanitized safety glasses with side shields or safety goggles during lab setup, hands-on-activity, and takedown.

LAB 11

- Wash your hands with soap and water when you are done collecting the data.

Investigation Proposal Required?
☐ Yes ☐ No

Getting Started

To answer the guiding question, you will need to design and carry out an experiment to determine the relationship between the distance between two magnets and the force each magnet feels from the other. Figure L11.3 shows one possible way to set up the equipment that you will be provided. Before you can design your experiment, however, you must determine what type of data you need to collect, how you will collect it, and how you will analyze it.

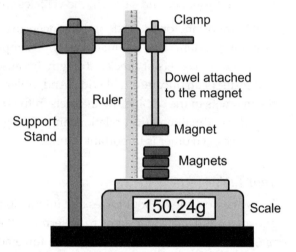

FIGURE L11.3

One possible way to set up the equipment in this investigation

To determine *what type of data you need to collect,* think about the following questions:

- What units are required for measuring each variable?
- What is the independent variable and what is the dependent variable?
- What is going on at the unobservable level that could cause the things that you observe?
- How could you keep track of changes in this system quantitatively?
- What could be the underlying cause of this phenomenon?

To determine *how you will collect your data,* think about the following questions:

- What other factors will you need to control during each experiment?
- How will you make sure that your data are of high quality (i.e., how will you reduce error)?
- How will you keep track of and organize the data you collect?
- What conditions need to be satisfied to establish a cause-and-effect relationship?

To determine *how you will analyze your data,* think about the following questions:

- What type of calculations will you need to make?
- What types of patterns might you look for as you analyze your data?

Magnetic Forces

What Is the Mathematical Relationship Between the Distance Between Two Magnets and the Strength of the Force Acting on Them?

- What type of table or graph could you create to help make sense of your data?
- How will you mathematically model your data?

Connections to the Nature of Science Knowledge and Scientific Inquiry

As you work through your investigation, be sure to think about

- the difference between observations and inferences in science, and
- the nature and role of experiments in science.

Argument presentation on a whiteboard

The Guiding Question:	
Our Claim:	
Our Evidence:	Our Justification of the Evidence:

Initial Argument

Once your group has finished collecting and analyzing your data, your group will need to develop an initial argument. Your initial argument needs to include a claim, evidence to support your claim, and a justification of the evidence. The *claim* is your group's answer the guiding question. The *evidence* is an analysis and interpretation of your data. Finally, the *justification* of the evidence is why your group thinks the evidence matters. The justification of the evidence is important because scientists can use different kinds of evidence to support their claims. You group will create your initial argument on a whiteboard. Your whiteboard should include all the information shown in Figure L11.4.

Argumentation Session

The argumentation session allows all of the groups to share their arguments. One or two members of each group will stay at the lab station to share that group's argument, while the other members of the group go to the other lab stations to listen to and critique the other arguments. This is similar to what scientists do when they propose, support, evaluate, and refine new ideas during a poster session at a conference. If you are presenting your group's argument, your goal is to share your ideas and answer questions. You should also keep a record of the critiques and suggestions made by your classmates so you can use this feedback to make your initial argument stronger. You can keep track of specific critiques and suggestions for improvement that your classmates mention in the space below.

Critiques about our initial argument and suggestions for improvement:

LAB 11

If you are critiquing your classmates' arguments, your goal is to look for mistakes in their arguments and offer suggestions for improvement so these mistakes can be fixed. You should look for ways you to make your initial argument stronger by looking for things that the other groups did well. You can keep track of interesting ideas that you see and hear during the argumentation in the space below. You can also use this space to keep track of any questions that you will need to discuss with your team.

Interesting ideas from other groups or questions to take back to my group:

Once the argumentation session is complete, you will have a chance to meet with your group and revise your initial argument. Your group might need to gather more data or design a way to test one or more alternative claims as part of this process. Remember, your goal at this stage of the investigation is to develop the best argument possible.

Report

Once you have completed your research, you will need to prepare an investigation report that consists of three sections. Each section should provide an answer for the following questions:

1. What question were you trying to answer and why?
2. What did you do to answer your question and why?
3. What is your argument?

Your report should answer these questions in two pages or less. This report must be typed, and any diagrams, figures, or tables should be embedded into the document. Be sure to write in a persuasive style; you are trying to convince others that your claim is acceptable or valid!

Checkout Questions

Lab 11. Magnetic Forces: What Is the Mathematical Relationship Between the Distance Between Two Magnets and the Strength of the Force Acting on Them?

1. Sketch the graph of force, **F**, as a function of the distance, **r**, the magnets are from one another. Label and scale your graph.

2. Using your mathematical model, extrapolate the force of the magnets when they are

   a. 2 m apart
   b. 20 m apart
   c. 2 km apart
   d. For those students with calculus background, find $\lim_{d \to \infty} F$.

3. Using your mathematical model, extrapolate the force of the magnets when they are

   a. 1 cm apart
   b. 1 mm apart
   c. 0.1 mm apart
   d. For those students with calculus background, find $\lim\limits_{d \to \infty} F$.

4. There is always a repulsion force between two magnets no matter how far away the two magnets are from one another.

   a. I agree with this statement.
   b. I disagree with this statement.

   Explain your answer, using information from question 2 and your experience during the investigation on magnetic force.

Magnetic Forces

What Is the Mathematical Relationship Between the Distance Between Two Magnets and the Strength of the Force Acting on Them?

5. There is a maximum repulsive force between two magnets, and this occurs when the magnets are touching one another with like poles facing each other (e.g., north pole facing north pole).

   a. I agree with this statement.
   b. I disagree with this statement.

   Explain your answer, using information from question 3 and your experience during the investigation on magnetic force.

6. The difference between observations and inferences is that observations are based on data and inferences are one's opinion.

   a. I agree with this statement.
   b. I disagree with this statement.

   Explain your answer, using examples from this investigation and at least one other investigation you have conducted.

LAB 11

7. What patterns did you identify during your investigation of magnetic force? Are these patterns similar to any patterns you know of regarding other forces?

8. Why is determining cause-and-effect relationships an important part of many scientific investigations? In your answer, be sure to use examples from this investigation and at least one other investigation you have conducted.

9. What conditions are necessary for an investigation to be considered an experiment? In your answer, be sure to use examples from this investigation and at least one other investigation you have conducted.

LAB 12

Teacher Notes

Lab 12. Magnetic Fields Around Current-Carrying Wires: How Does Changing the Magnitude and Direction of a Current in Two Parallel Wires Affect the Magnetic Field Around the Two-Wire System?

Purpose

The purpose of this lab is to *introduce* students to the disciplinary core idea (DCI) of Motion and Stability: Forces and Interactions (PS2) from the *NGSS* by having them investigate the factors influencing the strength of the magnetic field around a two-wire system. In addition, this lab can be used to help students understand two big ideas from AP Physics: (a) fields existing in space can be used to explain interactions and (b) the interactions of an object with other objects can be described by forces. This lab also gives students an opportunity to learn about the crosscutting concepts (CCs) of (a) Cause and Effect: Mechanism and Explanation; and (b) Systems and System Models from the NGSS. As part of the explicit and reflective discussion, students will also learn about (a) the difference between data and evidence in science and (b) how scientists use different methods to answer different types of questions.

FIGURE 12.1
A current-carrying wire

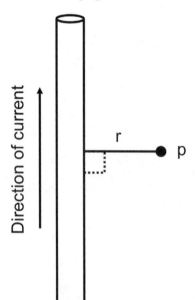

Underlying Physics Concepts

Ampère's law describes the strength of the magnetic field surrounding a current-carrying conductor. When applied to a long, straight current-carrying wire, Ampère's law suggests the magnetic field will be directly proportional to the strength of the current and inversely proportional to the distance from the wire measured perpendicular to the surface of the wire. Figure 12.1 shows a long, straight current-carrying wire and point p some distance **r** from the wire.

The precise mathematical relationship is shown in Equation 12.1, where **B** is the strength of the magnetic field, I is the current through the wire, **r** is the distance from the wire, and μ_o is the permeability of free space and is a constant. In SI units, magnetic field is measured in teslas (T), current is measured in amperes (A), distance is measured in meters (m), and the permeability of free space has a value of $4\pi \times 10^{-7}$ tesla-meters per ampere (T·m/A).

Magnetic Fields Around Current-Carrying Wires
How Does Changing the Magnitude and Direction of a Current in Two Parallel Wires Affect the Magnetic Field Around the Two-Wire System?

$$\text{(Equation 12.1)} \quad \mathbf{B} = \frac{\mu_0 I}{2\pi \mathbf{r}}$$

To determine the direction of the magnetic field, we can use a right-hand rule, as shown in Figure 12.2. This right-hand rule works by wrapping your fingers around the wires with your thumb pointing in the direction the current is flowing, curling your fingers. It is important to note that the magnetic field encircles the wire—the field forms a continuous loop around the wire.

Returning to Figure 12.1, if we want to know the direction of the magnetic field at point p, we can use this right-hand rule to determine that the magnetic field points into the page at point p. When representing magnetic fields around a wire, we use the symbol ⊗ when the field points into the page (or, more generally, into a plane), and we use the symbol ⊙ when the magnetic field points out of the page (or, more generally, out of a plane). Figure 12.3 shows the direction of the magnetic field at point p due to the current-carrying wire.

Another important thing to note is that magnetic fields obey properties of vector addition. If two wires are near each other, we can add the magnetic field from each wire to find the total magnetic field at a given point.

Figure 12.4 (p. 270) shows two parallel wires with current running through them in the same direction. Using the right-hand rule, we can determine the magnetic field at point p due to both wires; the magnetic field at point p due to wire A is into the page, and the magnetic field due to wire B is out of the page. If the current through the two wires is the same, then the magnetic fields will add to zero because point p is equidistant from wire A and wire B. At that point in space, there will be no magnetic field due to the wires. We can also ask what the magnetic field is at point q, to the left of wire A, and point s, to the right of wire B. Using the right-hand rule at point q, the magnetic field due to wire A and wire B will both be out of the page, and

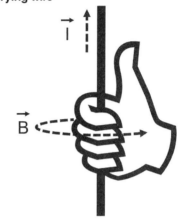

FIGURE 12.2
The magnetic field around a current-carrying wire

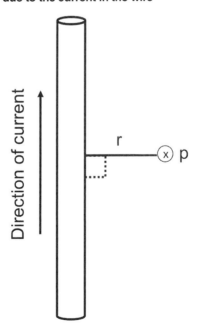

FIGURE 12.3
The magnetic field direction at point p due to the current in the wire

FIGURE 12.4

Two parallel wires with current (I) running in the same direction

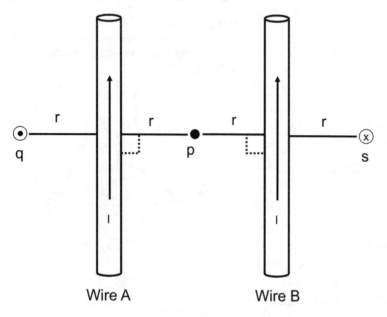

FIGURE 12.5

Forces on two wires carrying current in the same direction

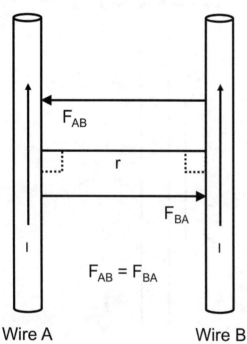

the magnitude of the magnetic field will be the sum of the strength of the two individual magnetic fields at that point. At point s, the right-hand rule shows that the magnetic field from both wires is pointing into the page and will be the sum of the two individual magnetic field strengths at point s.

We can also ask about the force on each individual wire from its interaction with the magnetic field created by the other wire (e.g., the force on wire A due to the magnetic field from wire B, and vice versa). In this case, the two magnetic fields interact, producing a force on each wire. Because the magnetic field from wire A is in the opposite direction as the magnetic field from wire B in the space between them, the wires feel an attractive force. This is shown in Figure 12.5, where F_{AB} is the force from wire A acting on wire B and F_{BA} is the force from wire B acting on wire A. According to Newton's third law, these forces will be equal in magnitude.

If you want to determine the force acting on each wire, Equation 12.2 shows the mathematical relationship between the force and the current through each wire, where **F** is the force acting on each wire, I_A is the current in wire A, I_B is the current in wire B, L is the length of wire parallel to each other, **r** is the distance between the two wires, and μ_o is the permeability of free space and is a constant. In SI units, force is measured in newtons (N), current is measured in amperes (A), length and distance are measured in meters

Magnetic Fields Around Current-Carrying Wires
How Does Changing the Magnitude and Direction of a Current in Two Parallel Wires Affect the Magnetic Field Around the Two-Wire System?

(m), and the permeability of free space is equal to $4\pi \times 10^{-7}$ T·m/A.

(Equation 12.2) $F = \dfrac{\mu_0 I_A I_B L}{2\pi r}$

Figures 12.4 and 12.5 show the effect of the currents moving in the same direction. Figure 12.6 illustrates what happens when the currents move in opposite directions (antiparallel currents). In this case, using the right-hand rule, the magnetic fields due to both wire A and wire B are pointing into the page at point p. At point q, the magnetic field is pointing out of the page due to wire A and into the page due to wire B. If we assume that the current in each wire is the same, the magnetic field strength due to wire A is greater than the magnetic field strength due to wire B at point q. Subsequently, the resultant magnetic field at point q is pointing out of the page. Similarly, at point s, the magnetic field is pointing out of the page because the magnetic field from wire B (out of the page) is greater than the magnetic field from wire A (into the page).

Finally, we can ask what the force is on each wire due to the currents running in opposite directions. Because the fields point in the same direction at point p and because like magnetic poles repel each other, the force on each wire is directed away from the other wire. Figure 12.7 shows the forces acting on each wire.

FIGURE 12.6
The magnetic field due to antiparallel currents

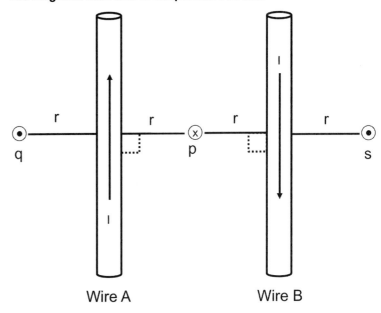

FIGURE 12.7
The forces acting on two wires with antiparallel currents

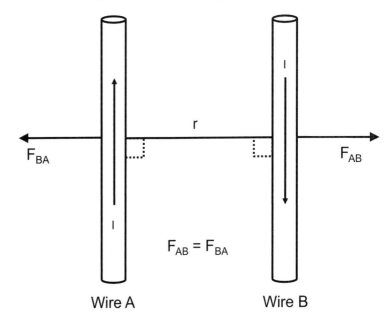

LAB 12

Timeline

The instructional time needed to complete this lab investigation is 170–230 minutes. Appendix 3 (p. 421) provides options for implementing this lab investigation over several class periods. Option C (230 minutes) should be used if students are unfamiliar with scientific writing, because this option provides extra instructional time for scaffolding the writing process. You can scaffold the writing process by modeling, providing examples, and providing hints as students write each section of the report. Option C can also be used if you are introducing students to digital interface sensors and/or the data analysis software. Option D (170 minutes) should be used if students are familiar with scientific writing and have developed the skills needed to write an investigation report on their own. In option D, students complete stage 6 (writing the investigation report) and stage 8 (revising the investigation report) as homework.

Materials and Preparation

The materials needed to implement this investigation are listed in Table 12.1. The equipment can be purchased from a science supply company such as Flinn Scientific, PASCO, Vernier, or Ward's Science.

TABLE 12.1

Materials list for Lab 12

Item	Quantity
Consumables	
D batteries	Several per group
9V batteries	Several per group
Tape	1 roll per group
String	As needed per group
Equipment and other materials	
Safety glasses or goggles	1 per student
Battery holders	2 per group
Wire with alligator clips on each end	4 segments per group
Magnetic field sensor	1 per group
Multimeters or ammeters	2 per group
Ruler	1 per group
Investigation Proposal C (optional)	1 per group

Continued

Magnetic Fields Around Current-Carrying Wires
How Does Changing the Magnitude and Direction of a Current in Two Parallel Wires Affect the Magnetic Field Around the Two-Wire System?

Table 12.1 (*continued*)

Item	Quantity
Whiteboard, 2' × 3'*	1 per group
Lab Handout	1 per student
Peer-review guide and teacher scoring rubric	1 per student
Checkout Questions	1 per student
Equipment for digital interface measurements (optional)	
Digital interface with USB or wireless connections	1 per group
Current measurement sensor	1 per group
Computer or tablet with appropriate data analysis software installed	1 per group

* As an alternative, students can use computer and presentation software such as Microsoft PowerPoint or Apple Keynote to create their arguments.

Be sure to use a set routine for distributing and collecting the materials during the lab investigation. One option is to set up the materials for each group at each group's lab station before class begins. This option works well when there is a dedicated section of the classroom for lab work and the materials are large and difficult to move. A second option is to have all the materials on a table or cart at a central location. You can then assign a member of each group to be the "materials manager." This individual is responsible for collecting all the materials his or her group needs from the table or cart during class and for returning all the materials at the end of the class. This option works well when the materials are small and easy to move (such as magnets, wire, and bulbs). It also makes it easy to inventory the materials at the end of the class before students leave for the day.

Safety Precautions and Laboratory Waste Disposal

Remind students to follow all normal lab safety rules. In addition, tell students to take the following safety precautions:

1. Wear sanitized safety glasses or goggles during lab setup, hands-on activity, and takedown.

2. Never put consumables in their mouth.

3. Wire and other metals with electric current flowing through them may get hot. Use caution when handling components of a closed circuit.

4. Handle electrical wires with caution. They have sharp ends, which can cut or puncture skin.

5. Wash their hands with soap and water when they are done collecting the data.

Batteries and wire may be stored for future use. When batteries need replacing, dispose of old batteries according to manufacturer's recommendations.

Topics for the Explicit and Reflective Discussion

Reflecting on the Use of Core Ideas and Crosscutting Concepts During the Investigation

Teachers should begin the explicit and reflective discussion by asking students to discuss what they know about the core ideas they used during the investigation. The following are some important concepts related to the core idea of forces and interactions that students need to use to determine the strength of the magnetic field around the two wires:

- A field associates a value of some physical quantity with every point in space. Fields are a model that physicists use to describe interactions that occur over a distance. Fields permeate space, and objects experience forces due to their interaction with a field.
- Vector fields are represented by field lines indicating direction and magnitude. The magnetic field is a vector field. Magnetic fields obey vector addition.
- A magnetic field is caused by a permanent magnet or a moving electric charge. Current is the flow of charge per unit time. The larger the magnitude of the current, the stronger the magnetic field produced by the current.

To help students reflect on what they know about magnetic fields, electromagnetism, and forces and interactions, we recommend showing them two or three images using presentation software that help illustrate these important ideas. You can then ask the students the following questions to encourage them to share how they are thinking about these important concepts:

1. What do we see going on in this image?
2. Does anyone have anything else to add?
3. What might be going on that we can't see?
4. What are some things that we are not sure about here?

You can then encourage students to think about how CCs played a role in their investigation. There are at least two CCs that students need to use to determine the factors influencing the magnetic field around the two-wire system: (a) Cause and Effect: Mechanism and Explanation and (b) Systems and System Models (see Appendix 2 [p. 417] for a brief description of these CCs). To help students reflect on what they know about these CCs, we recommend asking them the following questions:

1. What conditions do scientists need to create in order to determine a causal relationship during an investigation?

2. If you were to graph one of the causal relationships you determined in this investigation, would the graph be a function?

3. How did you define the system under study during this investigation? What parts made up your system?

4. Why is it advantageous to assume a closed system during an investigation?

You can then encourage the students to think about how they used all these different concepts to help answer the guiding question and why it is important to use these ideas to help justify their evidence for their final arguments. Be sure to remind your students to explain why they included the evidence in their arguments and make the assumptions underlying their analysis and interpretation of the data explicit in order to provide an adequate justification of their evidence.

Reflecting on Ways to Design Better Investigations

It is important for students to reflect on the strengths and weaknesses of the investigation they designed during the explicit and reflective discussion. Students should therefore be encouraged to discuss ways to eliminate potential flaws, measurement errors, or sources of uncertainty in their investigations. To help students be more reflective about the design of their investigation and what they can do to make their investigations more rigorous in the future, you can ask them the following questions:

1. What were some of the strengths of the way you planned and carried out your investigation? In other words, what made it scientific?

2. What were some of the weaknesses of the way you planned and carried out your investigation? In other words, what made it less scientific?

3. What rules can we make, as a class, to ensure that our next investigation is more scientific?

Reflecting on the Nature of Scientific Knowledge and Scientific Inquiry

This investigation can be used to illustrate two important concepts related to the nature of scientific knowledge and the nature of scientific inquiry: (a) the difference between data and evidence in science and (b) how scientists use different methods to answer different types of questions (see Appendix 2 [p. 417] for a brief description of these concepts). Be sure to review these concepts during and at the end of the explicit and reflective discussion. To help students think about these concepts in relation to what they did during the lab, you can ask them the following questions:

1. You had to talk about data and evidence during your investigation. Can you give me some examples of data and evidence from your investigation?

2. Can you work with your group to come up with a rule that you can use to decide if a piece of information is data or evidence? Be ready to share in a few minutes.

3. There is no universal step-by-step scientific method that all scientists follow. Why do you think there is no universal scientific method?

4. Think about what you did during this investigation. How would you describe the method you used to determine the magnetic field around the two-wire system? Why would you call it that?

You can also use presentation software or other techniques to encourage your students to think about these concepts. You can show examples of information from the investigation that are either data or evidence and ask students to classify each example and explain their thinking. You can also show one or more images of a "universal scientific method" that misrepresent the nature of scientific inquiry (see, e.g., *https://commons.wikimedia.org/wiki/File:The_Scientific_Method_as_an_Ongoing_Process.svg*) and ask students why each image is *not* a good representation of what scientists do to develop scientific knowledge. You can also ask students to suggest revisions to the image that would make it more consistent with the way scientists develop scientific knowledge. Be sure to remind your students that it is important for them to understand what counts as scientific knowledge and how that knowledge develops over time in order to be proficient in science.

Hints for Implementing the Lab

- Allowing students to design their own procedures for collecting data gives students an opportunity to try, to fail, and to learn from their mistakes. However, you can scaffold students as they develop their own procedure by having them fill out an investigation proposal. The proposals provide a way for you to offer students hints and suggestions without telling them how to do it. You can also check the proposals quickly during a class period. For this lab we suggest using Investigation Proposal C.

- Learn how to use the magnetic field sensor and the multimeter or ammeter before the lab begins. It is important for you to know how to use the equipment so you can help students when technical issues arise.

- Allow the students to become familiar with the magnetic field sensor and multimeter or ammeter as part of the tool talk before they begin to design their investigation. Give them 5–10 minutes to examine the equipment and materials before they begin designing their investigations. This gives students a chance to see what they can and cannot do with the equipment.

- To increase the current flow in each wire, students can place multiple D batteries in series.

Magnetic Fields Around Current-Carrying Wires
How Does Changing the Magnitude and Direction of a Current in Two Parallel Wires Affect the Magnetic Field Around the Two-Wire System?

- We suggest students use the same length wire for each of the two wires in the system. This will result in approximately equal internal resistances, which will aid students in creating similar currents through each wire.
- With the wires connected only to a battery, the batteries will drain quickly. We suggest giving each class a new set of batteries to use. Alternatively, you can have students wire a resistor in series with each of the circuits to moderate the flow of current.
- For the best results, minimize the distance between the two wires. We suggest no more than 5 cm distance between the two wires. This will produce the strongest magnetic field between the two wires.
- We also suggest having students tape the wires down to the table to maintain their parallel arrangement while collecting data.
- Be sure to allow students to go back and re-collect data at the end of the argumentation session. Students often realize that they made numerous mistakes when they were collecting data as a result of their discussions during the argumentation session. The students, as a result, will want a chance to re-collect data, and the re-collection of data should be encouraged when time allows. This also offers an opportunity to discuss what scientists do when they realize a mistake is made inside the lab.

If students use digital interface measurement equipment and analysis

- We suggest allowing students to familiarize themselves with the sensors and the data analysis software before they finalize the procedure for the investigation, especially if they have not used such software previously. This gives students an opportunity to learn how to work with the software and to improve the quality of the data they collect.
- Remind students to follow the user's guide to correctly connect any sensors to avoid damage to lab equipment.

Connections to Standards

Table 12.2 (p. 278) highlights how the investigation can be used to address specific performance expectations from the *NGSS*; learning objectives from AP Physics 1 and 2; learning objectives from AP Physics C: Electricity and Magnetism; *Common Core State Standards for English Language Arts* (*CCSS ELA*); and *Common Core State Standards for Mathematics* (*CCSS Mathematics*).

LAB 12

TABLE 12.2
Lab 12 alignment with standards

NGSS performance expectation	• HS-PS2-5: Plan and conduct an investigation to provide evidence that an electric current can produce a magnetic field and that a changing magnetic field can produce an electric current.
AP Physics 1 and AP Physics 2 learning objectives	• 2.D.2.1: Create a verbal or visual representation of a magnetic field around a straight wire or a pair of parallel wires.
AP Physics C: Electricity and Magnetism learning objectives	• FIE-5.A: Calculate the magnitude and direction of a magnetic field produced at a point near a long, straight, current-carrying wire. • FIE-5.B.a: Describe the direction of a magnetic-field vector at various points near multiple long, straight, current-carrying wires. • FIE-5.B.b: Calculate the magnitude of a magnetic field at various points near multiple long, straight, current-carrying wires. • FIE-5.C.a: Calculate the force of attraction or repulsion between two long, straight, current-carrying wires. • FIE-5.C.b: Describe the consequence (attract or repel) when two long, straight, current-carrying wires have known current directions. • CNV-8.D: Describe the relationship of the magnetic field as a function of distance for various configurations of current-carrying cylindrical conductors with either a single current or multiple currents, at points inside and outside of the conductors. • CNV-8.E.a: Describe the direction of a magnetic field at a point in space due to various combinations of conductors, wires, cylindrical conductors, or loops. • CNV-8.E.b: Calculate the magnitude of a magnetic field at a point in space due to various combinations of conductors, wires, cylindrical conductors, or loops.
Literacy connections (CCSS ELA)	• *Reading*: Key ideas and details, craft and structure, integration of knowledge and ideas • *Writing*: Text types and purposes, production and distribution of writing, research to build and present knowledge, range of writing • *Speaking and listening*: Comprehension and collaboration, presentation of knowledge and ideas

Continued

Table 12.2 (continued)

Mathematics connections (*CCSS Mathematics*)	• *Mathematical practices*: Make sense of problems and persevere in solving them, reason abstractly and quantitatively, construct viable arguments and critique the reasoning of others, model with mathematics, use appropriate tools strategically, attend to precision • *Number and quantity*: Reason quantitatively and use units to solve problems, represent and model with vector quantities, perform operations on vectors • *Algebra*: Interpret the structure of expressions, understand solving equations as a process of reasoning and explain the reasoning, solve equations and inequalities in one variable, represent and solve equations and inequalities graphically • *Functions*: Understand the concept of a function and use function notation; interpret functions that arise in applications in terms of the context; analyze functions using different representations; construct and compare linear, quadratic, and exponential models and solve problems; interpret expressions for functions in terms of the situation they model • *Statistics and probability*: Summarize, represent, and interpret data on two categorical and quantitative variables; interpret linear models; make inferences and justify conclusions from sample surveys, experiments, and observational studies

LAB 12

Lab Handout

Lab 12. Magnetic Fields Around Current-Carrying Wires: How Does Changing the Magnitude and Direction of a Current in Two Parallel Wires Affect the Magnetic Field Around the Two-Wire System?

Introduction

Many of the important historical experiments that students learn about during a high school physics class were conducted by one or two physicists in a lab. For example, the Nobel Prize–winning physicist Marie Curie (see Figure L12.1) did most of her investigations on her own even though she shared a lab with her husband, Pierre. In comparison, many investigations in modern physics involve large teams of people working together to design an investigation, build the equipment needed to carry it out, and then collect the actual data.

FIGURE L12.1

Marie Curie (right) in her lab with her husband Pierre (center) and another man. Marie Curie won the 1903 Nobel Prize in Physics (shared with her husband) and the 1911 Nobel Prize in Chemistry, making her the first person to win two Nobel Prizes.

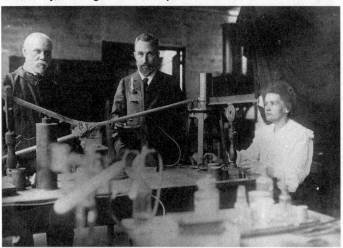

Figure L12.2 shows some of the equipment that was built for an experiment exploring the components of an atom conducted at Fermi National Accelerator Laboratory (Fermilab; *www.fnal.gov*), located outside of Chicago. Researchers at Fermilab and at similar labs around the world use electromagnets to accelerate electrons and protons to over 90% the speed of light. To achieve these high velocities, the particles are accelerated in a circular track until they reach a high enough velocity that they are smashed into each other. Physicists then study the result of the collision to more fully understand the workings of the universe at scales smaller than an atom.

Electromagnets work because of an important relationship between moving charges and magnetic fields. When an electric charge, like an electron, is moving, it creates a *magnetic field*. The magnetic field is directly proportional to the velocity of the charge—the faster

Magnetic Fields Around Current-Carrying Wires

How Does Changing the Magnitude and Direction of a Current in Two Parallel Wires Affect the Magnetic Field Around the Two-Wire System?

the charged particle moves, the stronger the magnetic field it will create. An electromagnet takes advantage of this by running a current through a wire, because an electric current is the flow of electrons in a wire. If the current is steady (i.e., unchanging), then a stable magnetic field will be created around the wire.

Similar to other fields, when an object with certain properties is present in the field created by the current-carrying wire, it feels a *force*. This force can be used to accelerate another charged particle, such as in experiments at Fermilab. It can also be used to produce the sound in loudspeakers or headphones, or in medical equipment such as a magnetic resonance imaging (MRI) machine.

FIGURE L12.2

The particle detector at Fermilab. Notice the people working in the bottom right-hand corner—this gives you an idea of the size of the particle detector.

Note: A full-color version of this figure is available on the book's Extras page at *www.nsta.org/adi-physics2*.

In many of these devices, a specific magnetic field must be precisely maintained. In an MRI, for example, too strong a magnetic field can be harmful to the people inside the MRI machine but too weak of a magnetic field will not provide important test results. In the experiments at Fermilab, if the magnetic fields are not at the exact value needed, the experiment will fail. Maintaining a specific magnetic field is trivial when the device only has one wire carrying current at a time. But, as can be seen in Figure L12.2, there can be hundreds of wires carrying current at the same time in one experiment at Fermilab. Physicists and engineers must be able to predict how a system of current-carrying wires interact to produce a magnetic field around the system.

LAB 12

Your Task

Use what you know about magnetic fields, causal relationships, and systems and system models to design and carry out an investigation to determine how the magnitude and direction of the current in a two-wire system affect the magnetic field around the system.

The guiding question of this investigation is, *How does changing the magnitude and direction of a current in two parallel wires affect the magnetic field around the two-wire system?*

Materials

You may use any of the following materials during your investigation:

Consumables
- D batteries
- 9V batteries
- Tape
- String

Equipment
- Safety glasses or goggles (required)
- Battery holders
- Wire with alligator clips on each end
- Magnetic field sensor
- Multimeters or ammeters
- Ruler

If you have access to the following equipment, you may also consider using a digital current sensor with an accompanying interface and a computer or tablet.

Safety Precautions

Follow all normal lab safety rules. In addition, take the following safety precautions:

1. Wear sanitized safety glasses or goggles during lab setup, hands-on activity, and takedown.

2. Never put consumables in your mouth.

3. Wire and other metals with electric current flowing through them may get hot. Use caution when handling components of a closed circuit.

4. Handle electrical wires with caution. They have sharp ends, which can cut or puncture skin.

5. Wash your hands with soap and water when you are done collecting the data.

Investigation Proposal Required?   ☐ Yes   ☐ No

Getting Started

To answer the guiding question, you will need to design and carry out one or more experiments to determine the effect of changing the magnitude and direction of current on the magnetic field around the two-wire system. For the best results, we suggest keeping the

Magnetic Fields Around Current-Carrying Wires
How Does Changing the Magnitude and Direction of a Current in Two Parallel Wires Affect the Magnetic Field Around the Two-Wire System?

distance between the two wires at around 5 cm. You should be able to identify the effect of the magnetic field both outside the system and in between the two wires. Before you can design your experiments, however, you must determine what type of data you need to collect, how you will collect it, and how you will analyze it.

To determine *what type of data you need to collect*, think about the following questions:

- What are the boundaries and components of the system?
- How do the components of the system interact with each other?
- When is this system stable and under which conditions does it change?
- How could you keep track of changes in this system quantitatively?
- What is going on at the unobservable level that could cause the things that you observe?

To determine *how you will collect the data*, think about the following questions:

- What is the independent variable and what is the dependent variable?
- What other factors will you need to control during your experiment or experiments?
- What scale or scales should you use when you take your measurements?
- How will you make sure that your data are of high quality (i.e., how will you reduce error)?
- How will you keep track of and organize the data you collect?
- What type of research design needs to be used to establish a cause-and-effect relationship?
- What are the components of this phenomenon or system and how do they interact?

To determine *how you will analyze the data*, think about the following questions:

- What type of calculations, if any, will you need to make?
- What types of patterns might you look for as you analyze your data?
- What type of table or graph could you create to help make sense of your data?
- How could you use mathematics to describe a relationship between variables?

Connections to the Nature of Scientific Knowledge and Scientific Inquiry

As you work through your investigation, you may want to consider

- the difference between data and evidence in science, and
- how scientists use different methods to answer different types of questions.

LAB 12

Initial Argument

Once your group has finished collecting and analyzing your data, your group will need to develop an initial argument. Your initial argument needs to include a claim, evidence to support your claim, and a justification of the evidence. The *claim* is your group's answer to the guiding question. The *evidence* is an analysis and interpretation of your data. Finally, the *justification* of the evidence is why your group thinks the evidence matters. The justification of the evidence is important because scientists can use different kinds of evidence to support their claims. Your group will create your initial argument on a whiteboard. Your whiteboard should include all the information shown in Figure L12.3.

FIGURE L12.3
Argument presentation on a whiteboard

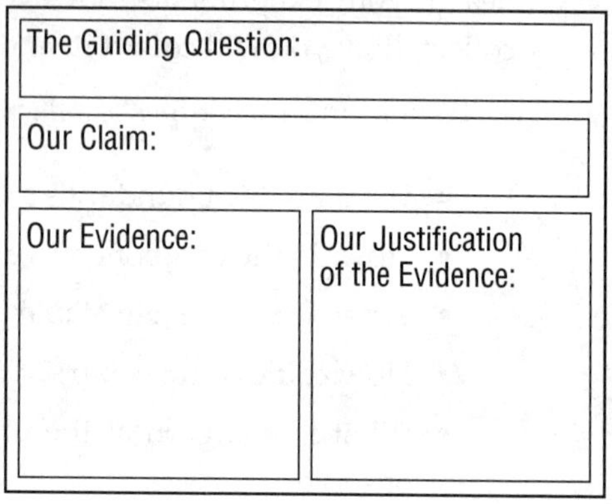

The Guiding Question:	
Our Claim:	
Our Evidence:	Our Justification of the Evidence:

Argumentation Session

The argumentation session allows all of the groups to share their arguments. One or two members of each group will stay at the lab station to share that group's argument, while the other members of the group go to the other lab stations to listen to and critique the other arguments. This is similar to what scientists do when they propose, support, evaluate, and refine new ideas during a poster session at a conference. If you are presenting your group's argument, your goal is to share your ideas and answer questions. You should also keep a record of the critiques and suggestions made by your classmates so you can use this feedback to make your initial argument stronger. You can keep track of specific critiques and suggestions for improvement that your classmates mention in the space below.

Critiques about our initial argument and suggestions for improvement:

Magnetic Fields Around Current-Carrying Wires

How Does Changing the Magnitude and Direction of a Current in Two Parallel Wires Affect the Magnetic Field Around the Two-Wire System?

If you are critiquing your classmates' arguments, your goal is to look for mistakes in their arguments and offer suggestions for improvement so these mistakes can be fixed. You should look for ways to make your initial argument stronger by looking for things that the other groups did well. You can keep track of interesting ideas that you see and hear during the argumentation in the space below. You can also use this space to keep track of any questions that you will need to discuss with your team.

Interesting ideas from other groups or questions to take back to my group:

Once the argumentation session is complete, you will have a chance to meet with your group and revise your initial argument. Your group might need to gather more data or design a way to test one or more alternative claims as part of this process. Remember, your goal at this stage of the investigation is to develop the best argument possible.

Report

Once you have completed your research, you will need to prepare an *investigation report* that consists of three sections. Each section should provide an answer to the following questions:

1. What question were you trying to answer and why?
2. What did you do to answer your question and why?
3. What is your argument?

Your report should answer these questions in two pages or less. This report must be typed, and any diagrams, figures, or tables should be embedded into the document. Be sure to write in a persuasive style; you are trying to convince others that your claim is acceptable or valid!

Checkout Questions

Lab 12. Magnetic Fields Around Current-Carrying Wires: How Does Changing the Magnitude and Direction of a Current in Two Parallel Wires Affect the Magnetic Field Around the Two-Wire System?

1. The picture below shows two wires with equal current running through them but in opposite directions.

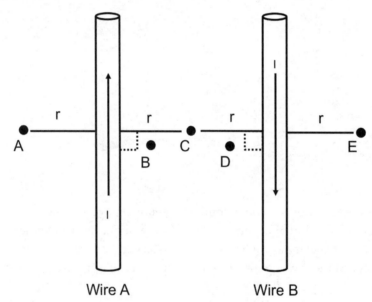

   a. Place the points in order from the point with the strongest magnetic field to the point with the weakest magnetic field. For each point, also indicate the direction of the magnetic field at that point.

   b. Explain your reasoning.

286

National Science Teaching Association

Magnetic Fields Around Current-Carrying Wires
How Does Changing the Magnitude and Direction of a Current in Two Parallel Wires Affect the Magnetic Field Around the Two-Wire System?

2. The picture below shows two wires with equal current running through them in the same direction.

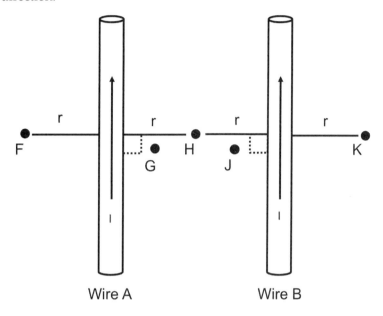

a. Place the points in order from the point with the strongest magnetic field to the point with the weakest magnetic field. For each point, also indicate the direction of the magnetic field at that point.

b. Explain your reasoning.

LAB 12

3. Data and evidence are different things. Data is all the information collected during an investigation, and evidence is analyzed data used to support a claim.

    a. I agree with this statement.
    b. I disagree with this statement.

    Explain your answer, using examples from this investigation and at least one other investigation you have conducted.

4. Scientists have always used the same method for investigating questions regarding magnetic fields.

    a. I agree with this statement.
    b. I disagree with this statement.

    Explain your answer, using an example from your investigation about the magnetic field around two wires.

Magnetic Fields Around Current-Carrying Wires
How Does Changing the Magnitude and Direction of a Current in Two Parallel Wires Affect the Magnetic Field Around the Two-Wire System?

5. Why is it useful to understand identify causal relationships in science? In your answer, be sure to use examples from this investigation and at least one other investigation you have conducted.

6. Why is it useful to assume that you are studying a closed system during an investigation? In your answer, be sure to use examples from this investigation and at least one other investigation you have conducted.

LAB 13

Teacher Notes

Lab 13. Electromagnets: What Variables Affect the Strength of the Electromagnet?

Purpose

The purpose of this lab is to *introduce* students to the disciplinary core idea (DCI) of Types of Interactions (PS2.B) from the *NGSS* and to Ampère's law and the relationship between electric current and magnetic fields by having them investigate the variables that affect the strength of an electric field inside a solenoid. In addition, this lab can be used to help students understand two big ideas from AP Physics: (a) fields existing in space can be used to explain interactions and (b) interactions between systems can result in changes in those systems. This lab also gives students an opportunity to learn about the crosscutting concepts (CCs) of (a) Cause and Effect: Mechanism and Explanation and (b) Scale, Proportion, and Quantity from the *NGSS*. As part of the explicit and reflective discussion, students will also learn about (a) the difference between laws and theories in science and (b) how scientists investigate questions about the natural or material world.

Underlying Physics Concepts

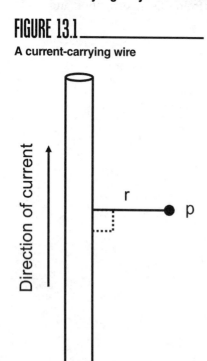

FIGURE 13.1

A current-carrying wire

Ampère's law describes the strength of the magnetic field surrounding a current-carrying conductor. When applied to a long, straight current-carrying wire, Ampère's law suggests that the magnetic field will be directly proportional to the strength of the current and inversely proportional to the distance from the wire measured perpendicular to the surface of the wire. Figure 13.1 shows a long, straight current-carrying wire and point p some distance *r* from the wire.

The mathematical form of Ampère's law, when applied to a long, straight current-carrying wire, is shown in Equation 13.1, where **B** is the strength of the magnetic field, *I* is the current through the wire, **r** is the distance from the wire, and μ_o is the permeability of free space and is a constant. In SI units, magnetic field is measured in teslas (T), current is measured in amperes (A), distance is measured in meters (m), and the permeability of free space has a value of $4\pi \times 10^{-7}$ tesla-meters per ampere (T·m/A).

(Equation 13.1) $\mathbf{B} = \dfrac{\mu_o I}{2\pi \mathbf{r}}$

Electromagnets
What Variables Affect the Strength of the Electromagnet?

Ampère's law can also be applied to a wire of any shape, including a solenoid such as in this lab. Figure 13.2 shows the magnetic field lines due to the current (*I*) flowing through the solenoid. There are a few things to notice. First, the magnetic field lines are all parallel inside the solenoid, but they fan out at the ends of the solenoid. Second, there are no magnetic field lines drawn outside the solenoid. This is because the magnetic field outside the solenoid is negligible compared with the magnetic field inside the solenoid, particularly for the currents that students will use in the lab.

Equation 13.2 relates the strength of the magnetic field to properties of the solenoid, where *N* is the number of loops of the solenoid and *l* is the length of the solenoid from end to end. In SI units, length is measured in meters (m).

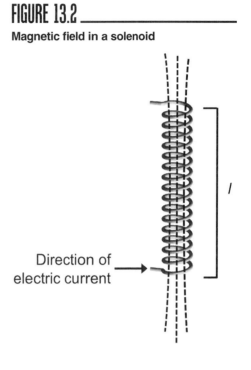

FIGURE 13.2
Magnetic field in a solenoid

(Equation 13.2) $$B = \frac{\mu_0 IN}{l}$$

The magnetic field inside the solenoid is a function of the current flowing through the solenoid, the number of turns of coil, and the length of the solenoid itself. Underlying this equation is the assumption that the coils are evenly spaced. This is important, because the total magnetic field is a function of the magnetic field created by each coil. If the coils are spaced evenly, then the magnetic field inside the solenoid is uniform. If the coils are not spaced evenly, the magnetic field at any given point must be calculated in a more complex manner.

To determine the direction of the magnetic field, we can use a right-hand rule, as shown in Figure 13.3. This right-hand rule works by wrapping your fingers around the solenoid in the direction the current is flowing. The thumb then points toward the north pole of the magnetic field.

An even spacing of the coils allows us to think about the shape of the solenoid in terms of the coil density (i.e., *N/l*). That is, it is the

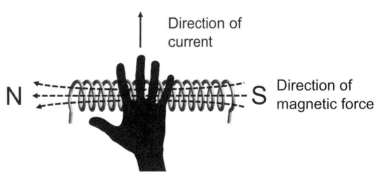

FIGURE 13.3
The right-hand rule for a solenoid

number of turns per unit length that determines the strength of the magnetic field inside the solenoid. Thus, the magnetic field strength inside a solenoid with 10 turns in 5 cm of length will be equal to the magnetic field strength inside a solenoid with 20 turns in 10 cm of length (assuming the current in the wire is the same in each case).

The magnetic field inside the solenoid is also a function of the current flowing through the solenoid wire. It turns out, however, that the current itself is a function of other variables as well—the voltage source and the resistance of the wire. Ohm's law, shown in Equation 13.3, describes the relationship between the current in a wire, the voltage source, and the total resistance of the circuit, where I is the current in the wire, V is the voltage source, and R is the resistance of the circuit. In SI units, voltage is measured in volts (V) and resistance is measured in ohms (Ω).

(Equation 13.3) $I = V/R$

Thus, if the voltage source of the solenoid is increased, the strength of the magnetic field will also increase. And if the resistance of the circuit is increased, the strength of the magnetic field will decrease.

For most students, these are the three relationships they will test: the number of coils in the solenoid, the spacing of the coils, and the voltage attached to the solenoid. It is important to stress that (a) voltage does not create a magnetic field, but a moving charge does; and (b) current is the movement of charge in a closed circuit.

Finally, the resistance of the circuit for the solenoid will affect the strength of the magnetic field in the electromagnet. For most solenoids the resistance in the circuit comes from the internal resistance of the wire (however, you could place other resistors either in series or in parallel with the solenoid; this would be a good way to demonstrate that the magnetic field is a function of current, and not voltage per se). The formula for determining the internal resistance (R) of a wire is shown in Equation 13.4, where ρ is the resistivity of the material the wire is made of, L is the length of the wire, and A is the cross-sectional area of the wire. In SI units, resistivity is measured in ohm-meters ($\Omega \cdot m$), length is measured in meters (m), and area is measured in meters squared (m²).

(Equation 13.4) $R = \rho \dfrac{L}{A}$

Thus, for materials with low resistivity, such as copper, the strength of the magnetic field is likely to be high. As the length of wire used to create the solenoid increases, the strength of the magnetic field will decrease. As the cross-sectional area (gauge) of the wire increases, the strength of the magnetic field will increase. The effect of increasing the length of the wire in this investigation is likely to be negligible if students decide they need a longer wire to increase the number of turns of the coil in the solenoid.

Electromagnets
What Variables Affect the Strength of the Electromagnet?

Timeline

The instructional time needed to complete this lab investigation is 170–230 minutes. Appendix 3 (p. 421) provides options for implementing this lab investigation over several class periods. Option C (230 minutes) should be used if students are unfamiliar with scientific writing, because this option provides extra instructional time for scaffolding the writing process. You can scaffold the writing process by modeling, providing examples, and providing hints as students write each section of the report. Option C can also be used if you are introducing students to the digital interface sensors and/or the data analysis software. Option D (170 minutes) should be used if students are familiar with scientific writing and have developed the skills needed to write an investigation report on their own. In option D, students complete stage 6 (writing the investigation report) and stage 8 (revising the investigation report) as homework.

Materials and Preparation

The materials needed to implement this investigation are listed in Table 13.1. The equipment can be purchased from a science supply company such as Flinn Scientific, PASCO, Vernier, or Ward's Science.

TABLE 13.1

Materials list for Lab 13

Item	Quantity
Consumables	
D batteries	3 per group
Electrical tape or duct tape	As needed
Equipment and other materials	
Safety glasses or goggles	1 per student
Iron nail	1 per group
Electrical wire	As needed
Paper clips	As needed
Battery holders	3 per group
Electronic or triple beam balance	1 per group
Ammeter	1 per group
Voltmeter	1 per group
Resistor (approximately 30–100 Ω)	1 per group

Continued

Table 13.1 (*continued*)

Item	Quantity
Investigation Proposal C (optional)	1 per group
Whiteboard, 2' × 3'*	1 per group
Lab Handout	1 per student
Peer-review guide and teacher scoring rubric	1 per student
Checkout Questions	1 per student
Equipment for digital interface measurements (optional)	
Digital interface with USB or wireless connections	1 per group
Magnetic field sensor†	1 per group
Current measurement sensor	1 per group
Voltage measurement sensor	1 per group
Computer or tablet with appropriate data analysis software installed	1 per group

* As an alternative, students can use computer and presentation software such as Microsoft PowerPoint or Apple Keynote to create their arguments.

† We recommend using the magnetic field sensor because it is more precise than the paper clips (see "Hints" section), but we list it as an option because the lab can be conducted without using the magnetic field sensor. The current measurement sensor and voltage measurement sensor are optional because any voltmeter and ammeter will suffice.

Be sure to use a set routine for distributing and collecting the materials during the lab investigation. One option is to set up the materials for each group at each group's lab station before class begins. This option works well when there is a dedicated section of the classroom for lab work and the materials are large and difficult to move. A second option is to have all the materials on a table or cart at a central location. You can then assign a member of each group to be the "materials manager." This individual is responsible for collecting all the materials his or her group needs from the table or cart during class and for returning all the materials at the end of the class. This option works well when the materials are small and easy to move (such as magnets, wire, and bulbs). It also makes it easy to inventory the materials at the end of the class before students leave for the day.

Safety Precautions and Laboratory Waste Disposal

Remind students to follow all normal lab safety rules. In addition, tell students to take the following safety precautions:

1. Wear sanitized safety glasses or goggles during lab setup, hands-on activity, and takedown.

2. Never put consumables in their mouth.

3. Wire and other metals with electric current flowing through them may get hot. Use caution when handling components of a closed circuit.

4. Handle electrical wires with caution. They have sharp ends, which can cut or puncture skin.

5. Wash their hands with soap and water when they are done collecting the data.

Batteries and wire may be stored for future use. When batteries need replacing, dispose of old batteries according to manufacturer's recommendations.

Topics for the Explicit and Reflective Discussion
Reflecting on the Use of Core Ideas and Crosscutting Concepts During the Investigation
Teachers should begin the explicit and reflective discussion by asking students to discuss what they know about the core ideas they used during the investigation. The following are some important concepts related to the core idea of Ampère's law that students need to use to explain the effects of the variables they tested on the strength of the electromagnet:

- According to Ampère's law, the magnitude of the magnetic field produced by a current-carrying wire is proportional to the magnitude of the current in a long, straight wire. For a solenoid, Ampère's law also shows that the magnitude of the magnetic field is directly proportional to the number of coils and inversely proportional to the length of the solenoid.

- The magnetic properties of materials can be affected by magnetic fields surrounding them. In an electromagnet, the ferromagnetic core becomes magnetized as a result of the magnetic field created by the solenoid.

- Ohm's law is expressed mathematically as $I = V/R$. Increasing voltage will lead to an increased current.

- A field associates a value of some physical quantity with every point in space. Fields are a model that physicists use to describe interactions that occur over a distance. Fields permeate space, and objects experience forces due to their interaction with a field.

To help students reflect on what they know about electromagnets, we recommend showing them two or three images using presentation software that help illustrate these important ideas. You can then ask the students the following questions to encourage them to share how they are thinking about these important concepts:

1. What do we see going on in this image?

2. Does anyone have anything else to add?

3. What might be going on that we can't see?

4. What are some things that we are not sure about here?

You can then encourage students to think about how CCs played a role in their investigation. There are at least two CCs that students need to use to determine variables that affect the strength of the magnetic field inside the solenoid: (a) Cause and Effect: Mechanism and Explanation and (b) Scale, Proportion, and Quantity (see Appendix 2 [p. 417] for a brief description of these CCs). To help students reflect on what they know about these CCs, we recommend asking them the following questions:

1. What kinds of proportional relationships did you identify in this investigation? How are they different? How are they the same?

2. Can two variables that are inversely proportional exist in a cause-and-effect relationship?

3. Why is it important to identify cause-and-effect relationships in science?

4. What measured quantities from this investigation were vectors and what measured quantities were scalars?

You can then encourage the students to think about how they used all these different concepts to help answer the guiding question and why it is important to use these ideas to help justify their evidence for their final arguments. Be sure to remind your students to explain why they included the evidence in their arguments and make the assumptions underlying their analysis and interpretation of the data explicit in order to provide an adequate justification of their evidence.

Reflecting on Ways to Design Better Investigations

It is important for students to reflect on the strengths and weaknesses of the investigation they designed during the explicit and reflective discussion. Students should therefore be encouraged to discuss ways to eliminate potential flaws, measurement errors, or sources of uncertainty in their investigations. To help students be more reflective about the design of their investigation and what they can do to make their investigations more rigorous in the future, you can ask them the following questions:

1. What were some of the strengths of the way you planned and carried out your investigation? In other words, what made it scientific?

2. What were some of the weaknesses of the way you planned and carried out your investigation? In other words, what made it less scientific?

3. What rules can we make, as a class, to ensure that our next investigation is more scientific?

Electromagnets
What Variables Affect the Strength of the Electromagnet?

Reflecting on the Nature of Scientific Knowledge and Scientific Inquiry

This investigation can be used to illustrate two important concepts related to the nature of scientific knowledge and the nature of scientific inquiry: (a) the difference between laws and theories in science and (b) how scientists investigate questions about the natural or material world (see Appendix 2 [p. 417] for a brief description of these two concepts). Be sure to review these concepts during and at the end of the explicit and reflective discussion. To help students think about these concepts in relation to what they did during the lab, you can ask them the following questions:

1. Laws and theories are different in science. What laws did you use as part of your justification, and what theories did you use as part of your justification?

2. Can you work with your group to come up with a rule that you can use to decide if something is a law or a theory? Be ready to share in a few minutes.

3. Not all questions can be answered by science. Can you give me some examples of questions related to this investigation that can and cannot be answered by science?

4. Can you work with your group to come up with a rule that you can use to decide if a question can or cannot be answered by science?

You can also use presentation software or other techniques to encourage your students to think about these concepts. You can show examples of laws (such as Ampère's law or Ohm's law) and theories (such as Maxwell's equations) and ask students to indicate if they think each example is a law or a theory and then explain their thinking. You can also show one or more examples of questions that can be answered by science (e.g., How does the current affect the strength of the electromagnet?) and questions that cannot be answered by science (e.g., Are there too many electromagnetic devices?) and then ask students why each example is or is not a question that can be answered by science. Be sure to remind your students that it is important for them to understand what counts as scientific knowledge and how that knowledge develops over time in order to be proficient in science.

Hints for Implementing the Lab

- Allowing students to design their own procedures for collecting data gives students an opportunity to try, to fail, and to learn from their mistakes. However, you can scaffold students as they develop their own procedure by having them fill out an investigation proposal. The proposals provide a way for you to offer students hints and suggestions without telling them how to do it. You can also check the proposals quickly during a class period. For this lab we suggest using Investigation Proposal C.

LAB 13

- Learn how to create the electromagnet and how to use the magnetic field sensor, the ammeter, and the voltmeter before the lab begins. It is important for you to know how to use the equipment so you can help students when technical issues arise.
- Allow the students to become familiar with the electromagnet, the magnetic field sensor, the ammeter, and the voltmeter as part of the tool talk before they begin to design their investigation. Give them 5–10 minutes to examine the equipment and materials before they begin designing their investigations. This gives students a chance to see what they can and cannot do with the equipment.
- Because the resistance of the wire is minimal, the battery may not last very long. We suggest using new batteries for each class.
- Although use of the magnetic field sensor is highly recommended because it is more precise than paper clips, the lab can be implemented without using the magnetic field sensor. In this case, students can measure the strength of the magnetic field by using the iron core to attract paper clips. Students can then choose a number of ways in which the attraction of paper clips is related to the magnetic field strength. The two ways that we anticipate most students using are the number of paper clips and the mass of paper clips.
- If you want to make the lab more challenging (this may be appropriate for students in AP physics), you can also give them wire of different gauges and different materials. This will affect the internal resistance of the wire and, as a result, change the current flowing through the solenoid. For the mathematics behind how the gauge of the wire and the material of the wire affect the current through a wire, see the Lab 7 Teacher Notes.
- To demonstrate that the magnetic field strength is a function of the current through the wire, you can also have students insert small resistors in series with the battery such that the total resistance of the circuit increases. This will cause a decrease in the electromagnet field strength despite a constant voltage.
- Be sure to allow students to go back and re-collect data at the end of the argumentation session. Students often realize that they made numerous mistakes when they were collecting data as a result of their discussions during the argumentation session. The students, as a result, will want a chance to re-collect data, and the re-collection of data should be encouraged when time allows. This also offers an opportunity to discuss what scientists do when they realize a mistake is made inside the lab.
- As mentioned earlier, the spacing of the coils is important. Students may not attend to this issue. As a result, they may get results that are not in agreement with other groups' findings or with the exact mathematical description of the relationship (see Equation 13.2). If this happens, you will have an opportunity to reinforce the importance of identifying and controlling variables during the explicit and reflective discussion. You can then have students re-collect data if you want.

Electromagnets
What Variables Affect the Strength of the Electromagnet?

If students use digital interface measurement equipment and analysis

- We suggest allowing students to familiarize themselves with the sensors and the data analysis software before they finalize the procedure for the investigation, especially if they have not used such software previously. This gives students an opportunity to learn how to work with the software and to improve the quality of the data they collect.
- Remind students to begin recording data just before they close the circuits.
- Remind students to follow the user's guide to correctly connect any sensors to avoid damage to lab equipment.

Connections to Standards

Table 13.2 highlights how the investigation can be used to address specific performance expectations from the *NGSS*; learning objectives from AP Physics 1 and 2; learning objectives from AP Physics C: Electricity and Magnetism; *Common Core State Standards for English Language Arts* (*CCSS ELA*); and *Common Core State Standards for Mathematics* (*CCSS Mathematics*).

TABLE 13.2

Lab 13 alignment with standards

NGSS performance expectation	• HS-PS2-5: Plan and conduct an investigation to provide evidence that an electric current can produce a magnetic field and that a changing magnetic field can produce an electric current.
AP Physics 1 and AP Physics 2 learning objectives	• 4.E.1.1: Use representations and models to qualitatively describe the magnetic properties of some materials that can be affected by magnetic properties of other objects in the system.
AP Physics C: Electricity & Magnetism Learning Objectives	• CNV-8.B.a: Derive the expression for the magnitude of magnetic field on the axis of a circular loop of current or a segment of a circular loop. • CNV-8.C.c: Derive the expression for the magnetic field of an ideal solenoid (length dimension is much larger than the radius of the solenoid) using Ampère's Law.
Literacy connections (*CCSS ELA*)	• *Reading*: Key ideas and details, craft and structure, integration of knowledge and ideas • *Writing*: Text types and purposes, production and distribution of writing, research to build and present knowledge, range of writing • *Speaking and listening*: Comprehension and collaboration, presentation of knowledge and ideas

Continued

Table 13.2 (continued)

Mathematics connections (*CCSS Mathematics*)	<ul><li>*Mathematical practices*: Make sense of problems and persevere in solving them, reason abstractly and quantitatively, construct viable arguments and critique the reasoning of others, model with mathematics, use appropriate tools strategically, attend to precision</li><li>*Number and quantity*: Reason quantitatively and use units to solve problems, represent and model with vector quantities, perform operations on vectors</li><li>*Algebra*: Interpret the structure of expressions, create equations that describe numbers or relationships, understand solving equations as a process of reasoning and explain the reasoning, solve equations and inequalities in one variable, represent and solve equations and inequalities graphically</li><li>*Functions*: Understand the concept of a function and use function notation; interpret functions that arise in applications in terms of the context; analyze functions using different representations; build a function that models a relationship between two quantities; construct and compare linear, quadratic, and exponential models and solve problems; interpret expressions for functions in terms of the situation they model,</li><li>*Statistics and probability*: Summarize, represent, and interpret data on two categorical and quantitative variables; interpret linear models; make inferences and justify conclusions from sample surveys, experiments, and observational studies</li></ul>

Lab Handout

Lab 13. Electromagnets: What Variables Affect the Strength of the Electromagnet?

Introduction

The first electromagnet was invented by the English scientist William Sturgeon in 1825 (see the article on Sturgeon in the online *Oxford Dictionary of National Biography* at *www.oxforddnb.com/search?q=sturgeon&searchBtn=Search&isQuickSearch=true*). By the end of the 19th century, electromagnets had become prevalent in a number of household devices. One of the common devices that use an electromagnet is the doorbell. Figure L13.1 shows a 1904 patent from W. T. Wheeler for a doorbell with an illuminated button, to help people ring the bell at night. Component 15 in the figure is the electromagnet, which is a coil of wire wrapped around two iron bars. When the button is pressed (component 44), an electric circuit is closed and current flows through the wire in the electromagnet. This creates a magnetic field which exerts a force on component 14, pulling it down. When component 14 is pulled down, it strikes the bell (component 16) and the bell rings. Today, electromagnets are used in many electronic devices, such as medical devices, speakers, headphones, and locking mechanisms.

FIGURE L13.1

Figure from W. T. Wheeler's patent for an illuminated electric doorbell

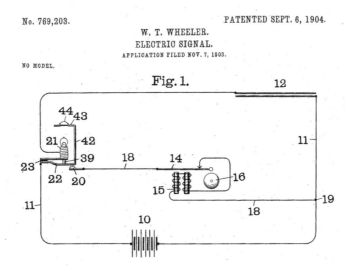

The most common type of electromagnet is a solenoid wrapped around a ferromagnetic core. A *solenoid* is the technical term for a coil of conducting wire with a length of wire greater than the diameter of the coil. The solenoid is then connected to a voltage source,

LAB 13

and a current flows through the solenoid. According to *Ampère's law*, the current through a wire will create a magnetic field around the wire. When the wire is coiled into a solenoid, this concentrates the magnetic field inside the solenoid (the magnetic field outside the solenoid is negligible). The core of the electromagnet is made from a ferromagnetic material—a material that has properties such that its individual atoms have a north pole and a south pole. Most often, the ferromagnetic core is made from iron. When a magnetic field is applied to a ferromagnetic material, the individual atoms line up such that their magnetic poles align. In a solenoid, this causes the ferromagnetic core to act like a permanent magnet and attract other ferromagnetic materials.

Unlike permanent magnets, electromagnets can be turned off and on. This allows users to control when a magnetic field will be created. A second advantage is that the user can control the strength of the magnet. Some electromagnets are relatively weak—like those inside your headphones. Others are so strong, such as electromagnets used in magnetic resonance imaging (MRI) machines, that people's exposure to the magnetic field must be limited. When designing electromagnets, it is important for scientists and engineers to understand the factors that influence the strength of electromagnets.

Your Task

Use what you know about electromagnetism, cause and effect, and scale, proportion, and quantity to design and carry out one or more experiments to determine what variables affect the strength of the electromagnet and how they affect the strength of the electromagnet. In other words, if you determine that a variable affects the strength of the magnetic field, is this relationship directly proportional (i.e., $y \propto x$), inversely proportional (i.e., $y \propto 1/x$), or another relationship?

The guiding question of this investigation is, **What variables affect the strength of the electromagnet?**

Materials

You may use any of the following materials during your investigation:

Consumables
- D batteries
- Electrical tape or duct tape

Equipment
- Safety glasses or goggles (required)
- Iron nail
- Electrical wire
- Paper clips

- Battery holders
- Electronic or triple beam balance
- Ammeter
- Voltmeter
- Resistor

If you have access to the following equipment, you may also consider using a digital magnetic field sensor, a digital current sensor, and a digital voltage sensor with an accompanying interface and a computer or tablet.

Electromagnets
What Variables Affect the Strength of the Electromagnet?

Safety Precautions

Follow all normal lab safety rules. In addition, take the following safety precautions:

1. Wear sanitized safety glasses or goggles during lab setup, hands-on activity, and takedown.
2. Never put consumables in your mouth.
3. Wire and other metals with electric current flowing through them may get hot. Use caution when handling components of a closed circuit.
4. Handle electrical wires with caution. They have sharp ends, which can cut or puncture skin.
5. Wash your hands with soap and water when you are done collecting the data.

Investigation Proposal Required?   ☐ Yes   ☐ No

Getting Started

To answer the guiding question, you will need to design and carry out one or more experiments to determine what variables affect the strength of the electromagnet. Figure L13.2 illustrates how you can set up the wire, battery, and iron core to create your electromagnet. There are several potential variables, so you will need to develop an experiment for each potential variable. Before you can design your investigation, however, you must determine what type of data you need to collect, how you will collect it, and how you will analyze it.

To determine *what type of data you need to collect*, think about the following questions:

FIGURE L13.2

An electromagnet made from an iron nail and insulated copper wire connected to a D battery

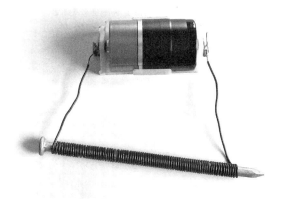

- What variables might affect the strength of the electromagnet?
- What are the boundaries and components of the system?
- How do the components of the system interact with each other?
- When is this system stable and under which conditions does it change?
- How could you keep track of changes in this system quantitatively?
- How will you measure the strength of the electromagnet?
- What is going on at the unobservable level that could cause the things that you observe?

To determine *how you will collect the data*, think about the following questions:

- What is the independent variable and what is the dependent variable for each experiment?
- What other factors will you need to control during each experiment?
- What scale or scales should you use when you take your measurements?
- How will you make sure that your data are of high quality (i.e., how will you reduce error)?
- How will you keep track of and organize the data you collect?
- What type of research design needs to be used to establish a cause-and-effect relationship?

To determine *how you will analyze the data*, think about the following questions:

- What type of calculations, if any, will you need to make?
- What types of patterns might you look for as you analyze your data?
- What type of table or graph could you create to help make sense of your data?
- Are there any proportional relationships that you can identify?

Connections to the Nature of Scientific Knowledge and Scientific Inquiry

As you work through your investigation, you may want to consider

- the difference between laws and theories in science, and
- how scientists investigate questions about the natural or material world.

Initial Argument

Once your group has finished collecting and analyzing your data, your group will need to develop an initial argument. Your initial argument needs to include a claim, evidence to support your claim, and a justification of the evidence. The *claim* is your group's answer to the guiding question. The *evidence* is an analysis and interpretation of your data. Finally, the justification of the evidence is why your group thinks the evidence matters. The *justification* of the evidence is important because scientists can use different kinds of evidence to support their claims. Your group will create your initial argument on a whiteboard. Your whiteboard should include all the information shown in Figure L13.3.

FIGURE L13.3

Argument presentation on a whiteboard

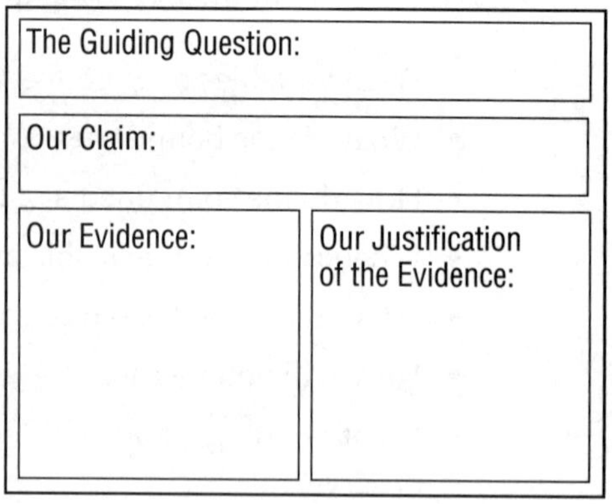

The Guiding Question:	
Our Claim:	
Our Evidence:	Our Justification of the Evidence:

Electromagnets
What Variables Affect the Strength of the Electromagnet?

Argumentation Session

The argumentation session allows all of the groups to share their arguments. One or two members of each group will stay at the lab station to share that group's argument, while the other members of the group go to the other lab stations to listen to and critique the other arguments. This is similar to what scientists do when they propose, support, evaluate, and refine new ideas during a poster session at a conference. If you are presenting your group's argument, your goal is to share your ideas and answer questions. You should also keep a record of the critiques and suggestions made by your classmates so you can use this feedback to make your initial argument stronger. You can keep track of specific critiques and suggestions for improvement that your classmates mention in the space below.

Critiques about our initial argument and suggestions for improvement:

If you are critiquing your classmates' arguments, your goal is to look for mistakes in their arguments and offer suggestions for improvement so these mistakes can be fixed. You should look for ways to make your initial argument stronger by looking for things that the other groups did well. You can keep track of interesting ideas that you see and hear during the argumentation in the space below. You can also use this space to keep track of any questions that you will need to discuss with your team.

Interesting ideas from other groups or questions to take back to my group:

LAB 13

Once the argumentation session is complete, you will have a chance to meet with your group and revise your initial argument. Your group might need to gather more data or design a way to test one or more alternative claims as part of this process. Remember, your goal at this stage of the investigation is to develop the best argument possible.

Report

Once you have completed your research, you will need to prepare an *investigation report* that consists of three sections. Each section should provide an answer to the following questions:

1. What question were you trying to answer and why?
2. What did you do to answer your question and why?
3. What is your argument?

Your report should answer these questions in two pages or less. This report must be typed, and any diagrams, figures, or tables should be embedded into the document. Be sure to write in a persuasive style; you are trying to convince others that your claim is acceptable or valid!

Reference

Wheeler, W. T. 1904. Electric signal. US Patent No. 769,203. Washington, DC: US Patent and Trademark Office. *https://pdfpiw.uspto.gov/.piw?Docid=769203&idkey=NONE&homeurl= http%3A%252F%252Fpatft.uspto.gov%252Fnetahtml%252FPTO%252Fpatimg.htm.*

Checkout Questions

Lab 13. Electromagnets: What Variables Affect the Strength of the Electromagnet?

1. A magnetic field B exists inside a solenoid. Assuming the number of coils is doubled and the length of the solenoid is doubled, what will happen to the strength of the magnetic field inside the solenoid?

   How do you know?

2. A magnetic field B exists inside a solenoid when the solenoid is connected in series with a battery of voltage V. What happens to the field strength inside the solenoid if the voltage source is doubled?

   How do you know?

3. A magnetic field B exists inside a solenoid when the solenoid is connected in series with a voltage source V and another resistor R. What happens to the field strength inside the solenoid if the resistance is doubled?

How do you know?

4. A theory turns into a law once it has been proven to be true.

    a. I agree with this statement.
    b. I disagree with this statement.

   Explain your answer, using examples from this investigation and at least one other investigation you have conducted.

5. Science can provide a solution to all of the problems that exist in the world.

    a. I agree with this statement.
    b. I disagree with this statement.

   Explain your answer, using examples from this investigation and at least one other investigation you have conducted.

6. All proportional relationships create a straight line if they are graphed.

    a. I agree with this statement.
    b. I disagree with this statement.

Electromagnets
What Variables Affect the Strength of the Electromagnet?

Explain your answer, using examples from this investigation and at least one other investigation you have conducted.

7. Why is it useful to identify causal relationships in science? In your answer, be sure to use examples from this investigation and at least one other investigation you have conducted.

8. Why is it useful to assume that you are studying a closed system during an investigation? In your answer, be sure to use examples from this investigation and at least one other investigation you have conducted.

LAB 14

Teacher Notes

Lab 14. Wire in a Magnetic Field: What Variables Affect the Strength of the Force Acting on a Wire in the Magnetic Field?

Purpose

The purpose of this lab is to *introduce* students to the disciplinary core idea (DCI) of Types of Interactions (PS2.B) from the *NGSS* and to the relationship between electricity and magnetism by having them identify the variables affecting the strength of the force acting on a current-carrying wire inside a magnetic field. In addition, this lab can be used to help students understand two big ideas from AP Physics: (a) fields existing in space can be used to explain interactions and (b) the interactions of an object with other objects can be described by forces. This lab also gives students an opportunity to learn about the cross-cutting concepts (CCs) of (a) Structure and Function and (b) Stability and Change from the *NGSS*. As part of the explicit and reflective discussion, students will also learn about (a) the difference between observations and inferences in science and (b) the assumptions made by scientists about order and consistency in nature.

Underlying Physics Concepts

There are two potential approaches to understanding the physics underlying this lab. The first approach we describe is the more direct approach and, therefore, more common. However, we describe the second approach because some students may think to take this approach, depending on where in a unit on electricity and magnetism you choose to conduct this lab.

The first approach starts by recognizing that current in a wire is the flow of a large number of electrons over time. We also know that a moving charge in a magnetic field experiences a force due to the magnetic field. Equation 14.1 shows this relationship, where **F** is the force on the moving charge, q is the charge on the particle, **v** is the velocity of the particle, **B** is the magnetic field, and θ is the angle between the direction of the magnetic field and the direction of the velocity. In SI units, force is measured in newtons (N), charge is measured in coulombs (C), velocity is measured in meters per second (m/s), magnetic field strength is measured in teslas (T), and the angle is measured in degrees (°) or radians (rad).

$$\text{(Equation 14.1)} \quad F = q\mathbf{vB}\sin\theta$$

We can rewrite velocity as a displacement or length traveled through a wire over time, which would give us Equation 14.2, where **l** is the length traveled and t is the time in

Wire in a Magnetic Field
What Variables Affect the Strength of the Force Acting on a Wire in the Magnetic Field?

which the particle traveled length l. In SI units, length is measured in meters (m) and time is measured in seconds (s).

$$\text{(Equation 14.2)} \quad F = \frac{q\mathbf{l}\mathbf{B}\sin\theta}{t}$$

Notice that in Equation 14.2 we now have the term q/t, which is charge per unit time. This is the definition of the current in a wire—the total charge passing a point in the wire per unit time. Thus, we can rewrite Equation 14.2 as Equation 14.3 to give us the force acting on a current-carrying wire in a magnetic field. In Equation 14.3, I is the current in the wire. In SI units, current is measured in amperes (A).

$$\text{(Equation 14.3)} \quad F = IlB\sin\theta$$

There are a few important things to note regarding Equation 14.3. First, the force on the wire can be thought of as the sum of the forces on all the individual particles (in most cases, electrons) moving through the wire. Second, the length of the wire must be in the magnetic field. That is, if the wire is 0.5 m long but only 0.25 m of wire is inside the magnetic field, then only the 0.25 m length of wire will contribute to the force due to the magnetic field. The other section of the wire will not feel a force due to the magnetic field because it is not present in the magnetic field. Third, the direction of the current in wire must be perpendicular to the magnetic field. The term $\sin\theta$ accounts for this because only the perpendicular component of the current will contribute to the force acting on the wire. Figure 14.1a–d (p. 312) shows a graphical representation of this, with the solid arrows representing the magnetic field and the dashed arrows representing the direction of current flow. The force will be maximum for the orientation of Figure 14.1a because the wire is perpendicular to the magnetic field (i.e., $\theta = 90°$ and $\sin\theta = 1$). In Figure 14.1c, the wire is parallel to the magnetic field and results in no force acting on the wire. Figure 14.1b shows the wire at an angle relative to the magnetic field, and the force acting on the wire will be between 0 N and the maximum force. Finally, in Figure 14.1d, the direction of the current is antiparallel to the direction of the magnetic field—the magnetic field points up while the current is oriented 180° from the magnetic field and flows down. Because there is no perpendicular component of the current to the magnetic field, the force on the wire is zero.

FIGURE 14.1

Different orientations of a current-carrying wire in a magnetic field

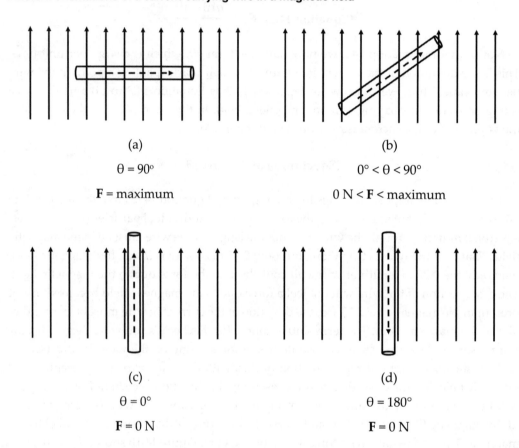

Based on our discussion thus far, the strength of the force on a current-carrying wire in a magnetic field is directly proportional to (1) the strength of the magnetic field, (2) the current strength, and (3) the length of the wire in the magnetic field.

The direction of the force on the wire can be determined using a right-hand rule because current multiplied by the magnetic field is a cross product. Figure 14.2 shows the result of the cross product of the magnetic field and the current using the right-hand rule. Remember that even though current is the flow of electrons, we assume current flows from the positive terminal. Equation 14.3 accounts for the cross product by the inclusion of the $\sin\theta$ term.

Finally, the orientation of the current with respect to the magnetic field is important. In Figure 14.1a, using the right-hand rule, we can see that the force acting on the wire in the magnetic field would be out of the page. Often, we can define "out of the page" as a positive direction (thus, the force would be positive). If we reverse the direction of the current, the force acting on the wire would be "into the page"—a negative force.

Wire in a Magnetic Field
What Variables Affect the Strength of the Force Acting on a Wire in the Magnetic Field?

As mentioned earlier in this section, there is a second approach to viewing the interaction of the current-carrying wire with the magnetic field. As we have discussed in prior labs (and you may have covered previously in your physics class), a magnetic field will be established around a current-carrying wire. The strength of the magnetic field is a function of the current flowing in the wire. As the current increases, the strength of the magnetic field also increases.

Students may use ideas surrounding the interaction of two magnetic fields to understand the force acting on the wire. In this case, the current through the wire creates a magnetic field around the wire that interacts with the external magnetic field. The force on the wire will be directly proportional to the strength of each field. We do not present an equation in this circumstance, because the mathematics are beyond the scope of even a first-year physics course. However, we point this out because students are likely to be familiar with and potentially use as part of their justification the concepts of (a) forces resulting from the interaction of two fields and (b) the creation of a magnetic field due to a current-carrying wire.

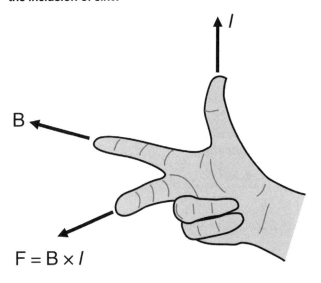

FIGURE 14.2

The right-hand rule showing the direction of the magnetic field, the current, and the resulting force acting on the wire. The × in the equation B × *I* indicates a cross product. In Equation 14.3, the cross product is accounted for by the inclusion of sinθ.

$F = B \times I$

Timeline

The instructional time needed to complete this lab investigation is 170–230 minutes. Appendix 3 (p. 421) provides options for implementing this lab investigation over several class periods. Option C (230 minutes) should be used if students are unfamiliar with scientific writing, because this option provides extra instructional time for scaffolding the writing process. You can scaffold the writing process by modeling, providing examples, and providing hints as students write each section of the report. Option C can also be used if you are introducing students to the magnetic field sensor, the data analysis software, and/or the video analysis software. Option D (170 minutes) should be used if students are familiar with scientific writing and have developed the skills needed to write an investigation report on their own. In option D, students complete stage 6 (writing the investigation report) and stage 8 (revising the investigation report) as homework.

LAB 14

Materials and Preparation

The materials needed to implement this investigation are listed in Table 14.1. The equipment can be purchased from a science supply company such as Flinn Scientific, PASCO, Vernier, or Ward's Science. Video analysis software can be purchased from Vernier (Logger *Pro*) or PASCO (SPARKvue or Capstone). These companies also have apps that can be used on Apple- or Android-based tablets and cell phones. We recommend consulting with your school's information technology coordinator to determine the best option for your students.

TABLE 14.1

Materials list for Lab 14

Item	Quantity
Consumables	
D batteries	As needed per group
Tape	1 roll per group
Equipment and other materials	
Safety glasses with side shields or goggles	1 per student
Permanent magnets of different strength	1 per strength per group
Copper wire	As needed per group
Alligator clips	As needed per group
Battery holders	Several per group
Multimeter	1 per group
Other equipment (see text below table)	
Investigation Proposal C (optional)	1 per group
Whiteboard, 2'× 3'*	1 per group
Lab Handout	1 per student
Peer-review guide and teacher scoring rubric	1 per student
Checkout Questions	1 per student
Equipment for digital interface measurements and video analysis (optional)	
Digital interface with USB or wireless connections	1 per group
Magnetic field sensor	1 per group

Continued

Table 14.1 (*continued*)

Item	Quantity
Current measurement sensor	1 per group
Video camera	1 per group
Computer or tablet with appropriate data analysis and video analysis software installed	1 per group

* As an alternative, students can use computer and presentation software such as Microsoft PowerPoint or Apple Keynote to create their arguments.

Science lab supply companies, such as those mentioned above, sell equipment that can be used to explore the force on a current-carrying wire in a magnetic field. Each supplier sells a slightly different setup to investigate the same concepts. To provide flexibility for teachers in purchasing, we have not committed to any specific set of equipment and instead have listed "other equipment" in Table 14.1. We also want to make it clear that specialized equipment is not necessary—the lab can be conducted with materials purchased individually.

This lab includes optional use of video analysis software. Using this software will allow students to more precisely measure the force on the wire. The wire will accelerate briefly upon closing the circuit due to its presence in the magnetic field. The video analysis will make it easier to measure this movement.

Be sure to use a set routine for distributing and collecting the materials during the lab investigation. One option is to set up the materials for each group at each group's lab station before class begins. This option works well when there is a dedicated section of the classroom for lab work and the materials are large and difficult to move. A second option is to have all the materials on a table or cart at a central location. You can then assign a member of each group to be the "materials manager." This individual is responsible for collecting all the materials his or her group needs from the table or cart during class and for returning all the materials at the end of the class. This option works well when the materials are small and easy to move (such as magnets, wire, and bulbs). It also makes it easy to inventory the materials at the end of the class before students leave for the day.

Safety Precautions and Laboratory Waste Disposal

Remind students to follow all normal lab safety rules. In addition, tell students to take the following safety precautions:

1. Wear sanitized safety glasses with side shields or goggles during lab setup, hands-on activity, and takedown.

2. Never put consumables in their mouth.

LAB 14

3. Keep their fingers and toes out of the way of the moving objects.

4. Wire and other metals with electric current flowing through them may get hot. Use caution when handling components of a closed circuit.

5. Handle electrical wires with caution. They have sharp ends, which can cut or puncture skin.

6. Wash their hands with soap and water when they are done collecting the data.

Batteries and wire may be stored for future use. When batteries need replacing, dispose of old batteries according to manufacturer's recommendations.

Topics for the Explicit and Reflective Discussion
Reflecting on the Use of Core Ideas and Crosscutting Concepts During the Investigation
Teachers should begin the explicit and reflective discussion by asking students to discuss what they know about the core ideas they used during the investigation. The following are some important concepts related to the core idea of magnetic fields that students need to use to explain the force acting on the wire:

- A field associates a value of some physical quantity with every point in space. Fields are a model that physicists use to describe interactions that occur over a distance. Fields permeate space, and objects experience forces due to their interaction with a field.

- Moving charges produce a magnetic field, which will subsequently interact with other magnetic fields produced by external objects or systems. This interaction is a force.

- A magnetic field exerts a force on a moving electrically charged object. For the electrons that make up the flow of current in a wire, the force is directly proportional to the current, the magnetic field, and the length of wire in the magnetic field. The equation describing this relationship is $\mathbf{F} = I\mathbf{B}\sin\theta$. The force is perpendicular to the direction of velocity of the object and to the magnetic field. Thus, the force also depends on the angle between the velocity and the magnetic field.

- Cross products can be modeled via a right-hand rule for determining the direction of the cross product. The direction of the force on the wire is perpendicular to the plane created by the magnetic field and the direction of the current.

- Current-carrying wires also produce a magnetic field. The interaction of two magnetic fields can result in an unbalanced force acting on the sources of each magnetic field.

To help students reflect on what they know about forces and the relationship between current and magnetic fields, we recommend showing them two or three images using presentation software that help illustrate these important ideas. You can then ask the students the following questions to encourage them to share how they are thinking about these important concepts:

1. What do we see going on in this image?
2. Does anyone have anything else to add?
3. What might be going on that we can't see?
4. What are some things that we are not sure about here?

You can then encourage students to think about how CCs played a role in their investigation. There are at least two CCs that students need to use to determine the variables affecting the force on the wire in the magnetic field: (a) Structure and Function and (b) Stability and Change (see Appendix 2 [p. 417] for a brief description of these CCs). To help students reflect on what they know about these CCs, we recommend asking them the following questions:

1. How does the structure of the wire and magnetic field affect how the system behaves?
2. Why is this knowledge important for designed systems?
3. What rates of change, if any, did you measure in your investigation?
4. What rates of change, if any, did you calculate in your investigation?

You can then encourage the students to think about how they used all these different concepts to help answer the guiding question and why it is important to use these ideas to help justify their evidence for their final arguments. Be sure to remind your students to explain why they included the evidence in their arguments and make the assumptions underlying their analysis and interpretation of the data explicit in order to provide an adequate justification of their evidence.

Reflecting on Ways to Design Better Investigations

It is important for students to reflect on the strengths and weaknesses of the investigation they designed during the explicit and reflective discussion. Students should therefore be encouraged to discuss ways to eliminate potential flaws, measurement errors, or sources of uncertainty in their investigations. To help students be more reflective about the design of their investigation and what they can do to make their investigations more rigorous in the future, you can ask them the following questions:

1. What were some of the strengths of the way you planned and carried out your investigation? In other words, what made it scientific?
2. What were some of the weaknesses of the way you planned and carried out your investigation? In other words, what made it less scientific?

3. What rules can we make, as a class, to ensure that our next investigation is more scientific?

Reflecting on the Nature of Scientific Knowledge and Scientific Inquiry

This investigation can be used to illustrate two important concepts related to the nature of scientific knowledge and the nature of scientific inquiry: (a) the difference between observations and inferences in science and (b) the assumptions made by scientists about order and consistency in nature (see Appendix 2 [p. 417] for a brief description of these concepts). Be sure to review these concepts during and at the end of the explicit and reflective discussion. To help students think about these concepts in relation to what they did during the lab, you can ask them the following questions:

1. You made observations and inferences during your investigation. Can you give me some examples of these observations and inferences?

2. Can you work with your group to come up with a rule that you can use to tell the difference between an observation and an inference? Be ready to share in a few minutes.

3. Scientists assume that natural laws operate today as they did in the past and that they will continue to do so in the future. Why do you think this assumption is important?

4. Think about what you were trying to do during this investigation. What would you have had to do differently if you could not assume natural laws operate today as they have in the past?

You can also use presentation software or other techniques to encourage your students to think about these concepts. You can show examples of information from the investigation that are either observations or inferences and ask students to classify each example and explain their thinking. You can also show images of different scientific laws (such as the Biot-Savart law or Ohm's law) and ask students if they think these laws have been the same throughout the universe's history. Be sure to remind your students that it is important for them to understand what counts as scientific knowledge and how that knowledge develops over time in order to be proficient in science.

Hints for Implementing the Lab

- Allowing students to design their own procedures for collecting data gives students an opportunity to try, to fail, and to learn from their mistakes. However, you can scaffold students as they develop their own procedure by having them fill out an investigation proposal. The proposals provide a way for you to offer students hints and suggestions without telling them how to do it. You can also check the proposals quickly during a class period. For this lab we suggest using Investigation Proposal C.

- Learn how to use the equipment, including the multimeter and the magnetic field sensor, before the lab begins. It is important for you to know how to use the equipment so you can help students when technical issues arise.

- Allow the students to become familiar with the equipment, especially the multimeter and magnetic field sensor, as part of the tool talk before they begin to design their investigation. Give them 5–10 minutes to examine the equipment and materials before they begin designing their investigations. This gives students a chance to see what they can and cannot do with the equipment.

- Although the force on the wire is directly proportional to the current through the wire, students must indirectly alter the current by changing the voltage. Thus, students may develop a relationship between the force acting on the wire and the voltage or resistance. Using Ohm's law, and assuming the resistance of the circuit remains constant, as the voltage increases, the current will also increase. We suggest pointing this out during the explicit and reflective discussion.

- If students are using a strong magnet and copper wire (as opposed to a specific kit for the concepts in this investigation), the forces on the wire may be small. We suggest pointing this out for students, because they will need to figure out how to identify and measure the small forces. One option is to measure the movement of the wire (or the magnet). According to Newton's second law, an unbalanced force produces an acceleration. Thus, the wire and/or magnet will move as a result from the interaction once the current begins to flow.

- Be sure to allow students to go back and re-collect data at the end of the argumentation session. Students often realize that they made numerous mistakes when they were collecting data as a result of their discussions during the argumentation session. The students, as a result, will want a chance to re-collect data, and the re-collection of data should be encouraged when time allows. This also offers an opportunity to discuss what scientists do when they realize a mistake is made inside the lab.

If students use digital interface measurement equipment and video analysis

- We suggest allowing students to familiarize themselves with the sensors, data analysis software, and video analysis software before they finalize the procedure for the investigation, especially if they have not used such software previously. This gives students an opportunity to learn how to work with the software and to improve the quality of the data they collect and the video they take.

- Remind students to begin recording data just before they close the circuits.

- Remind students to follow the user's guide to correctly connect any sensors to avoid damage to lab equipment.

LAB 14

- Remind students to hold the video camera as still as possible. Any movement of the camera will introduce error into their analysis. If using actual camcorders, we recommend using a tripod to hold the camera steady. If students are using a camera on a cell phone or tablet, we recommend using a table to help steady the camera.
- Remind students to place a meterstick in the same field of view as the motion they are capturing with the video camera. Also, the meterstick should be approximately the same distance from the camera as the motion. Most video analysis software requires the user to define a scale in the video (this allows the software to establish distances and, subsequently, other variables dependent on distance and displacement).

Connections to Standards

Table 14.2 highlights how the investigation can be used to address specific performance expectations from the *NGSS*; learning objectives from AP Physics 1 and 2; learning objectives from AP Physics C: Electricity and Magnetism; *Common Core State Standards for English Language Arts* (*CCSS ELA*); and *Common Core State Standards for Mathematics* (*CCSS Mathematics*).

TABLE 14.2

Lab 14 alignment with standards

NGSS performance expectation	• HS-PS2-5: Plan and conduct an investigation to provide evidence that an electric current can produce a magnetic field and that a changing magnetic field can produce an electric current.
AP Physics 1 and AP Physics 2 learning objectives	• 2.D.1.1: Apply mathematical routines to express the force exerted on a moving charged object by a magnetic field. • 3.A.3.1: Analyze a scenario and make claims (develop arguments, justify assertions) about the forces exerted on an object by other objects for different types of forces or components of forces. • 3.G.2.1: Connect the strength of electromagnetic forces with the spatial scale of the situation, the magnitude of the electric charges, and the motion of the electrically charged objects involved.
AP Physics C: Electricity and Magnetism learning objectives	• FIE-4.A.a: Calculate the magnitude of the magnetic force acting on a straight-line segment of a conductor with current in a uniform magnetic field. • FIE-4.A.b: Describe the direction of the magnetic force of interaction on a segment of a straight current-carrying conductor in a specified uniform magnetic field.

Continued

Table 14.2 (*continued*)

Literacy connections (*CCSS ELA*)	• *Reading*: Key ideas and details, craft and structure, integration of knowledge and ideas • *Writing*: Text types and purposes, production and distribution of writing, research to build and present knowledge, range of writing • *Speaking and listening:* Comprehension and collaboration, presentation of knowledge and ideas
Mathematics connections (*CCSS Mathematics*)	• *Mathematical practices*: Make sense of problems and persevere in solving them, reason abstractly and quantitatively, construct viable arguments and critique the reasoning of others, model with mathematics, use appropriate tools strategically, attend to precision • *Number and quantity*: Reason quantitatively and use units to solve problems, represent and model with vector quantities, perform operations on vectors • *Algebra*: Interpret the structure of expressions, create equations that describe numbers or relationships, understand solving equations as a process of reasoning and explain the reasoning, solve equations and inequalities in one variable, represent and solve equations and inequalities graphically • *Functions*: Understand the concept of a function and use function notation; interpret functions that arise in applications in terms of the context; analyze functions using different representations; build a function that models a relationship between two quantities; construct and compare linear, quadratic, and exponential models and solve problems; interpret expressions for functions in terms of the situation they model • *Statistics and probability*: Summarize, represent, and interpret data on two categorical and quantitative variables; interpret linear models; make inferences and justify conclusions from sample surveys, experiments, and observational studies

LAB 14

Lab Handout

Lab 14. Wire in a Magnetic Field: What Variables Affect the Strength of the Force Acting on a Wire in the Magnetic Field?

Introduction

All modern electronic speakers work by converting the energy carried by an electric current into sound waves. The approach to converting the electrical energy into sound waves was jointly invented by engineers Chester W. Rice and Edward W. Kellogg in 1924 and 1925 while working for General Electric. Figure L14.1 shows a diagram from the patent application filed by Rice in 1925. For comparison, Figure L14.2 shows a modern speaker, produced in 2015, that uses the same basic principles as the one invented by Rice and Kellogg (it looks quite similar as well). The same approach to converting electrical energy into sound energy underlies the speaker systems in televisions, headphones, cell phones, and loudspeaker systems.

FIGURE L14.1

Diagrams from U.S. patent application for the dynamic loudspeaker filed by Rice in 1925

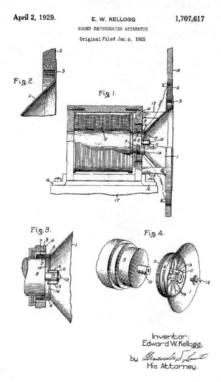

FIGURE L14.2

A modern speaker system produced in 2015

Note: A full-color version of this figure is available on the book's Extras page at *www.nsta.org/adi-physics2*.

Wire in a Magnetic Field
What Variables Affect the Strength of the Force Acting on a Wire in the Magnetic Field?

Loudspeakers work by placing a coil of wire in a magnetic field created by a permanent magnet. The wire is also placed near, or connected to, the back of a cone. The cone is covered by a diaphragm made of a special type of paper or cloth. Figure L14.3 shows a cross-sectional diagram of a loudspeaker. The wire is placed inside the ring magnet. When the wire is connected to a voltage source and the circuit is closed, a current will flow through the wire coil. According to the Biot-Savart law, a current-carrying wire will generate a magnetic field around the wire. The magnetic field generated by the current will interact with the field created by the permanent ring magnet surrounding the wire. This will produce a force acting on the wire, causing the wire to move. Because the wire is connected to the cone, the cone will move with the wire. This movement is fast enough that it causes the air to vibrate, producing a sound wave.

To produce sounds such as voices or music, the current passing through the wire must be controlled such that the wire vibrates very quickly. When designing speakers or other electronic devices, it is important for engineers and inventors to know what factors influence the force acting on the wire. If they are to precisely control the pattern of vibration to make sound, they need to know how to change the force acting on the wire. Thus, it is important to identify the variables that affect the force on the wire.

FIGURE L14.3
Cross-sectional diagram of a loudspeaker

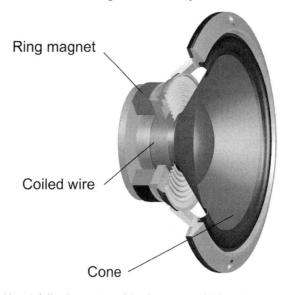

Note: A full-color version of this figure is available on the book's Extras page at *www.nsta.org/adi-physics2*.

Your Task

Use what you know about electric currents, magnetic fields, the relationship between structure and function, and what controls rates of change in systems to design and carry out one or more experiments to determine what variables affect the strength of the force acting on the wire when it is placed in the magnetic field. You should develop a model that allows you to make predictions about both the strength and direction of the force given a change to one of the variables you identify as affecting the strength of the force acting on the wire. Your model can be conceptual (e.g., an increase in x leads to an increase in y) or mathematical (e.g., $y = kx$, where k is a constant).

The guiding question of this investigation is, *What variables affect the strength of the force acting on a wire in the magnetic field?*

LAB 14

Materials
You may use any of the following materials during your investigation:

Consumables
- D batteries
- Tape

Equipment
- Safety glasses with side shields or goggles (required)
- Permanent magnets of different strength
- Copper wire
- Alligator clips
- Battery holders
- Multimeter

If you have access to the following equipment, you may also consider using a video camera, a digital magnetic field sensor, and a digital current sensor with an accompanying interface and a computer or tablet.

Safety Precautions
Follow all normal lab safety rules. In addition, take the following safety precautions:

1. Wear sanitized safety glasses with side shields or goggles during lab setup, hands-on activity, and takedown.
2. Never put consumables in your mouth.
3. Keep fingers and toes out of the way of the moving objects.
4. Wire and other metals with electric current flowing through them may get hot. Use caution when handling components of a closed circuit.
5. Handle electrical wires with caution. They have sharp ends, which can cut or puncture skin.
6. Wash your hands with soap and water when you are done collecting the data.

Investigation Proposal Required? ☐ Yes ☐ No

Getting Started
To answer the guiding question, you will need to design and carry out an investigation to determine what variables affect the strength of the force acting on the wire. Before you can design your investigation, however, you must determine what type of data you need to collect, how you will collect it, and how you will analyze it.

To determine *what type of data you need to collect*, think about the following questions:

- What variables might affect the strength of the force acting on the wire?
- What are the boundaries and components of the system?
- How do the components of the system interact with each other?

- When is this system stable and under which conditions does it change?
- Which factor(s) might control the rate of change in this system?
- How might the structure of what you are studying relate to its function?
- How could you keep track of changes in this system quantitatively?

To determine *how you will collect the data*, think about the following questions:

- What is the independent variable and what is the dependent variable for each experiment?
- What other factors will you need to control during each experiment?
- What scale or scales should you use when you take your measurements?
- How will you make sure that your data are of high quality (i.e., how will you reduce error)?
- How will you keep track of and organize the data you collect?
- How will you measure change over time during your investigation?

To determine *how you will analyze the data*, think about the following questions:

- What type of calculations, if any, will you need to make?
- What types of patterns might you look for as you analyze your data?
- What type of table or graph could you create to help make sense of your data?
- How could you use mathematics to describe a change over time?
- How could you use mathematics to describe a relationship between variables?

Connections to the Nature of Scientific Knowledge and Scientific Inquiry

As you work through your investigation, you may want to consider

- the difference between observations and inferences in science, and
- the assumptions made by scientists about order and consistency in nature.

Initial Argument

Once your group has finished collecting and analyzing your data, your group will need to develop an initial argument. Your initial argument needs to include a claim, evidence to support your claim, and a justification of the evidence. The *claim* is your group's answer to the guiding question. The *evidence* is an analysis and interpretation of your data. Finally, the *justification* of the evidence is why your group thinks the evidence matters. The justification of the evidence is important because scientists can use different kinds of evidence to support their claims. Your group will create your initial argument on a whiteboard. Your whiteboard should include all the information shown in Figure L14.4 (p. 326).

LAB 14

Argumentation Session

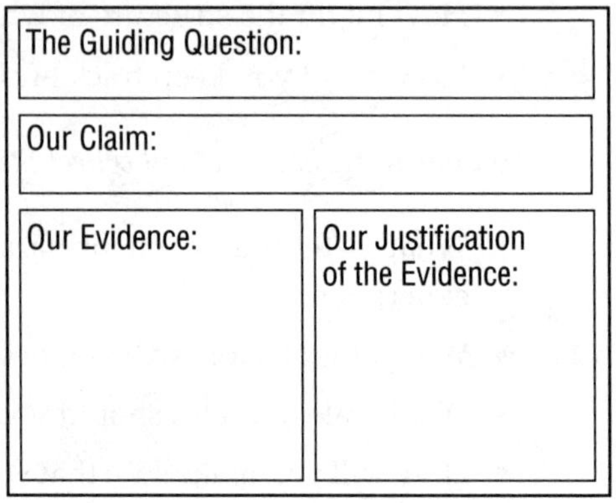

The argumentation session allows all of the groups to share their arguments. One or two members of each group will stay at the lab station to share that group's argument, while the other members of the group go to the other lab stations to listen to and critique the other arguments. This is similar to what scientists do when they propose, support, evaluate, and refine new ideas during a poster session at a conference. If you are presenting your group's argument, your goal is to share your ideas and answer questions. You should also keep a record of the critiques and suggestions made by your classmates so you can use this feedback to make your initial argument stronger. You can keep track of specific critiques and suggestions for improvement that your classmates mention in the space below.

FIGURE L14.4
Argument presentation on a whiteboard

The Guiding Question:	
Our Claim:	
Our Evidence:	Our Justification of the Evidence:

Critiques about our initial argument and suggestions for improvement:

If you are critiquing your classmates' arguments, your goal is to look for mistakes in their arguments and offer suggestions for improvement so these mistakes can be fixed. You should look for ways to make your initial argument stronger by looking for things that the other groups did well. You can keep track of interesting ideas that you see and hear during the argumentation in the space below. You can also use this space to keep track of any questions that you will need to discuss with your team.

Interesting ideas from other groups or questions to take back to my group:

Once the argumentation session is complete, you will have a chance to meet with your group and revise your initial argument. Your group might need to gather more data or design a way to test one or more alternative claims as part of this process. Remember, your goal at this stage of the investigation is to develop the best argument possible.

Report

Once you have completed your research, you will need to prepare an *investigation report* that consists of three sections. Each section should provide an answer to the following questions:

1. What question were you trying to answer and why?
2. What did you do to answer your question and why?
3. What is your argument?

Your report should answer these questions in two pages or less. This report must be typed, and any diagrams, figures, or tables should be embedded into the document. Be sure to write in a persuasive style; you are trying to convince others that your claim is acceptable or valid!

References

Audio Engineering Society. 1960. Dr. Edward W. Kellogg. [Obituary]. *Journal of the Audio Engineering Society* 8 (4): 283. *www.aes.org/aeshc/jaes.obit/JAES_V8_4_PG283.pdf.*

MIX Staff. 2017. 1925 Chester Rice & Edward Kellogg, General Electric Co. modern dynamic loudspeaker. *www.mixonline.com/technology/1925-chester-rice-edward-kellogg-general-electric-co-modern-dynamic-loudspeaker-377962.*

LAB 14

Checkout Questions

Lab 14. Wire in a Magnetic Field: What Variables Affect the Strength of the Force Acting on a Wire in the Magnetic Field?

1. The images below show a current-carrying wire in a magnetic field. The solid arrows represent the magnetic field. The dashed arrows represent the direction of the current in the wire. Assume the current through each is the same.

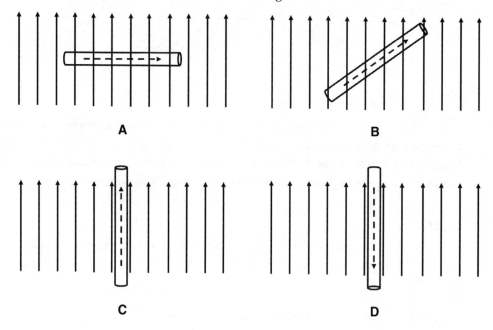

a. Rank the four images from greatest to smallest in terms of the force felt by the current-carrying wire in the magnetic field.

_____   _____   _____   _____
Smallest   Greatest

b. How do you know?

Wire in a Magnetic Field
What Variables Affect the Strength of the Force Acting on a Wire in the Magnetic Field?

2. A wire connected in series to a resistor R and a voltage source V is placed in a magnetic field of strength **B** such that a segment of the wire feels a force **F** applied to it. Assuming that the resistance and the voltage source are both doubled, what will happen to the force on the wire?

    a. It will increase.
    b. It will decrease.
    c. It will stay the same.

    How do you know?

3. A wire connected in series to a resistor R and a voltage source V is placed in a magnetic field of strength **B** such that a segment of the wire feels a force **F** applied to it. Assuming that the magnetic field strength is decreased by half, what will happen to the force on the wire?

    a. It will increase.
    b. It will decrease.
    c. It will stay the same.

    How do you know?

LAB 14

4. Understanding what factors affect the rates of change in systems does not apply to electronics because the wires are not moving.

    a. I agree with this statement.
    b. I disagree with this statement.

    Explain your answer, using an example from your investigation about magnetic fields and current-carrying wires.

5. Scientists assume that the laws governing the universe remain constant throughout time.

    a. I agree with this statement.
    b. I disagree with this statement.

    Explain your answer, using examples from this investigation and at least one other investigation you have conducted.

6. Observations and inferences are not the same thing in science. What is the difference between observations and inferences in science? In your answer, be sure to use examples from this investigation and at least one other investigation you have conducted.

7. Why is it useful to understand the relationship between structure and function of designed systems? In your answer, be sure to use examples from this investigation and at least one other investigation you have conducted.

LAB 15

Teacher Notes

Lab 15. Electromagnetic Induction: What Factors Influence the Induced Voltage in a Loop of Wire Placed in a Changing Magnetic Field?

Purpose

The purpose of this lab is to *introduce* students to the disciplinary core ideas (DCI) of Types of Interactions (PS2.B) and Conservation of Energy and Energy Transfer (PS3.B) from the *NGSS* and to electromagnetic induction by having them investigate the factors influencing the induced electromotive force (emf) in a coil of wire due to a changing magnetic field. In addition, this lab can be used to help students understand two big ideas from AP Physics: (a) fields existing in space can be used to explain interactions and (b) changes that occur as a result of interactions are constrained by conservation laws. This lab also gives students an opportunity to learn about the crosscutting concepts (CCs) of (a) Energy and Matter: Flows, Cycles, and Conservation and (b) Stability and Change from the *NGSS*. As part of the explicit and reflective discussion, students will also learn about (1) the difference between data and evidence in science and (b) the nature and role of experiments in science.

Underlying Physics Concepts

Two related laws lie at the heart of this lab. The first law is Faraday's law of induction, which states that a changing magnetic flux will induce an emf in a loop of wire resulting in the flow of current. Faraday's law of induction is shown in Equation 15.1, where ε is the emf, Δ is the symbol for change in the subsequent quantity, Φ is the magnetic flux, and t is time. In SI units, emf is measured in volts (V), magnetic flux is measured in webers (Wb), and time is measured in seconds.

$$\text{(Equation 15.1)} \quad \varepsilon = -\frac{\Delta \Phi}{t}$$

The second law relevant to this lab is Lenz's law, which states that a current resulting from an induced emf created by changing magnetic flux will flow in a direction such that the magnetic field created by the current will oppose the change in magnetic flux. Lenz's law is included within Faraday's law, represented as the negative sign. The negative sign tells us that the induced emf will produce a magnetic field to oppose that of the changing magnetic flux.

The reason the induced magnetic field will oppose the changing magnetic flux is a result of the law of conservation of energy. The changing magnetic flux can be thought of as doing work on the individual electrons (hence the term *electromotive force*, or the force on each electron giving rise to a voltage and current) causing their kinetic energy to increase—a flow

Electromagnetic Induction
What Factors Influence the Induced Voltage in a Loop of Wire Placed in a Changing Magnetic Field?

of current. Because the total energy of the system must be conserved, the induced emf does work back on the source of the changing magnetic field. In the case of a permanent magnet moving through a loop of wire (this is one option for setting up the equipment, as shown in Figure 15.2), that means inducing a magnetic field to oppose the change in magnetic flux.

It is important to define the term *magnetic flux*. Magnetic flux is the strength of the magnetic field passing through a surface. In other words, the magnetic flux is analogous to the number of field lines that pass through a surface. More precisely, the magnetic flux is the product of the magnetic field strength parallel to the normal of the surface and the area of the surface it is passing through. The *normal of the surface* is defined as a line perpendicular to the surface. This relationship is shown mathematically in Equation 15.2, where Φ is the magnetic flux, **B** is the magnetic field strength, A is the area of the surface, and θ is the angle between the magnetic field and the normal to the surface. In SI units, area is measured in meters squared (m^2) and the angle is measured in degrees.

(Equation 15.2) $\Phi = BA\cos\theta$

Figure 15.1 shows how magnetic flux inside of the loop changes with a change in the loop's orientation in the magnetic field. In Figure 15.1a, the magnetic field lines are parallel to the normal of the loop of wire. Thus, the angle between the magnetic field and the normal of the loop is 0°. The cosine of 0° is equal to 1, so the magnetic flux in Figure 15.1a is equal to the magnetic field strength multiplied by the area of the loop. This is also the maximum value for the magnetic flux. In Figure 15.1c, the magnetic field strength is perpendicular to the normal of the loop. In this case, the angle between the magnetic field and the normal is 90°. The cosine of 90° is equal to 0, so there is no magnetic flux through the loop in Figure 15.1c. Finally, Figure 15.1b shows the magnetic field at an angle relative to the normal of the surface. The magnetic flux will be between the maximum value and zero.

FIGURE 15.1

Changes in magnetic flux through the loop of wire corresponding to the loop's orientation in the magnetic field

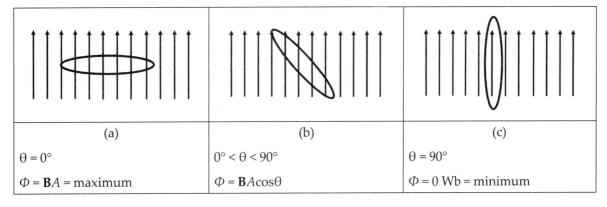

(a)	(b)	(c)
$\theta = 0°$	$0° < \theta < 90°$	$\theta = 90°$
$\Phi = BA$ = maximum	$\Phi = BA\cos\theta$	$\Phi = 0$ Wb = minimum

LAB 15

We can combine Equations 15.1 and 15.2 to give us the factors that influence the strength of the induced emf in a loop of wire by changing the flux, shown in Equation 15.3. The greater the change in the magnetic field, the area of the loop of wire, or the angle between the magnetic field and loop of wire, the stronger the emf will be and the larger the induced current.

$$\text{(Equation 15.3)} \quad \varepsilon = -\frac{\Delta B A \cos\theta}{t}$$

In this lab, students will identify the factors influencing the emf in a coil of wire by changing the flux passing through the loop of wire. There are a number of different ways to do this. One possible lab setup, also shown in the Lab Handout, is shown in Figure 15.2. In this setup, the permanent magnet is fastened to a spring and allowed to oscillate within the loop of wire. Students can change the strength of the permanent magnet, with a stronger magnet leading to a greater emf. They can also change the spring such that the period of oscillation changes. Students will find that the smaller the period, the greater the induced emf.

FIGURE 15.2

Equipment setup for changing the magnetic field strength through the coil of wire

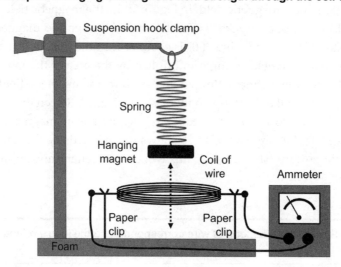

A second possible way to set up the equipment is shown in Figure 15.3. In this setup, a permanent magnet is placed below a coil of wire and the wire is laid on top of a stand such that it can rotate on the stand. The students can then spin the coil to induce an emf in the coil. Students can use magnets of different strengths to identify how a change in magnetic strength will affect the induced emf. Students can also investigate the effects of changing the rate of spin of the loop of wire. In this case, flux will change because the angle between the loop and the magnet will change. The faster the coil spins, the greater the induced emf.

Electromagnetic Induction
What Factors Influence the Induced Voltage in a Loop of Wire Placed in a Changing Magnetic Field?

FIGURE 15.3

Equipment setup for changing the angle between the magnetic field and the surface of the loop

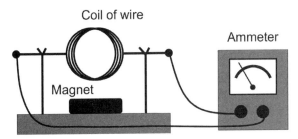

Students can also explore the effect of changing the number of loops of the coil wire. Assuming the distance between the loops of wire in a coil is negligible, the magnetic induction occurring through each loop is the same and adds together. This gives us Equation 15.4, where N is the number of full loops in the coil.

$$\text{(Equation 15.4)} \quad \varepsilon = -N\frac{\Delta\Phi}{t}$$

It is also possible to think about the number of loops with respect to the area of the coil. Because the magnetic flux is proportional to the area, if the number of loops of wire increases, the area that the magnetic field passes through also increases (assuming the radius of the coil is the same for all loops). Thus, if there are three loops of the same radius, then the area through which the magnetic field passes is three times the area of a single loop.

If students use an ammeter to collect their data, they also need to use Ohm's law to infer the induced emf (or voltage) from their measurement of current. Ohm's law is shown below in Equation 15.5, where V is the voltage or emf, I is the current flowing through the wire, and R is the resistance of a closed circuit. In SI units, current is measured in amperes (A) and resistance is measured in ohms (Ω). Assuming the resistance of the circuit remains constant, the induced voltage is directly proportional to the induced current.

$$\text{(Equation 15.5)} \quad V = IR$$

Finally, it is important to note that in both experimental setups (as shown in Figures 15.2 and 15.3), the current will oscillate between a positive maximum value and a negative maximum value that are equal in magnitude. This oscillation occurs due to the periodic nature of the experimental design. For example, in Figure 15.2, when the magnet is moving down through the loop, the current will be in a direction to oppose the changing magnetic flux and does work up against the magnet. When the magnet is moving back up through the loop, the current will again oppose the changing magnetic flux. To do this, the magnetic field must do work down against the magnet—this can only result if the current is in a different direction than when the magnet was moving down.

LAB 15

Timeline

The instructional time needed to complete this lab investigation is 200–280 minutes. Appendix 3 (p. 421) provides options for implementing this lab investigation over several class periods. Option E (280 minutes) should be used if students are unfamiliar with scientific writing, because this option provides extra instructional time for scaffolding the writing process. You can scaffold the writing process by modeling, providing examples, and providing hints as students write each section of the report. Option E can also be used if you are introducing students to the digital interface sensors and/or data analysis software. Option F (200 minutes) should be used if students are familiar with scientific writing and have developed the skills needed to write an investigation report on their own. In option F, students complete stage 6 (writing the investigation report) and stage 8 (revising the investigation report) as homework.

Materials and Preparation

The materials needed to implement this investigation are listed in Table 15.1. The equipment can be purchased from a science supply company such as Flinn Scientific, PASCO, Vernier, or Ward's Science.

TABLE 15.1

Materials list for Lab 15

Item	Quantity
Consumables	
Tape	As needed
Paper towel tube	2 per group
String	As needed
Paper clips (metal)	5 per group
Foam pieces	2 per group
Equipment and other materials	
Safety glasses with side shields or safety goggles	1 per student
Ring stand	1 per group
Clamps	As needed
Permanent magnets of different strengths	Several per group
Copper wire	As needed
Ammeter	1 per group
Alligator clips	4 per group

Continued

Electromagnetic Induction
What Factors Influence the Induced Voltage in a Loop of Wire Placed in a Changing Magnetic Field?

Table 15.1 (*continued*)

Item	Quantity
Wire cutter	1 per group
Ruler	2 per group
Springs of different spring constants	Several per group
Other equipment (see text below table)	
Investigation Proposal C (optional)	1 per group
Whiteboard, 2' × 3'*	1 per group
Lab Handout	1 per student
Peer-review guide and teacher scoring rubric	1 per student
Checkout Questions	1 per student
Equipment for digital interface measurements (optional)	
Digital interface with USB or wireless connections	1 per group
Magnetic field sensor	1 per group
Current measurement sensor	1 per group
Computer or tablet with appropriate data analysis software installed	1 per group

* As an alternative, students can use computer and presentation software such as Microsoft PowerPoint or Apple Keynote to create their arguments.

Science lab supply companies, such as those mentioned above, also make magnetic induction kits and electromotor kits that students can use to more quickly set up the equipment as shown in Figures 15.2 and 15.3. Each supplier sells a slightly different setup to investigate the same concepts. To provide flexibility for teachers in purchasing, we have not committed to any specific set of equipment and instead have listed "other equipment" in Table 15.1. We also want to make it clear that specialized equipment is not necessary—the lab can be conducted with the materials listed in Table 15.1.

Be sure to use a set routine for distributing and collecting the materials during the lab investigation. One option is to set up the materials for each group at each group's lab station before class begins. This option works well when there is a dedicated section of the classroom for lab work and the materials are large and difficult to move. A second option is to have all the materials on a table or cart at a central location. You can then assign a member of each group to be the "materials manager." This individual is responsible for collecting all the materials his or her group needs from the table or cart during class and for returning all the materials at the end of the class. This option works well when the materials are small and easy to move (such as magnets, wire, and bulbs). It also makes it easy to inventory the materials at the end of the class before students leave for the day.

LAB 15

Safety Precautions and Laboratory Waste Disposal

Remind students to follow all normal lab safety rules. In addition, tell students to take the following safety precautions:

1. Wear sanitized safety glasses with side shields or goggles during lab setup, hands-on activity, and takedown.

2. Wire and other metals with electric current flowing through them may get hot. Use caution when handling components of a closed circuit.

3. Wire cutters are very sharp. Make sure your fingers are out of the way when cutting wire.

4. Use caution when working with sharp objects (e.g., wires, clips) because they can cut or puncture skin.

5. Wash their hands with soap and water when they are done collecting the data.

There is no laboratory waste associated with this activity.

Topics for the Explicit and Reflective Discussion

Reflecting on the Use of Core Ideas and Crosscutting Concepts During the Investigation

Teachers should begin the explicit and reflective discussion by asking students to discuss what they know about the core ideas they used during the investigation. The following are some important concepts related to the core idea of magnetic induction that students need to use to identify the factors contributing to magnetic induction:

- According to Faraday's law, a changing magnetic flux induces an electromotive force (emf) in a system, giving rise to a current. The magnitude of the induced emf is equal to the rate of change in magnetic flux.

- The magnetic flux is the product of the magnetic field strength parallel to the normal of the surface and the area of the surface it is passing through. The normal of the surface is defined as a line perpendicular to the surface.

- Lenz's law states that a current resulting from an induced emf created by changing magnetic flux will flow in a direction such that the magnetic field created by the current will oppose the change in magnetic flux. The law of conservation of energy determines the direction of the induced emf, current, and magnetic field relative to the change in the magnetic flux.

- Work is done on an object or system by applying a force to the object or system through a displacement of the object or system. In this lab, work is done on individual electrons as they move through the wire. Work also changes the type or amount of energy of the object or system.

- Ohm's law is expressed mathematically as $V = IR$. We can infer the effect of the changing magnetic field on the voltage if we measure the current in the system.

To help students reflect on what they know about electromagnetic induction, we recommend showing them two or three images using presentation software that help illustrate these important ideas. You can then ask the students the following questions to encourage them to share how they are thinking about these important concepts:

1. What do we see going on in this image?
2. Does anyone have anything else to add?
3. What might be going on that we can't see?
4. What are some things that we are not sure about here?

You can then encourage students to think about how CCs played a role in their investigation. There are at least two CCs that students need to use to determine what factors influence the induced current in the wire: (a) Energy and Matter: Flows, Cycles, and Conservation and (b) Stability and Change (see Appendix 2 [p. 417] for a brief description of these CCs). To help students reflect on what they know about these CCs, we recommend asking them the following questions:

1. The law of conservation of energy states that energy cannot be created or destroyed, yet you created an electric current in this investigation. Did you violate the law of conservation of energy?
2. If the magnet does work on the electrons in the wire to generate a current, what work was done to conserve energy?
3. Why do you think the induced current kept changing values?
4. What factors control the rate of change of the current in your investigation?

You can then encourage the students to think about how they used all these different concepts to help answer the guiding question and why it is important to use these ideas to help justify their evidence for their final arguments. Be sure to remind your students to explain why they included the evidence in their arguments and make the assumptions underlying their analysis and interpretation of the data explicit in order to provide an adequate justification of their evidence.

Reflecting on Ways to Design Better Investigations

It is important for students to reflect on the strengths and weaknesses of the investigation they designed during the explicit and reflective discussion. Students should therefore be encouraged to discuss ways to eliminate potential flaws, measurement errors, or sources

LAB 15

of uncertainty in their investigations. To help students be more reflective about the design of their investigation and what they can do to make their investigations more rigorous in the future, you can ask them the following questions:

1. What were some of the strengths of the way you planned and carried out your investigation? In other words, what made it scientific?

2. What were some of the weaknesses of the way you planned and carried out your investigation? In other words, what made it less scientific?

3. What rules can we make, as a class, to ensure that our next investigation is more scientific?

Reflecting on the Nature of Scientific Knowledge and Scientific Inquiry

This investigation can be used to illustrate two important concepts related to the nature of scientific knowledge and the nature of scientific inquiry: (a) the difference between data and evidence in science and (b) the nature and role of experiments in science (see Appendix 2 [p. 417] for a brief description of these concepts). Be sure to review these concepts during and at the end of the explicit and reflective discussion. To help students think about these concepts in relation to what they did during the lab, you can ask them the following questions:

1. You had to talk about data and evidence during your investigation. Can you give me some examples of data and evidence from your investigation?

2. Can you work with your group to come up with a rule that you can use to decide if a piece of information is data or evidence? Be ready to share in a few minutes.

3. I asked you to design an experiment as part of your investigation. Are all investigations in science experiments? Why or why not?

4. Can you work with your group to come up with a rule that you can use to decide if an investigation is an experiment or not? Be ready to share in a few minutes.

You can also use presentation software or other techniques to encourage your students to think about these concepts. You can show examples of information from the investigation that are either data or evidence and ask students to classify each example and explain their thinking. You can also show images of different types of investigations (such as an astronomer using a telescope to observe the motion of Jupiter, a person working on a computer to analyze an existing data set, and an actual experiment) and ask students to indicate if they think each image represents an experiment and why or why not. Be sure to remind your students that it is important for them to understand what counts as scientific knowledge and how that knowledge develops over time in order to be proficient in science.

Electromagnetic Induction
What Factors Influence the Induced Voltage in a Loop of Wire Placed in a Changing Magnetic Field?

Hints for Implementing the Lab

- Allowing students to design their own procedures for collecting data gives students an opportunity to try, to fail, and to learn from their mistakes. However, you can scaffold students as they develop their own procedure by having them fill out an investigation proposal. The proposals provide a way for you to offer students hints and suggestions without telling them how to do it. You can also check the proposals quickly during a class period. For this lab we suggest using Investigation Proposal C.

- Learn how to use the equipment before the lab begins. It is important for you to know how to use the equipment so you can help students when technical issues arise. You may want to make sure you can get an induced emf when you set up the lab so you can help any students who struggle

- Allow the students to become familiar with the equipment as part of the tool talk before they begin to design their investigation. Give them 5–10 minutes to examine the equipment and materials before they begin designing their investigations. This gives students a chance to see what they can and cannot do with the equipment.

- For this lab, we strongly suggest using probeware that allows you to collect data at several measurements per second and that will easily graph the data for students. This is because the current will not reach a steady value—instead, it will consistently change as the magnetic flux changes. Analog measurement devices will work, but they will not provide precise results.

- To capture the effect of changing the angle of orientation on magnetic induction, we like to have students build the apparatus in Figure 15.3. To support the coil, you can bend a paper clip and stick the bent paper clip into a piece of foam. Then, attach the alligator clips on the end of the wire to the paper clip to make a closed circuit with the ammeter. Alternatively, science supply companies do sell a kit that includes such an apparatus.

- We suggest using neodymium magnets of different strengths. These are typically strong magnets and will allow students to see more pronounced results.

- If students measure the current through the circuit, they will need to infer the effect of the changing magnetic field on the voltage through their measurement of current. Because current and voltage are directly proportional, when a larger current is produced, it follows that a larger emf is produced as well.

- Be sure to allow students to go back and re-collect data at the end of the argumentation session. Students often realize that they made numerous mistakes when they were collecting data as a result of their discussions during the argumentation session. The students, as a result, will want a chance to re-collect data, and the re-collection of data should be encouraged when time allows. This also offers an opportunity to discuss what scientists do when they realize a mistake is made inside the lab.

LAB 15

If students use digital interface measurement equipment and analysis

- We suggest allowing students to familiarize themselves with the sensors and the data analysis software before they finalize the procedure for the investigation, especially if they have not used such software previously. This gives students an opportunity to learn how to work with the software and to improve the quality of the data they collect.
- Remind students to begin recording data just before they close the circuits.
- Remind students to follow the user's guide to correctly connect any sensors to avoid damage to lab equipment.

Connections to Standards

Table 15.2 highlights how the investigation can be used to address specific performance expectations from the *NGSS*; learning objectives from AP Physics 1 and 2; learning objectives from AP Physics C: Electricity and Magnetism; *Common Core State Standards for English Language Arts* (*CCSS ELA*); and *Common Core State Standards for Mathematics* (*CCSS Mathematics*).

TABLE 15.2

Lab 15 alignment with standards

NGSS performance expectations	• HS-PS2-5: Plan and conduct an investigation to provide evidence that an electric current can produce a magnetic field and that a changing magnetic field can produce an electric current. • HS-PS3-5: Develop and use a model of two objects interacting through electric or magnetic fields to illustrate the forces between objects and the changes in energy of the objects due to the interaction.
AP Physics 1 and AP Physics 2 learning objectives	• 2.C.4.1: Distinguish the characteristics that differ between monopole fields (gravitational field of spherical mass and electrical field due to single-point charge) and dipole fields (electric dipole field and magnetic field) and make claims about the spatial behavior of the fields using qualitative or semiquantitative arguments based on vector addition of fields due to each point source, including identifying the locations and signs of sources from a vector diagram of the field. • 4.E.2.1: Construct an explanation of the function of a simple electromagnetic device in which an induced emf is produced by a changing magnetic flux through an area defined by a current loop (i.e., a simple microphone or generator) or of the effect on behavior of a device in which an induced emf is produced by a constant magnetic field through a changing area.

Continued

Table 15.2 (continued)

AP Physics C: Electricity and Magnetism learning objectives	• FIE-6.A.a: Describe which physical situations with a changing magnetic field and a conductive loop will create an induced current in the loop. • FIE-6.A.b: Describe the direction of an induced current in a conductive loop that is placed in a changing magnetic field. • FIE-6.A.e: Calculate the magnitude and direction of induced EMF and induced current in a conductive loop (or conductive bar) when a physical quantity related to magnetic field or area is changing with a specified non-linear function of time.
Literacy connections (CCSS ELA)	• *Reading*: Key ideas and details, craft and structure, integration of knowledge and ideas • *Writing*: Text types and purposes, production and distribution of writing, research to build and present knowledge, range of writing • *Speaking and listening:* Comprehension and collaboration, presentation of knowledge and ideas.
Mathematics connections (CCSS Mathematics)	• *Mathematical practices*: Make sense of problems and persevere in solving them, reason abstractly and quantitatively, construct viable arguments and critique the reasoning of others, model with mathematics, use appropriate tools strategically, attend to precision • *Number and quantity*: Reason quantitatively and use units to solve problems, represent and model with vector quantities, perform operations on vectors • *Algebra*: Interpret the structure of expressions, create equations that describe numbers or relationships, understand solving equations as a process of reasoning and explain the reasoning, solve equations and inequalities in one variable, represent and solve equations and inequalities graphically • *Functions:* Understand the concept of a function and use function notation; interpret functions that arise in applications in terms of the context; analyze functions using different representations; build a function that models a relationship between two quantities; construct and compare linear, quadratic, and exponential models and solve problems; interpret expressions for functions in terms of the situation they model • *Statistics and probability*: Summarize, represent, and interpret data on two categorical and quantitative variables; interpret linear models; make inferences and justify conclusions from sample surveys, experiments, and observational studies

LAB 15

Lab Handout

Lab 15. Electromagnetic Induction: What Factors Influence the Induced Voltage in a Loop of Wire Placed in a Changing Magnetic Field?

Introduction

The early 1830s saw a number of important scientific and engineering breakthroughs related to electromagnetism (Tipler 1999). One of the scientists working in electromagnetism at the time was Michael Faraday (1791–1867). Among Faraday's many accomplishments was the invention of the first electric generator, a drawing of which is shown in Figure L15.1. The generator worked by placing a spinning metal disk (labeled component D in the figure) inside a horseshoe magnet (component A). When the disk was spun, a current was generated outward toward the edge of the disk. Component m is a contact that allowed the current to flow out from the disk toward the terminals (components B and B', respectively). The Faraday disk, as the invention has come to be known, was highly inefficient and not much energy could be harnessed from it. The importance of the Faraday disk was that it demonstrated the ability to use electromagnetic principles to engineer a device that could produce an electric current. This ability would give rise to further technological development and the electrification of modern society.

FIGURE L15.1

The Faraday disk, an electric generator, invented by Michael Faraday in 1831

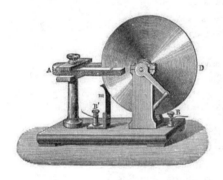

The physical mechanisms underlying the Faraday disk are the same as in modern electric generators. Faraday's law of electromagnetic induction states that a changing magnetic flux will induce an electric current in a conducting material. Magnetic flux is the strength of the magnetic field passing through a surface. This can be modeled by thinking about the magnetic flux as analogous to the number of field lines that pass through a surface. The field must be perpendicular to the surface—as you can see in Figure L15.1, the magnetic field of the horseshoe magnet is perpendicular to the surface of the metal disk. For cases where the field is not perpendicular to the surface, only the perpendicular component of the field contributes to the flux. Although the Faraday disk uses a solid metal surface, the flux can also be due to a magnetic field passing inside a wire loop. Modern electromagnetic generators use a loop of conducting wire instead of a solid metal disk. Mathematically, the magnetic flux is shown in Equation L.15.1, where Φ is the magnetic flux, **B** is the magnetic field strength, A is the area of the surface, and θ is the angle between the magnetic field and the normal to the surface. In SI units, magnetic flux is measured in webers (Wb), magnetic field is measured in teslas (T), area is measured in meters squared (m^2) and the angle is measured in degrees.

Electromagnetic Induction

What Factors Influence the Induced Voltage in a Loop of Wire Placed in a Changing Magnetic Field?

(Equation L15.1) $\Phi = BA\cos\theta$

Faraday's law can be applied to any situation with a magnetic field. This means that it can be used to understand how a magnetic field created by a permanent magnet (e.g., a bar magnet), a moving point charge, or an electric current can induce a current in a wire loop. To effectively harness the energy produced via an application of Faraday's law, it is important to understand what factors influence the induced voltage and subsequent current flow in a coil of wire. This knowledge allows engineers to design electric generators that provide electrical energy for a variety of uses.

Your Task

Use what you know about magnetic fields, forces and interaction, the law of conservation of energy, and what controls rates of change in a system to design and carry out one or more experiments to determine the effect of a number of variables on the induced current in a loop of wire. In this lab, you will study a system comprised of a coil of wire and permanent magnet. You must determine not only what variables influence the induced current in the loop of wire, but also the type of relationship (e.g. linear, inversely proportional).

The guiding question of this investigation is, **What factors influence the induced voltage in a loop of wire placed in a changing magnetic field?**

Materials

You may use any of the following materials during your investigation:

Consumables
- Tape
- Paper towel tubes
- String
- Paper clips
- Foam pieces

Equipment
- Safety glasses with side shields or goggles (required)
- Ring stand
- Clamps
- Permanent magnets of different strengths
- Copper wire

- Ammeter
- Alligator clips
- Wire cutter
- Ruler
- Springs of different spring constants

If you have access to the following equipment, you may also consider using a digital magnetic field sensor and digital current sensor with an accompanying interface and a computer or tablet.

Safety Precautions

Follow all normal lab safety rules. In addition, take the following safety precautions:

1. Wear sanitized safety glasses with side shields or goggles during lab setup, hands-on activity, and takedown.

LAB 15

2. Wire and other metals with electric current flowing through them may get hot. Use caution when handling components of a closed circuit.

3. Wire cutters are very sharp. Make sure your fingers are out of the way when cutting wire.

4. Use caution when working with sharp objects (e.g., wires, clips) because they can cut or puncture skin.

5. Wash your hands with soap and water when you are done collecting the data.

Investigation Proposal Required? ☐ Yes   ☐ No

Getting Started

To answer the guiding question, you will need to design and carry out an investigation to determine what variables affect electromagnetic induction. Figure L15.2 illustrates how you can use the available equipment to study the electromagnetic induction of the coil of wire. Your teacher may have additional equipment for you to use—if so, he or she will also demonstrate how to use the additional equipment. Before you can design your investigation, however, you must determine what type of data you need to collect, how you will collect it, and how you will analyze it.

FIGURE L15.2

One way to set up your equipment

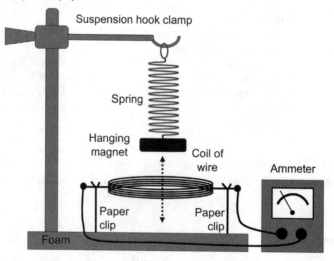

To determine *what type of data you need to collect*, think about the following questions:

- What variables might affect the induced current?
- What are the boundaries and components of the system?

- How do the components of the system interact with each other?
- When is this system stable and under which conditions does it change?
- How could you keep track of changes in this system quantitatively?
- How useful is it to track how energy flows into, out of, or within this system?
- Which factor(s) might control the rate of change in this system?

To determine *how you will collect the data*, think about the following questions:

- What is the independent variable and what is the dependent variable in each experiment?
- What other factors will you need to control during each experiment?
- What scale or scales should you use when you take your measurements?
- How will you make sure that your data are of high quality (i.e., how will you reduce error)?
- How will you keep track of and organize the data you collect?
- How can you track how energy flows into, out of, or within this system?
- How will you measure change over time during your investigation?

To determine *how you will analyze the data*, think about the following questions:

- What type of calculations, if any, will you need to make?
- What types of patterns might you look for as you analyze your data?
- What type of table or graph could you create to help make sense of your data?
- How could you use mathematics to describe a change over time?

Connections to the Nature of Scientific Knowledge and Scientific Inquiry

As you work through your investigation, you may want to consider

- the difference between data and evidence in science, and
- the nature and role of experiments in science.

Initial Argument

Once your group has finished collecting and analyzing your data, your group will need to develop an initial argument. Your initial argument needs to include a claim, evidence to support your claim, and a justification of the evidence. The *claim* is your group's answer to the guiding question. The *evidence* is an analysis and interpretation of your data. Finally, the *justification* of the evidence is why your group thinks the evidence matters. The justification of the evidence is important because scientists can use different kinds of evidence to support their claims. Your group will create your initial argument on a whiteboard.

LAB 15

FIGURE L15.3
Argument presentation on a whiteboard

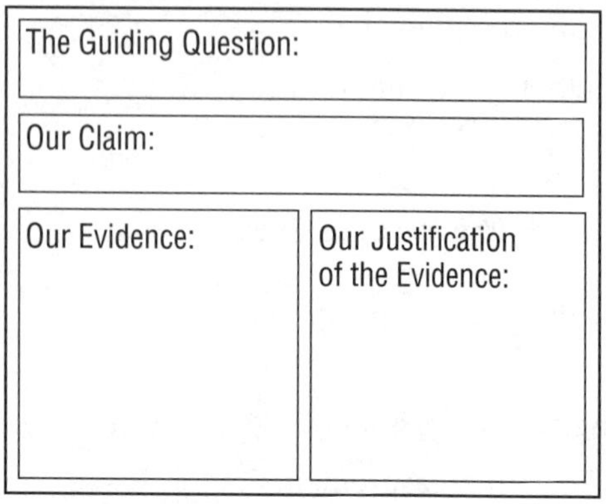

The Guiding Question:	
Our Claim:	
Our Evidence:	Our Justification of the Evidence:

Your whiteboard should include all the information shown in Figure L15.3.

Argumentation Session

The argumentation session allows all of the groups to share their arguments. One or two members of each group will stay at the lab station to share that group's argument, while the other members of the group go to the other lab stations to listen to and critique the other arguments. This is similar to what scientists do when they propose, support, evaluate, and refine new ideas during a poster session at a conference. If you are presenting your group's argument, your goal is to share your ideas and answer questions. You should also keep a record of the critiques and suggestions made by your classmates so you can use this feedback to make your initial argument stronger. You can keep track of specific critiques and suggestions for improvement that your classmates mention in the space below.

Critiques about our initial argument and suggestions for improvement:

If you are critiquing your classmates' arguments, your goal is to look for mistakes in their arguments and offer suggestions for improvement so these mistakes can be fixed. You should look for ways to make your initial argument stronger by looking for things that the other groups did well. You can keep track of interesting ideas that you see and hear during the argumentation in the space below. You can also use this space to keep track of any questions that you will need to discuss with your team.

Electromagnetic Induction

What Factors Influence the Induced Voltage in a Loop of Wire Placed in a Changing Magnetic Field?

Interesting ideas from other groups or questions to take back to my group:

Once the argumentation session is complete, you will have a chance to meet with your group and revise your initial argument. Your group might need to gather more data or design a way to test one or more alternative claims as part of this process. Remember, your goal at this stage of the investigation is to develop the best argument possible.

Report

Once you have completed your research, you will need to prepare an *investigation report* that consists of three sections. Each section should provide an answer to the following questions:

1. What question were you trying to answer and why?
2. What did you do to answer your question and why?
3. What is your argument?

Your report should answer these questions in two pages or less. This report must be typed, and any diagrams, figures, or tables should be embedded into the document. Be sure to write in a persuasive style; you are trying to convince others that your claim is acceptable or valid!

Reference

Tipler, P. A. 1999. *Physics for scientists and engineers. Volume 2: Electricity and magnetism, light.* 4th ed. New York: W. H. Freeman.

LAB 15

Checkout Questions

Lab 15. Electromagnetic Induction: What Factors Influence the Induced Voltage in a Loop of Wire Placed in a Changing Magnetic Field?

Use the following picture to answer questions 1–3.

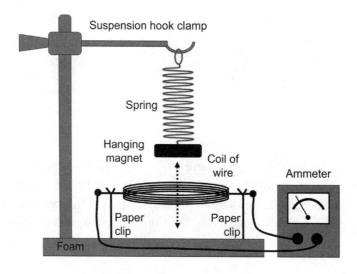

1. How does the strength of the magnet attached to the spring affect the current flowing through the loop of wire?

    a. The induced current increases as the strength of the magnet increases.

    b. The induced current decreases as the strength of the magnet increases.

    c. The induced current does not depend on the strength of the magnet.

    How do you know?

Electromagnetic Induction
What Factors Influence the Induced Voltage in a Loop of Wire Placed in a Changing Magnetic Field?

2. If you were to double the strength of the magnet, what would happen to the induced current?

   How do you know?

3. What would happen to the current induced in the coil of wire if the mass of the magnet on the end of the spring in the figure were doubled while the magnetic field created by the magnet remained the same?

   a. The induced current would increase as the mass of the magnet increases.
   b. The induced current would decrease as the mass of the magnet increases.
   c. The induced current does not depend on the mass of the magnet.

   How do you know?

4. The difference between data and evidence is that *data* is all the information collected during an investigation, whereas *evidence* is analyzed data used to support a claim.

   a. I agree with this statement.
   b. I disagree with this statement.

   Explain your answer, using examples from this investigation and at least one other investigation you have conducted.

LAB 15

5. All scientific investigations are experiments.

    a. I agree with this statement.

    b. I disagree with this statement.

    Explain your answer, using examples from this investigation and at least one other investigation you have conducted.

6. How does the wire and magnet system obey the law of conservation of energy when current is induced in the loop of wire?

7. Why is it useful to understand the factors that control rates of change in electromagnetic systems? In your answer, be sure to use examples from this investigation and at least one other investigation you have conducted.

Application Labs

LAB 16

Teacher Notes

Lab 16. Lenz's Law: Why Does the Magnet Fall Through the Metal Tube With an Acceleration That Is Not Equal to the Acceleration Due to Earth's Gravitational Field (−9.8 m/s²)?

Purpose

The purpose of this lab is for students to *apply* what they know about the disciplinary core idea (DCI) of Types of Interactions (PS2.B) from the *NGSS* and about fields, forces, and electromagnetic induction by having them develop a model to explain the motion of the magnet as it falls to the ground through a tube. In addition, this lab can be used to help students understand two big ideas from AP Physics: (a) fields existing in space can be used to explain interactions and (b) interactions between systems can result in changes in those systems. This lab also gives students an opportunity to learn about the crosscutting concepts (CCs) of (a) Scale, Proportion, and Quantity and (b) Systems and System Models from the *NGSS*. As part of the explicit and reflective discussion, students will also learn about (a) the difference between laws and theories in science and (b) the assumptions made by scientists about order and consistency in nature.

Underlying Physics Concepts

Three related laws lie at the heart of this lab. The first law is Faraday's law of induction, which states that a changing magnetic flux will induce an electromotive force (emf) in a loop of wire resulting in the flow of current. Faraday's law of induction is shown in Equation 16.1, where ε is the emf, Δ is the symbol for change in the subsequent quantity, Φ is the magnetic flux, and t is time. In SI units, emf is measured in volts (V), magnetic flux is measured in webers (Wb), and time is measured in seconds.

$$\text{(Equation 16.1)} \quad \varepsilon = -\frac{\Delta \Phi}{t}$$

The second law relevant to this lab is the Biot-Savart law, which describes the magnetic field produced by a current-carrying wire. When applied to a closed conducting loop (such as a copper ring), the magnetic field is directly proportional to the current in the loop and inversely proportional to the radius of the loop. The Biot-Savart law for a closed loop is shown mathematically in Equation 16.2, where **B** is the strength of the magnetic field, I is the current in the loop, **r** is the radius of the loop, and μ_o is the permeability of free space. In SI units, magnetic field is measured in teslas (T), current is measured in amperes (A), radius is measured in meters (m), and the permeability of free space is a constant with a value of $4\pi \times 10^{-7}$ tesla-meters per ampere (T·m/A).

Lenz's Law

Why Does the Magnet Fall Through the Metal Tube With an Acceleration That Is Not Equal to the Acceleration Due to Earth's Gravitational Field (–9.8 m/s²)?

(Equation 16.2) $$B = \frac{\mu_0 I}{2r}$$

The third, and final, law relevant to this lab is Lenz's law, which states that a current resulting from an induced emf created by changing magnetic flux will flow in a direction such that the magnetic field created by the current will oppose the change in magnetic flux. Lenz's law is included within Faraday's law, represented as the negative sign. The negative sign tells us that the induced emf will produce a magnetic field to oppose that of the changing magnetic flux.

The reason the induced magnetic field will oppose the changing magnetic flux is a result of the law of conservation of energy. The changing magnetic flux can be thought of as doing work on the individual electrons (hence the term *electromotive force*, or the force on each electron giving rise to a voltage and current) causing their kinetic energy to increase—a flow of current. Because the total energy of the system must be conserved, the induced emf does work back on the source of the changing magnetic field. In the case of a permanent magnet moving through a loop of wire (as in this lab), that means inducing a magnetic field to oppose the change in magnetic flux.

Finally, it is important to define the term *magnetic flux*. Magnetic flux is the strength of the magnetic field passing through a surface. We can model magnetic flux by thinking of it as analogous to the number of field lines that pass through a surface. More precisely, the magnetic flux is the product of the magnetic field strength parallel to the normal of the surface and the area of the surface it is passing through. This relationship is shown mathematically in Equation 16.3, where Φ is the magnetic flux, B is the magnetic field strength, A is the area of the surface, and θ is the angle between the magnetic field and the normal to the surface. In SI units, area is measured in meters squared (m²) and the angle is measured in degrees.

(Equation 16.3) $$\Phi = BA\cos\theta$$

Figure 16.1 (p. 356) shows how magnetic flux inside of the loop changes with a change in the loop's orientation in the magnetic field. In Figure 16.1a, the magnetic field lines are parallel to the normal of the loop of wire. Thus, the angle between the magnetic field and the normal of the loop is 0°. The cosine of 0° is equal to 1, so the magnetic flux in Figure 16.1a is equal to the magnetic field strength multiplied by the area of the loop. This is also the maximum value for the magnetic flux. In Figure 16.1c, the magnetic field is perpendicular to the normal of the loop. In this case, the angle between the magnetic field and the normal is 90°. The cosine of 90° is equal to 0, so there is no magnetic flux through the loop in Figure 16.1c. Finally, Figure 16.1b shows the magnetic field at an angle relative to the normal of the surface. The magnetic flux will be between the maximum value and zero.

LAB 16

FIGURE 16.1

Changes in magnetic flux through the loop of wire corresponding to the loop's orientation in the magnetic field

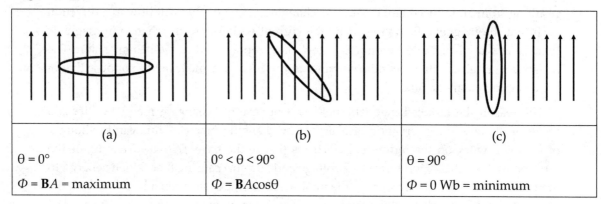

(a)	(b)	(c)
$\theta = 0°$	$0° < \theta < 90°$	$\theta = 90°$
$\Phi = BA$ = maximum	$\Phi = BA\cos\theta$	$\Phi = 0$ Wb = minimum

In this lab, students will drop a permanent magnet down a copper tube. We assume that the magnetic field near the permanent magnet is uniform and constant. We can use qualitative descriptions of Faraday's law, the Biot-Savart law, and Lenz's law to analyze the motion of the magnet. Each of the three equations above provides guidance on the influence of the change in one variable on the system.

To begin, we drop the magnet down the copper tube. As the magnet moves into the tube, the strength of the magnetic field through the surface of the tube will increase, producing a changing flux. According to Faraday's law, the changing magnetic flux will induce an emf in the copper tube and a resulting current. The Biot-Savart law states that a moving current will produce a magnetic field, and Lenz's law tells us the direction of magnetic field and the direction of the current that can produce the magnetic field. In this case, as the magnet falls into the tube, the magnetic flux is increasing. An increased flux will produce an increasing emf and a strengthening current.

The key question is to find what direction the induced magnetic field will be in. Lenz's law tells us that the induced magnetic field will oppose the change in the magnetic flux. If the magnetic flux is increasing, then using Lenz's law we can deduce that the magnetic field must be in a direction to oppose an increasing flux. Because the area remains constant, this means that the opposition to the change in flux must be due to the magnetic field created by the current. In other words, the induced magnetic field must oppose the strengthening magnetic field. If the field is increasing and pointing out of the magnet, the induced field must point into the magnet.

Figure 16.2 shows this step-by-step process. First, the magnet drops into the copper tube with the north pole facing down, and the flux increases as the magnet enters the tube (i.e., the magnetic field inside the tube increases from the initial value of 0 T). This leads to an induced emf and current in the tube that produces a magnetic field opposing

Lenz's Law

Why Does the Magnet Fall Through the Metal Tube With an Acceleration That Is Not Equal to the Acceleration Due to Earth's Gravitational Field (–9.8 m/s²)?

the change in flux. In this case, that means the magnetic field due to the current in tube will be pointed up. We used solid arrows to depict the permanent magnetic field due to the magnet, dashed lines to represent the induced current, and dashed arrows to represent the induced magnetic field.

It is helpful to think of the copper tube as being made of a series of very small copper rings. Thus, as the magnet falls through the copper tube, each "ring" of the tube experiences the same set of processes. We show this in Figure 16.3 for a smaller copper tube.

Note that the permanent magnet also has a south pole, with field lines pointed in toward the magnet (not shown in Figures 16.2 and 16.3). This will also cause a change in the magnetic flux through each copper "ring." Because the field remains in the down direction (relative to the copper tube and the motion of the magnet), the induced magnetic field due to the current flowing in the copper tube will again be pointed up.

Once the magnet begins to fall through the tube, it interacts with two fields. First, it is interacting with and moving in the Earth's gravitational field. Second, it is interacting with and moving through the magnetic field created by the induced emf due to the presence of a changing magnetic flux in the tube. Because the magnet has mass, it interacts with the Earth's gravitational field and feels a gravitational force. Because the magnet produces its own magnetic field, it also interacts with the induced magnetic field and feels a magnetic force.

To understand the motion of the magnet, we use Newton's second law, which states that the sum of the forces acting on an object causes that object to accelerate. Furthermore, the sum of the forces on the object is directly proportional to the acceleration of the object. Mathematically, Newton's second law is shown in Equation 16.4, where $\sum \mathbf{F}$ is the sum of the forces, or net force, acting on an object; m is the mass of the object; and $\mathbf{a}$ is the acceleration of the object. In SI units, force is measured in newtons (N), mass is measured in kilograms (kg), and acceleration is measured in meters per second squared (m/s²).

(Equation 16.4) $\sum \mathbf{F} = m\mathbf{a}$

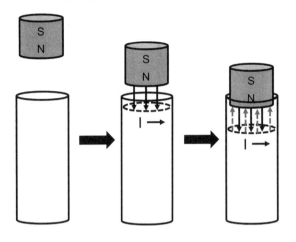

FIGURE 16.2

The result of placing the magnet into the tube. S = south pole; N = north pole.

FIGURE 16.3

The individual copper "rings" that make up the copper tube. S = south pole; N = north pole.

To identify the forces acting on the magnet, it is helpful to also draw a force diagram, shown in Figure 16.4. The diagram shows the gravitational force pointing down, which means it is a negative force, and the magnetic force pointing up, which means it is a positive force. Thus, we can rewrite Equation 16.4 to show all of the forces acting on the magnet as it falls through the copper tube. This is shown in Equation 16.5, where F_g is the force of gravity and F_B is the magnetic force.

FIGURE 16.4

The forces acting on the magnet as it falls through the copper tube

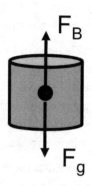

(**Equation 16.5**) $F_B - F_g = ma$

Initially, gravity causes the magnet to accelerate. This means the change in the magnetic flux per unit time is also increasing (see Equation 16.1). Eventually, the magnetic force acting on the magnet will reach the same magnitude as the gravitational force. When this occurs, the magnet reaches terminal velocity and falls through the tube at a constant rate. Thus, we can relate the strength of the magnetic force to the strength of the magnet—as the magnet's strength increases, the force due to the induced magnetic field will increase. The stronger the magnet, the quicker the magnet will reach terminal velocity inside the tube, and the longer the magnet will take to fall through the tube.

It is important to note that the strength of the current induced by the changing magnetic flux is also a function of material properties of the tube. One effect of the material properties of the tube is that the higher the electrical conductivity of the tube, the greater the induced current. As the induced current increases, the strength of the induced magnetic field increases, leading to a smaller terminal velocity and a longer time to fall through the tube. In this lab, students may test tubes made of copper, aluminum, and plastic. They will find that the magnet falls slowest through the copper tube and fastest through the plastic tube, because plastic is a very poor electrical conductor.

There are two ways that students can then find the average acceleration through the tube. First, they can assume a uniform acceleration through the tube and calculate the average acceleration of the magnet over the time interval it takes for the magnet to fall through the tube. Second, they can compare the time for the magnet to fall through the tube with the time it would take the magnet to fall the same height outside the tube.

Timeline

The instructional time needed to complete this lab investigation is 170–230 minutes. Appendix 3 (p. 421) provides options for implementing this lab investigation over several class periods. Option C (230 minutes) should be used if students are unfamiliar with scientific writing, because this option provides extra instructional time for scaffolding the

Lenz's Law

Why Does the Magnet Fall Through the Metal Tube With an Acceleration That Is Not Equal to the Acceleration Due to Earth's Gravitational Field (−9.8 m/s²)?

writing process. You can scaffold the writing process by modeling, providing examples, and providing hints as students write each section of the report. Option C can also be used if you are introducing students to the video analysis programs. Option D (170 minutes) should be used if students are familiar with scientific writing and have developed the skills needed to write an investigation report on their own. In option D, students complete stage 6 (writing the investigation report) and stage 8 (revising the investigation report) as homework.

Materials and Preparation

The materials needed to implement this investigation are listed in Table 16.1. The equipment can be purchased from a science supply company such as Flinn Scientific, PASCO, Vernier, or Ward's Science. Video analysis software can be purchased from Vernier (*Logger Pro*) or PASCO (SPARKvue or Capstone). These companies also have apps that can be used on Apple- or Android-based tablets and cell phones. We recommend consulting with your school's information technology coordinator to determine the best option for your students.

TABLE 16.1

Materials list for Lab 16

Item	Quantity
Safety glasses with side shields or safety goggles	1 per student
Cylindrical magnets of different strengths	1 per strength per group
Metal cylinders of different metals (in the same shape and size as the magnets)	1 per metal per group
Copper tube (1–1.5 m in length)	1 per group
Aluminum tube (1–1.5 m in length)	1 per group
Plastic tube (1–1.5 m in length)	1 per group
Magnetic field sensor	1 per group
Stopwatch	Several per group
Meterstick	1 per group
Electronic or triple beam balance	1 per group
Investigation Proposal A (optional)	1 per group
Whiteboard, 2' × 3'*	1 per group
Lab Handout	1 per student
Peer-review guide and teacher scoring rubric	1 per student
Checkout Questions	1 per student

Continued

Table 16.1 (continued)

Item	Quantity
Equipment for video analysis (optional)	
Video camera	1 per group
Computer or tablet with video analysis software	1 per group

* As an alternative, students can use computer and presentation software such as Microsoft PowerPoint or Apple Keynote to create their arguments.

You should conduct a demonstration of dropping a permanent magnet down a copper tube before students begin their investigation. Learn how to conduct the demonstration before the lab begins; the demonstration provides the context for this lab investigation, so you want to make sure that it works correctly.

Use of video analysis software is optional, but using the software will allow students to more precisely measure the rate at which the magnets and metal cylinders fall through each tube.

Be sure to use a set routine for distributing and collecting the materials during the lab investigation. One option is to set up the materials for each group at each group's lab station before class begins. This option works well when there is a dedicated section of the classroom for lab work and the materials are large and difficult to move. A second option is to have all the materials on a table or cart at a central location. You can then assign a member of each group to be the "materials manager." This individual is responsible for collecting all the materials his or her group needs from the table or cart during class and for returning all the materials at the end of the class. This option works well when the materials are small and easy to move (such as magnets, wire, and bulbs). It also makes it easy to inventory the materials at the end of the class before students leave for the day.

Safety Precautions and Laboratory Waste Disposal

Remind students to follow all normal lab safety rules. In addition, tell students to take the following safety precautions:

1. Wear sanitized safety glasses with side shields or goggles during lab setup, hands-on activity, and takedown

2. Keep their fingers and toes out of the way of the moving and falling objects.

3. Use caution in working with sharp objects (e.g., metal cylinders) because they may cut or puncture skin.

4. Wash their hands with soap and water when they are done collecting the data.

There is no laboratory waste associated with this activity.

Lenz's Law
Why Does the Magnet Fall Through the Metal Tube With an Acceleration That Is Not Equal to the Acceleration Due to Earth's Gravitational Field (–9.8 m/s²)?

Topics for the Explicit and Reflective Discussion

Reflecting on the Use of Core Ideas and Crosscutting Concepts During the Investigation

Teachers should begin the explicit and reflective discussion by asking students to discuss what they know about the core ideas they used during the investigation. The following are some important concepts related to the core idea of magnetic induction that students need to use to determine why the magnet does not fall with an acceleration equal to -9.8 m/s²:

- According to Faraday's law, a changing magnetic flux induces an electromotive force (emf) in a system, giving rise to a current. The magnitude of the induced emf is equal to the rate of change in magnetic flux.

- Lenz's law states that a current resulting from an induced emf created by changing magnetic flux will flow in a direction such that the magnetic field created by the current will oppose the change in magnetic flux. The law of conservation of energy determines the direction of the induced emf, current, and magnetic field relative to the change in the magnetic flux.

- Work is done on an object or system by applying a force to the object or system through a displacement of the object or system. Work also changes the type or amount of energy of the object or system.

- A field associates a value of some physical quantity with every point in space. Fields are a model that physicists use to describe interactions that occur over a distance. Fields permeate space, and objects experience forces due to their interaction with a field.

- Objects can interact with multiple fields at once, and the vector sum of the fields will determine the motion of the object.

- Newton's second law states that if multiple forces are acting on an object, the net force is the vector sum of the individual forces acting on the first object.

To help students reflect on what they know about fields, forces, and electromagnetic induction, we recommend showing them two or three images using presentation software that help illustrate these important ideas. You can then ask the students the following questions to encourage them to share how they are thinking about these important concepts:

1. What do we see going on in this image?
2. Does anyone have anything else to add?
3. What might be going on that we can't see?
4. What are some things that we are not sure about here?

You can then encourage students to think about how CCs played a role in their investigation. There are at least two CCs that students need to use to create their conceptual model:

LAB 16

(a) Scale, Proportion, and Quantity and (b) Systems and System Models (see Appendix 2 [p. 417] for a brief description of these CCs). To help students reflect on what they know about these CCs, we recommend asking them the following questions:

1. What vector quantities did you work with in this investigation? Why are vectors important in physics?

2. One of the goals that physicists share is to identify proportional relationships. Why is it important to identify proportional relationships in physics?

3. What are the advantages of creating models of complex systems in science?

4. What assumptions did you make about the magnet and tube system in order to model it?

You can then encourage the students to think about how they used all these different concepts to help answer the guiding question and why it is important to use these ideas to help justify their evidence for their final arguments. Be sure to remind your students to explain why they included the evidence in their arguments and make the assumptions underlying their analysis and interpretation of the data explicit in order to provide an adequate justification of their evidence.

Reflecting on Ways to Design Better Investigations
It is important for students to reflect on the strengths and weaknesses of the investigation they designed during the explicit and reflective discussion. Students should therefore be encouraged to discuss ways to eliminate potential flaws, measurement errors, or sources of uncertainty in their investigations. To help students be more reflective about the design of their investigation and what they can do to make their investigations more rigorous in the future, you can ask them the following questions:

1. What were some of the strengths of the way you planned and carried out your investigation? In other words, what made it scientific?

2. What were some of the weaknesses of the way you planned and carried out your investigation? In other words, what made it less scientific?

3. What rules can we make, as a class, to ensure that our next investigation is more scientific?

Reflecting on the Nature of Scientific Knowledge and Scientific Inquiry
This investigation can be used to illustrate two important concepts related to the nature of scientific knowledge and the nature of scientific inquiry: (a) the difference between laws and theories in science and (b) the assumptions made by scientists about order and consistency in nature (see Appendix 2 [p. 417] for a brief description of these concepts). Be sure

Lenz's Law

Why Does the Magnet Fall Through the Metal Tube With an Acceleration That Is Not Equal to the Acceleration Due to Earth's Gravitational Field (−9.8 m/s²)?

to review these concepts during and at the end of the explicit and reflective discussion. To help students think about these concepts in relation to what they did during the lab, you can ask them the following questions:

1. Laws and theories are different in science. In this lab, you used both laws and theories to plan your investigation and justify your evidence. What laws and theories did you use and how are they related to each other?

2. Can you work with your group to come up with a rule that you can use to decide if something is a law or a theory? Be ready to share in a few minutes.

3. Scientists assume that the universe is a vast single system in which basic laws are consistent. Why do you think this assumption is important?

4. Think about what you were trying to do during this investigation. What assumption did you make when planning your investigation?

You can also use presentation software or other techniques to encourage your students to think about these concepts. You can show examples of laws (such as Faraday's law or Lenz's law) and theories (such as Maxwell's equations) and ask students to indicate if they think each example is a law or a theory and explain their thinking. You can also show images of different scientific laws and ask students if these laws would be the same everywhere if the universe was not a single system. Then ask them to think about what scientists would need to do to be able to study the universe if it was made up of many different systems. Be sure to remind your students that it is important for them to understand what counts as scientific knowledge and how that knowledge develops over time in order to be proficient in science.

Hints for Implementing the Lab

- Allowing students to design their own procedures for collecting data gives students an opportunity to try, to fail, and to learn from their mistakes. However, you can scaffold students as they develop their own procedure by having them fill out an investigation proposal. The proposals provide a way for you to offer students hints and suggestions without telling them how to do it. You can also check the proposals quickly during a class period. For this lab we suggest using Investigation Proposal A.

- Learn how to use the equipment before the lab begins. It is important for you to know how to use the equipment so you can help students when technical issues arise.

- Allow the students to become familiar with the motion of the magnet as it falls through the copper tube as part of the tool talk before they begin to design their investigation. Give them 5–10 minutes to examine the equipment and materials

LAB 16

before they begin designing their investigations. This gives students a chance to see what they can and cannot do with the equipment.

- We suggest having students drop the magnets onto a soft surface. You can purchase foam to place underneath the tubes as the students drop the magnets. This way, students will not break the magnets when they fall out of the tubes.

- If you do not have access to magnetic field sensors for students, we suggest informing students of the strength of each magnet. This way, students can also test the effect of the strength of the magnet on how long it takes to fall through the tube.

- The magnet will reach a terminal velocity through the copper tube relatively quickly. We suggest that students calculate an average acceleration and compare this to the acceleration due to gravity of -9.8 m/s^2. As an alternative, students can measure the amount of time it takes each magnet/cylinder to fall through the tube and then calculate the expected amount of time for an object to fall the length of the tube.

- It is important to use cylindrical magnets that are slightly smaller in diameter than the tube. This allows you to assume a relatively uniform magnetic field inside the tube near the magnet. That is, as you move away from the magnet and toward the copper tube, the magnetic field remains constant. If the tube is much larger in diameter than the magnet, the field is no longer uniform, and the motion becomes a bit more complex. If you want to challenge students, you can also give students copper tubes with different diameters to see how the change in cross-sectional area influences the motion of the magnet as it falls.

- If students suggest that the reason the magnet slows down is because it is attracted to the metal tube, we suggest two options to push their thinking. First, you can purchase copper tubes with a small slit down the side. In this case, the loop is not closed, no current flows, and there is no magnetic field opposing the changing magnetic flux. Second, you can ask them to see if the magnet sticks to the copper tube. You can even have students let the magnet fall as it touches the outside of the copper tube. They will find the acceleration in this case is equal to the acceleration due to gravity.

- Be sure to allow students to go back and re-collect data at the end of the argumentation session. Students often realize that they made numerous mistakes when they were collecting data as a result of their discussions during the argumentation session. The students, as a result, will want a chance to re-collect data, and the re-collection of data should be encouraged when time allows. This also offers an opportunity to discuss what scientists do when they realize a mistake is made inside the lab.

If students use video analysis

- We suggest allowing students to familiarize themselves with the video analysis software before they finalize the procedure for the investigation, especially if they

Lenz's Law

Why Does the Magnet Fall Through the Metal Tube With an Acceleration That Is Not Equal to the Acceleration Due to Earth's Gravitational Field (–9.8 m/s²)?

have not used such software previously. This gives students an opportunity to learn how to work with the software and to improve the quality of the video they take.

- Remind students to hold the video camera as still as possible. Any movement of the camera will introduce error into their analysis. If using actual camcorders, we recommend using a tripod to hold the camera steady. If students are using a camera on a cell phone or tablet, we recommend using a table to help steady the camera.
- Remind students to place a meterstick in the same field of view as the motion they are capturing with the video camera. Also, the meterstick should be approximately the same distance from the camera as the motion. Most video analysis software requires the user to define a scale in the video (this allows the software to establish distances and, subsequently, other variables dependent on distance and displacement).

Connections to Standards

Table 16.2 highlights how the investigation can be used to address specific performance expectations from the *NGSS*; learning objectives from AP Physics 1 and 2; learning objectives from AP Physics C: Electricity and Magnetism; *Common Core State Standards for English Language Arts (CCSS ELA)*; and *Common Core State Standards for Mathematics (CCSS Mathematics)*.

TABLE 16.2

Lab 16 alignment with standards

NGSS performance expectations	• HS-PS2-5: Plan and conduct an investigation to provide evidence that an electric current can produce a magnetic field and that a changing magnetic field can produce an electric current. • HS-PS3-5: Develop and use a model of two objects interacting through electric or magnetic fields to illustrate the forces between objects and the changes in energy of the objects due to the interaction.
AP Physics 1 and AP Physics 2 learning objectives	• 2.C.4.1: Distinguish the characteristics that differ between monopole fields (gravitational field of spherical mass and electrical field due to single-point charge) and dipole fields (electric dipole field and magnetic field) and make claims about the spatial behavior of the fields using qualitative or semiquantitative arguments based on vector addition of fields due to each point source, including identifying the locations and signs of sources from a vector diagram of the field. • 3.A.1.1: Express the motion of an object using narrative, mathematical, and graphical representations. • 3.A.1.3: Analyze experimental data describing the motion of an object and is able to express the results of the analysis using narrative, mathematical, and graphical representations. • 3.A.2.1: Represent forces in diagrams or mathematically using appropriately labeled vectors with magnitude, direction, and units during the analysis of a situation.

Continued

LAB 16

Table 16.2 (continued)

AP Physics 1 and AP Physics 2 learning objectives (continued)	• 3.B.1.4: Predict the motion of an object subject to forces exerted by several objects using an application of Newton's second law in a variety of physical situations. • 4.E.2.1: Construct an explanation of the function of a simple electromagnetic device in which an induced emf is produced by a changing magnetic flux through an area defined by a current loop (i.e., a simple microphone or generator) or of the effect on behavior of a device in which an induced emf is produced by a constant magnetic field through a changing area. • 5.B.5.1: Design an experiment and analyze data to examine how a force exerted on an object or system does work on the object or system as it moves through a distance.
AP Physics C: Electricity and Magnetism learning objectives	• CNV-9.A.c: Calculate the magnetic flux of a non-uniform magnetic field that may have a magnitude that varies over one coordinate through a specified rectangular loop that is oriented perpendicularly to the field. • FIE-6.A.a: Describe which physical situations with a changing magnetic field and a conductive loop will create an induced current in the loop. • FIE-6.A.b: Describe the direction of an induced current in a conductive loop that is placed in a changing magnetic field.
Literacy connections (*CCSS ELA*)	• *Reading*: Key ideas and details, craft and structure, integration of knowledge and ideas • *Writing*: Text types and purposes, production and distribution of writing, research to build and present knowledge, range of writing • *Speaking and listening*: Comprehension and collaboration, presentation of knowledge and ideas
Mathematics connections (*CCSS Mathematics*)	• *Mathematical practices*: Make sense of problems and persevere in solving them, reason abstractly and quantitatively, construct viable arguments and critique the reasoning of others, model with mathematics, use appropriate tools strategically, attend to precision • *Number and quantity*: Reason quantitatively and use units to solve problems, represent and model with vector quantities, perform operations on vectors • *Algebra*: Interpret the structure of expressions, understand solving equations as a process of reasoning and explain the reasoning, solve equations and inequalities in one variable, represent and solve equations and inequalities graphically • *Functions*: Understand the concept of a function and use function notation; interpret functions that arise in applications in terms of the context; analyze functions using different representations; construct and compare linear, quadratic, and exponential models and solve problems; interpret expressions for functions in terms of the situation they model • *Statistics and probability*: Summarize, represent, and interpret data on two categorical and quantitative variables; interpret linear models; make inferences and justify conclusions from sample surveys, experiments, and observational studies

Lenz's Law
Why Does the Magnet Fall Through the Metal Tube With an Acceleration That Is Not Equal to the Acceleration Due to Earth's Gravitational Field (−9.8 m/s²)?

Lab Handout

Lab 16. Lenz's Law: Why Does the Magnet Fall Through the Metal Tube With an Acceleration That Is Not Equal to the Acceleration Due to Earth's Gravitational Field (−9.8 m/s²)?

Introduction

The 1800s saw a proliferation of research into electricity and magnetism, eventually giving rise to the unified theory of electromagnetism. The unified theory of electromagnetism allows scientists to understand a number of interesting phenomena related to the relationship between magnetic fields and moving electric charges. Your teacher will now demonstrate one of these phenomena by dropping a magnet down a copper tube.

Three important laws that are part of the unified theory of electromagnetism are Faraday's law, the Biot-Savart law, and Lenz's law. *Faraday's law* states that a changing magnetic flux will induce an electromotive force (emf) in a coil of wire resulting in the flow of current. The *Biot-Savart law* states that an electric current moving through a conductor produces a magnetic field. Finally, *Lenz's law* states that a current produced by a changing magnetic flux will flow in a direction such that the magnetic field produced by the current opposes the change in magnetic flux. *Magnetic flux* is the perpendicular component of a magnetic field passing through a particular surface, such as the surface formed by a loop of wire. Mathematically, magnetic flux (Φ) is related to the magnetic field strength (**B**) and the area of the loop (A) via the following equation: $\Phi = \mathbf{B}A\cos\theta$. In SI units, magnetic flux is measured in webers (Wb), magnetic field is measured in teslas (T), area is measured in meters squared (m²), and the angle is measured in degrees. Figure L16.1 (p. 368) is a visual representation of the relationship between the angle of the magnetic field relative to the wire and the flux.

Common to all three laws is the idea of a magnetic field. Fields are an important concept in physics and are used to describe how forces act on objects across space. A field is created by an object with specific properties—a magnetic field is established by either a permanent magnet or a moving electric charge. Similarly, a gravitational field (the field we are most familiar with) is established by an object with mass. When an object that establishes a field interacts with a field established by a similar object, it experiences a force. This is an important idea—forces exist between two objects interacting through a field. The force between two permanent magnets occurs because each magnet establishes a magnetic field. Similarly, the force of gravity one feels from Earth is an interaction with Earth's gravitational field. When an object is dropped, the acceleration due to gravity (−9.8 m/s²) is a measure of the strength of Earth's gravitational field.

Finally, an object can establish and interact with two fields at the same time. For example, Earth has both a gravitational field (due to Earth's mass) and a magnetic field (due to the convection currents in the molten iron core). This means that if you let a bar magnet

LAB 16

FIGURE L16.1

The magnetic flux through a single loop of wire. On the left, the magnetic field passes through the surface of the loop at an angle of 0° compared to perpendicular. This is the maximum flux. On the right, the magnetic field is parallel to the loop (i.e., an angle of 90° compared to perpendicular), resulting in a zero flux. The middle diagram shows the magnetic field at an angle between 0° and 90°, and the flux will be between the maximum flux and 0 Wb.

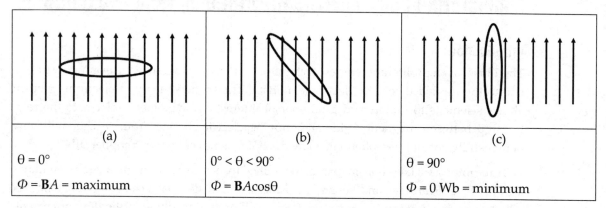

(a)　　　　　　　　　　　　(b)　　　　　　　　　　　　(c)

$\theta = 0°$　　　　　　　　　　$0° < \theta < 90°$　　　　　　　　$\theta = 90°$

$\Phi = \mathbf{B}A$ = maximum　　　　$\Phi = \mathbf{B}A\cos\theta$　　　　　　$\Phi = 0$ Wb = minimum

fall toward the ground, the motion can be described by the sum of the forces due to both the gravitational field and the magnetic field produced by Earth. For the majority of the types of motion you investigate in your physics class, the gravitational field is the most important factor influencing how an object moves; the influence of Earth's magnetic field is negligible and can be ignored.

It is also important to realize that other fields can exist for objects to interact with besides Earth's gravitational and magnetic fields. This is why you take off any metallic jewelry when you pass through a metal detector—the metal detector creates a magnetic field that will interact with any metal you carry with you through the metal detector. This interaction can then be measured by the metal detector as a change in the magnetic field of the metal detector. This is also why people with certain medical devices, such as a pacemaker, cannot walk through a metal detector—the magnetic field will disrupt the proper working of the pacemaker.

Your Task

Use what you know about electromagnetism, forces, fields, proportional relationships, and systems and system models to design and carry out an investigation to develop a model that explains the movement of the magnet as it falls through the copper tube. Your model should allow you to make predictions about variables of the system, such as the strength of the magnet and the composition of the tube. Your model should also describe the different fields acting on the magnet as it falls and how these fields are established. There may be other variables that influence the movement of the magnet that your model will want to account for as well.

Lenz's Law

Why Does the Magnet Fall Through the Metal Tube With an Acceleration That Is Not Equal to the Acceleration Due to Earth's Gravitational Field (–9.8 m/s²)?

The guiding question of this investigation is, **Why does the magnet fall through the metal tube with an acceleration that is not equal to the acceleration due to Earth's gravitational field (–9.8 m/s²)?**

Materials

You may use any of the following materials during your investigation:

- Safety glasses with side shields or goggles (required)
- Cylindrical magnets
- Metal cylinders
- Copper tube
- Aluminum tube
- Plastic tube
- Magnetic field sensor
- Stopwatch
- Meterstick
- Electronic or triple beam balance

If you have access to the following equipment, you may also consider using a video camera and a computer or tablet.

Safety Precautions

Follow all normal lab safety rules. In addition, take the following safety precautions:

1. Wear sanitized safety glasses with side shields or goggles during lab setup, hands-on activity, and takedown.
2. Keep fingers and toes out of the way of the moving and falling objects.
3. Use caution in working with sharp objects (e.g., metal cylinders) because they may cut or puncture skin.
4. Wash your hands with soap and water when you are done collecting the data.

Investigation Proposal Required? ☐ Yes   ☐ No

Getting Started

To answer the guiding question, you will need to design and carry out an investigation to determine why the magnet does not accelerate at –9.8 m/s² when dropped through the copper tube. Before you can design your investigation, however, you must determine what type of data you need to collect, how you will collect it, and how you will analyze it.

To determine *what type of data you need to collect*, think about the following questions:

- Can you directly measure the acceleration of the magnet, or will you need to calculate it based on other measurements?
- What are the boundaries and components of the system?
- How do the components of the system interact with each other?

LAB 16

- When is this system stable and under which conditions does it change?
- How could you keep track of changes in this system quantitatively?
- What quantities are vectors and what quantities are scalars?

To determine *how you will collect the data*, think about the following questions:

- What other factors will you need to control during your investigation?
- What scale or scales should you use when you take your measurements?
- How will you make sure that your data are of high quality (i.e., how will you reduce error)?
- How will you keep track of and organize the data you collect?

To determine *how you will analyze the data*, think about the following questions:

- What type of calculations, if any, will you need to make?
- What types of patterns might you look for as you analyze your data?
- What type of table or graph could you create to help make sense of your data?
- What types of proportional relationships might you look for as you analyze your data?

Connections to the Nature of Scientific Knowledge and Scientific Inquiry

As you work through your investigation, you may want to consider

- the difference between laws and theories, and
- the assumptions made by scientists about order and consistency in nature.

FIGURE L16.2

Argument presentation on a whiteboard

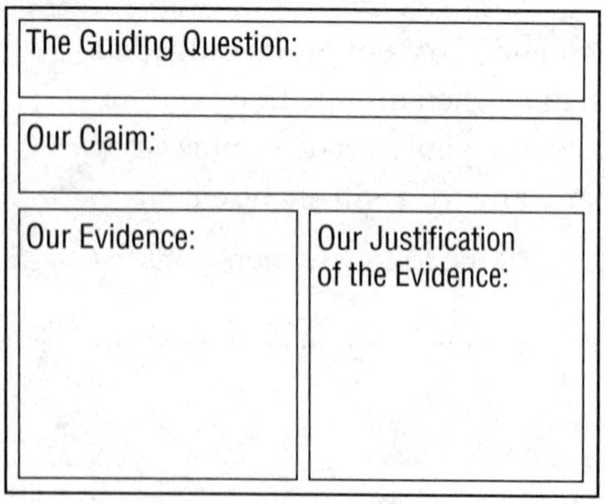

The Guiding Question:	
Our Claim:	
Our Evidence:	Our Justification of the Evidence:

Initial Argument

Once your group has finished collecting and analyzing your data, your group will need to develop an initial argument. Your initial argument needs to include a claim, evidence to support your claim, and a justification of the evidence. The *claim* is your group's answer to the guiding question. The *evidence* is an analysis and interpretation of your data. Finally, the *justification* of the evidence is why your group thinks the evidence matters. The justification of the evidence is important because scientists can use different kinds of evidence to support their claims. Your group will create your initial argument on a whiteboard. Your whiteboard should include all the information shown in Figure L16.2.

Lenz's Law

Why Does the Magnet Fall Through the Metal Tube With an Acceleration That Is Not Equal to the Acceleration Due to Earth's Gravitational Field (-9.8 m/s^2)?

Argumentation Session

The argumentation session allows all of the groups to share their arguments. One or two members of each group will stay at the lab station to share that group's argument, while the other members of the group go to the other lab stations to listen to and critique the other arguments. This is similar to what scientists do when they propose, support, evaluate, and refine new ideas during a poster session at a conference. If you are presenting your group's argument, your goal is to share your ideas and answer questions. You should also keep a record of the critiques and suggestions made by your classmates so you can use this feedback to make your initial argument stronger. You can keep track of specific critiques and suggestions for improvement that your classmates mention in the space below.

Critiques about our initial argument and suggestions for improvement:

If you are critiquing your classmates' arguments, your goal is to look for mistakes in their arguments and offer suggestions for improvement so these mistakes can be fixed. You should look for ways to make your initial argument stronger by looking for things that the other groups did well. You can keep track of interesting ideas that you see and hear during the argumentation in the space below. You can also use this space to keep track of any questions that you will need to discuss with your team.

Interesting ideas from other groups or questions to take back to my group:

LAB 16

Once the argumentation session is complete, you will have a chance to meet with your group and revise your initial argument. Your group might need to gather more data or design a way to test one or more alternative claims as part of this process. Remember, your goal at this stage of the investigation is to develop the best argument possible.

Report

Once you have completed your research, you will need to prepare an *investigation report* that consists of three sections. Each section should provide an answer to the following questions:

1. What question were you trying to answer and why?
2. What did you do to answer your question and why?
3. What is your argument?

Your report should answer these questions in two pages or less. This report must be typed, and any diagrams, figures, or tables should be embedded into the document. Be sure to write in a persuasive style; you are trying to convince others that your claim is acceptable or valid!

Checkout Questions

Lab 16. Lenz's Law: Why Does the Magnet Fall Through the Metal Tube With an Acceleration That Is Not Equal to the Acceleration Due to Earth's Gravitational Field (–9.8 m/s²)?

Use the pictures below to answer questions 1–3. The two rings are held up by a pin attached to the stand. The rings are free to rotate and tilt up and down. The side view shows the north end of the magnet being pushed forward through ring A.

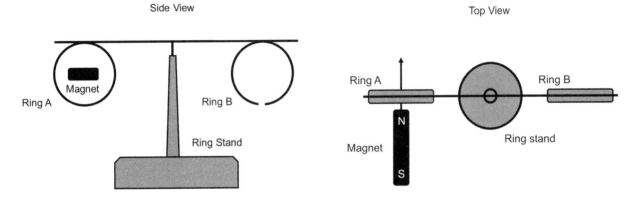

1. What will happen to the two-ring apparatus?

    a. Ring A will move in the same direction that the magnet is moving.

    b. Ring A will move in the opposite direction that the magnet is moving.

    c. Ring A will not move.

    How do you know?

LAB 16

2. The orientation of the magnet is changed such that the south end is pushed through ring A. What will happen to the two-ring apparatus?

    a. Ring A will move in the same direction that the magnet is moving.
    b. Ring A will move in the opposite direction that the magnet is moving.
    c. Ring A will not move.

    How do you know?

3. The north end of the magnet is pushed forward through ring B. What will happen to the two-ring apparatus?

    a. Ring B will move in the same direction that the magnet is moving.
    b. Ring B will move in the opposite direction that the magnet is moving.
    c. Ring B will not move.

    How do you know?

Lenz's Law

Why Does the Magnet Fall Through the Metal Tube With an Acceleration That Is Not Equal to the Acceleration Due to Earth's Gravitational Field (–9.8 m/s²)?

4. The difference between a law and a theory is that scientists are sure a law is true, but they think a theory is probably true but aren't sure yet.

    a. I agree with this statement.
    b. I disagree with this statement.

    Explain your answer, using examples from this investigation and at least one other investigation you have conducted.

5. Scientists make models because models can help them understand complex phenomena.

    a. I agree with this statement.
    b. I disagree with this statement.

    Explain your answer, using examples from this investigation and at least one other investigation you have conducted.

LAB 16

6. The distinction between vector quantities and scalar quantities is a unique distinction in physics. Why do physicists distinguish between scalar and vector quantities during an investigation? In your answer, be sure to use examples from this investigation and at least one other investigation you have conducted.

7. Why is it useful to assume that you are studying a closed system during an investigation? In your answer, be sure to use examples from this investigation and at least one other investigation you have conducted.

LAB 17

Teacher Notes

Lab 17. Electromagnetism: Why Does the Battery-and-Magnet "Car" Roll When It Is Placed on a Sheet of Aluminum Foil?

Purpose

The purpose of this lab is for students to *apply* what they know about the disciplinary core idea (DCI) of Types of Interactions (PS2.B) from the *NGSS* and about electromagnetism by having them develop a model to explain the movement of the battery-and-magnet "car." In addition, this lab can be used to help students understand two big ideas from AP Physics: (a) fields existing in space can be used to explain interactions and (b) the interactions of an object with other objects can be described by forces. This lab also gives students an opportunity to learn about the crosscutting concepts (CCs) of (a) Systems and System Models and (b) Stability and Change from the *NGSS*. As part of the explicit and reflective discussion, students will also learn about (a) how scientific knowledge changes over time and (b) how scientists use different methods to answer different types of questions.

Underlying Physics Concepts

To understand the physics underlying this lab, we start by recognizing that the magnets and the aluminum foil are electric conductors. Thus, when the battery-and-magnet car (henceforth referred to as just "the car") is placed on the aluminum foil, a closed circuit is formed. This leads to a current moving through the magnets and the foil. According to the Biot-Savart Law, the flow of current will create a magnetic field around the current. The Biot-Savart law also states that for a conducting loop carrying current, a magnetic field will be established inside the loop. Figure 17.1 shows the creation of the magnetic field created by the electric current flowing through the loop composed of the battery, magnets, and aluminum foil.

FIGURE 17.1

The magnetic field created by an electric current from the closed circuit formed when the battery-and-magnet car is placed on the aluminum foil

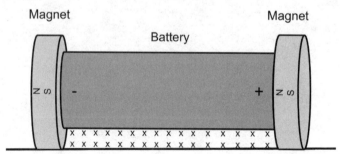

Electromagnetism
Why Does the Battery-and-Magnet "Car" Roll When It Is Placed on a Sheet of Aluminum Foil?

To determine the direction of the magnetic field, we can use a right-hand rule for the magnetic field created by a current-carrying wire to determine that the magnetic field created by the current will be pointing into the page. Notice that the magnetic field will only be established inside the loop of current. This is important for understanding the motion of the car.

Before placing the car on the aluminum foil, a magnetic field also existed around the car due to the presence of the two magnets on either end of the car. Figure 17.2 shows the magnetic field running from right to left, because the north pole of both magnets is pointing toward the left. We use dashed lines for the magnetic field due to the two permanent magnets in Figures 17.2 and 17.3 to avoid confusion with other lines in the figure. Because both magnets point in the same direction, a relatively uniform magnetic field is established around the battery.

FIGURE 17.2

The magnetic field around the battery due to the magnets on the ends of the battery. The north pole of both magnets faces to the left.

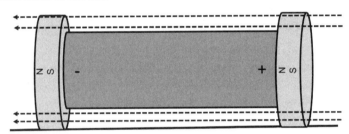

Thus, when we place the car down on the aluminum foil, we have two magnetic fields—one established by the magnets themselves (Figure 17.2) and one from the magnetic field created by the flow of the current (Figure 17.1). Figure 17.3 shows the combination of magnetic fields surrounding the battery when we place the car on the aluminum foil.

FIGURE 17.3

The magnetic fields when the car is placed on the aluminum foil

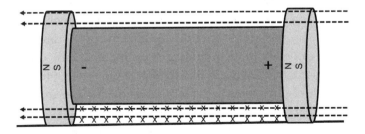

Notice how there is an additional magnetic field below the battery due to the flow of current but not above the battery. This field will interact with the field established by magnets themselves and create an unbalanced force. This is the force that causes the car to move.

It is important to note that the force exerted on the car exists only below the battery, not above the battery. This force then creates a torque on the magnets, as the force is exerted some radial distance away from their center. It is the torque on the lower half of the magnets that causes the car to move rotationally. It is also important to note that the direction the magnets face is the same. In Figure 17.3, the north pole of both magnets is to the left. This is important because it creates a uniform magnetic field from the magnets. If the magnets face in opposite directions, the magnetic field underneath the battery due to the magnets will not be uniform. This will result in different torques on each wheel. When students conduct the lab, it is OK to let them place the magnets in opposite directions, because this will lead to the car not moving and then they will need to establish why the car does not move under these circumstances.

Once the car starts moving, the physics becomes increasingly complex because the current will also be moving down the aluminum foil as the car moves and the wheels of the car will be rotating. Both magnetic fields will be moving, so students will need to account for additional factors. To mathematically represent this situation, complex sets of differential equations are necessary—mathematics that are beyond the scope of an introductory high school physics course and the AP physics courses. For this reason, we have chosen to present only a conceptual description of the underlying physics for this lab.

Students may establish quantitative relationships between the rotational acceleration of the car and the voltage of the battery and strength of the magnet. As the voltage of the battery increases, the angular acceleration of the car will also increase. This is because a higher voltage will create a larger current, leading to a greater magnetic field inside the loop. Similarly, strong magnets will create a larger magnetic field surrounding the battery. An increase in either magnetic field will lead to an increase in the torque acting on the wheels.

Timeline

The instructional time needed to complete this lab investigation is 170–230 minutes. Appendix 3 (p. 421) provides options for implementing this lab investigation over several class periods. Option C (230 minutes) should be used if students are unfamiliar with scientific writing, because this option provides extra instructional time for scaffolding the writing process. You can scaffold the writing process by modeling, providing examples, and providing hints as students write each section of the report. Option C should also be used if you are introducing students to digital sensors, the data analysis software, and/or the video analysis software. Option D (170 minutes) should be used if students are familiar with scientific writing and have developed the skills needed to write an investigation report on their own. In option D, students complete stage 6 (writing the investigation report) and stage 8 (revising the investigation report) as homework.

Materials and Preparation

The materials needed to implement this investigation are listed in Table 17.1. The equipment can be purchased from a science supply company such as Flinn Scientific, PASCO, Vernier, or Ward's Science. We also suggest companies that specialize in magnets, such as K&J Magnetics, as a source for the magnets for this lab. Video analysis software can be purchased from Vernier (Logger *Pro*) or PASCO (SPARKvue or Capstone). These companies also have apps that can be used on Apple- or Android-based tablets and cell phones. We recommend consulting with your school's information technology coordinator to determine the best option for your students.

TABLE 17.1

Materials list for Lab 17

Item	Quantity
Consumables	
AA batteries	2 per group
C batteries	2 per group
D batteries	2 per group
Duct tape	As needed
Aluminum foil	As needed
Wax paper	As needed
Butcher paper	As needed
Equipment and other materials	
Safety glasses with side shields or safety goggles	1 per student
Disc-shaped neodymium magnets	2 per group
Disc-shaped ceramic magnets (ideally, the same size as the neodymium magnets)	2 per group
Voltmeter	1 per group
Ammeter	1 per group
Multimeter	1 per group
Meterstick	1 per group

Continued

LAB 17

Table 17.1 (*continued*)

Item	Quantity
Stopwatch	1 per group
Electronic or triple beam balance	1 per group
Electronic pole identifier	1 per group
Investigation Proposal A (optional)	1 per group
Whiteboard, 2'× 3'*	1 per group
Lab Handout	1 per student
Peer-review guide and teacher scoring rubric	1 per student
Checkout Questions	1 per student
Equipment for digital interface measurements and video analysis (optional)	
Digital interface with USB or wireless connections	1 per group
Magnetic field sensor	1 per group
Current measurement sensor	1 per group
Voltage measurement sensor	1 per group
Video camera	1 per group
Computer or tablet with appropriate data analysis and video analysis software installed	1 per group

* As an alternative, students can use computer and presentation software such as Microsoft PowerPoint or Apple Keynote to create their arguments.

Use of video analysis software is optional, but using this software will allow students to more precisely measure the movement of the car.

You should conduct a demonstration with two magnets connected to a battery and then placed on a sheet of aluminum foil before students begin their investigation. Learn how to conduct the demonstration before the lab begins; the demonstration provides the context for this lab investigation, so you want to make sure it works correctly.

Be sure to use a set routine for distributing and collecting the materials during the lab investigation. One option is to set up the materials for each group at each group's lab station before class begins. This option works well when there is a dedicated section of the classroom for lab work and the materials are large and difficult to move. A second option is to have all the materials on a table or cart at a central location. You can then assign a member of each group to be the "materials manager." This individual is responsible for collecting all the materials his or her group needs from the table or cart during class and for returning all the materials at the end of the class. This option works well when the materials are small and easy to move (such as magnets, wire, and bulbs). It also makes it easy to inventory the materials at the end of the class before students leave for the day.

Safety Precautions and Laboratory Waste Disposal

Remind students to follow all normal lab safety rules. In addition, tell students to take the following safety precautions:

1. Wear sanitized safety glasses with side shields or safety goggles during lab setup, hands-on activity, and takedown.

2. Never put consumables in their mouth.

3. Wire and other metals with electric current flowing through them may get hot. Use caution when handling components of a closed circuit.

4. Us caution in working with sharp objects (e.g., wires) because they can cut or puncture skin.

5. Neodymium magnets should be at least 30 cm away from sensitive electronic and storage devices. These strong magnets could affect the functioning of pacemakers and implanted heart defibrillators.

6. Big magnets have a very strong attractive force. Unsafe handling could cause jamming of fingers or skin in between magnets. This may lead to contusions and bruises.

7. Neodymium magnets are brittle. Colliding magnets could crack, and sharp splinters could be catapulted away for several meters and injure eyes.

8. Wash their hands with soap and water when they are done collecting the data.

Batteries may be stored for future use. When batteries need replacing, dispose of old batteries according to manufacturer's recommendations.

Topics for the Explicit and Reflective Discussion

Reflecting on the Use of Core Ideas and Crosscutting Concepts During the Investigation

Teachers should begin the explicit and reflective discussion by asking students to discuss what they know about the core ideas they used during the investigation. The following are some important concepts related to the core ideas of types of interactions and electromagnetism that students need to use to explain the motion of the car:

- A field associates a value of some physical quantity with every point in space. Fields are a model that physicists use to describe interactions that occur over a distance. Fields permeate space, and objects experience forces due to their interaction with a field.

LAB 17

- Objects can interact with multiple fields at once, and the vector sum of the fields will determine the motion of the object.
- A current flowing through a conducting material will create a magnetic field around the material. If the current is flowing through a loop, a magnetic field is established inside the loop. The magnitude of the field is proportional to the magnitude of the current in the conducting material/loop. The direction of the magnetic field can be established using a right-hand rule.
- Torque is a measure of force applied perpendicular to a lever arm multiplied by the distance from the point of rotation. Torque is directly proportional to angular acceleration.

To help students reflect on what they know about electromagnetism, fields, and forces, we recommend showing them two or three images using presentation software that help illustrate these important ideas. You can then ask the students the following questions to encourage them to share how they are thinking about these important concepts:

1. What do we see going on in this image?
2. Does anyone have anything else to add?
3. What might be going on that we can't see?
4. What are some things that we are not sure about here?

You can then encourage students to think about how CCs played a role in their investigation. There are at least two CCs that students need to use to determine why the car moves: (a) Systems and System Models and (b) Stability and Change (see Appendix 2 [p. 417] for a brief description of these CCs). To help students reflect on what they know about these CCs, we recommend asking them the following questions:

1. In this investigation, you had to define your system under study. What assumptions did you have to make about the system in order to conduct your investigation?
2. You made models of your system in order to explain how it works. What types of models did you use during this investigation?
3. Your models allowed you to identify factors that affect the rates of change in your system. Why is it important to identify the factors that affect rates of change in systems?
4. What rates of change did you model during this investigation? What additional information would you have needed to model your system using a function?

You can then encourage the students to think about how they used all these different concepts to help answer the guiding question and why it is important to use these ideas to help justify their evidence for their final arguments. Be sure to remind your students to explain why they included the evidence in their arguments and make the assumptions underlying their analysis and interpretation of the data explicit in order to provide an adequate justification of their evidence.

Reflecting on Ways to Design Better Investigations

It is important for students to reflect on the strengths and weaknesses of the investigation they designed during the explicit and reflective discussion. Students should therefore be encouraged to discuss ways to eliminate potential flaws, measurement errors, or sources of uncertainty in their investigations. To help students be more reflective about the design of their investigation and what they can do to make their investigations more rigorous in the future, you can ask them the following questions:

1. What were some of the strengths of the way you planned and carried out your investigation? In other words, what made it scientific?

2. What were some of the weaknesses of the way you planned and carried out your investigation? In other words, what made it less scientific?

3. What rules can we make, as a class, to ensure that our next investigation is more scientific?

Reflecting on the Nature of Scientific Knowledge and Scientific Inquiry

This investigation can be used to illustrate two important concepts related to the nature of scientific knowledge and the nature of scientific inquiry: (a) how scientific knowledge changes over time and (b) how scientists use different methods to answer different types of questions (see Appendix 2 [p. 417] for a brief description of these concepts). Be sure to review these concepts during and at the end of the explicit and reflective discussion. To help students think about these concepts in relation to what they did during the lab, you can ask them the following questions:

1. Scientific knowledge can and does change over time. Can you tell me why it changes?

2. Can you work with your group to come up with some examples of how scientific knowledge related to electricity and magnetism has changed over time? Be ready to share in a few minutes.

3. There is no universal step-by-step scientific method that all scientists follow. Why do you think there is no universal scientific method?

4. Think about what you did during this investigation. How would you describe the method you used to understand why the battery-and-magnet car starts rolling? Why would you call it that?

LAB 17

You can also use presentation software or other techniques to encourage your students to think about these concepts. You can show examples of how our thinking about electricity and magnetism has changed over time (and continues to change as scientists search for a grand unified theory) and ask students to discuss what they think led to those changes. You can show one or more images of a "universal scientific method" that misrepresent the nature of scientific inquiry (see, e.g., *https://commons.wikimedia.org/wiki/ File:The_Scientific_Method_as_an_Ongoing_Process.svg*) and ask students why each image is *not* a good representation of what scientists do to develop scientific knowledge. You can also ask students to suggest revisions to the image that would make it more consistent with the way scientists develop scientific knowledge. Be sure to remind your students that it is important for them to understand what counts as scientific knowledge and how that knowledge develops over time in order to be proficient in science.

Hints for Implementing the Lab

- Allowing students to design their own procedures for collecting data gives students an opportunity to try, to fail, and to learn from their mistakes. However, you can scaffold students as they develop their own procedure by having them fill out an investigation proposal. The proposals provide a way for you to offer students hints and suggestions without telling them how to do it. You can also check the proposals quickly during a class period. For this lab we suggest using Investigation Proposal A.

- Learn how to set up the battery-and-magnet car and how to use the equipment before the lab begins. If you set up the magnets improperly, the car will not move. It is also important for you to know how to use the equipment so you can help students when technical issues arise.

- When setting up the demonstration to introduce the lab, make sure that the north poles of the magnets face in the same direction. This establishes a uniform magnetic field around the battery and will result in the car moving. If the magnets' north poles face in opposite directions, the car will not move.

- Allow the students to become familiar with the equipment as part of the tool talk before they begin to design their investigation. Give them 5–10 minutes to examine the equipment and materials before they begin designing their investigations. This gives students a chance to see what they can and cannot do with the equipment.

- The resistance in the circuit comprised of the battery, magnets, and aluminum foil is minimal, so the battery will deplete rather quickly. We suggest providing fresh batteries to each group as they begin to collect data. Because of this, we also suggest using the investigation proposal guide before providing groups with batteries as a way to minimize students' undirected use of the batteries.

- The performance of the magnet car will be improved by using flat aluminum foil. If the foil is wrinkled, this will cause the current to flow in random directions and not directly underneath the battery.
- If you do not have an electronic pole identifier for each group, you can identify the north pole on the magnets prior to class and then place a small sticker on the north pole of each magnet.
- In the lab materials, we have included butcher paper and wax paper, because students must show that the car is not just interacting with Earth's magnetic field to produce the motion. We anticipate that most groups will not initially think of this possibility. This is a good opportunity to discuss initial assumptions and how we often need to run experiments to rule out other possible explanations.
- If students want to test the relationship between the voltage of the battery and the angular acceleration, you can either give them batteries with higher voltage (e.g., 3V batteries, which are sold by Duracell and Energizer) or have them tape two 1.5V batteries together in series.
- Be sure to allow students to go back and re-collect data at the end of the argumentation session. Students often realize that they made numerous mistakes when they were collecting data as a result of their discussions during the argumentation session. The students, as a result, will want a chance to re-collect data, and the re-collection of data should be encouraged when time allows. This also offers an opportunity to discuss what scientists do when they realize a mistake is made inside the lab.

If students use digital interface measurement equipment and video analysis

- We suggest allowing students to familiarize themselves with the sensors, data analysis software, and video analysis software before they finalize the procedure for the investigation, especially if they have not used such software previously. This gives students an opportunity to learn how to work with the software and to improve the quality of the data they collect and the video they take.
- Remind students to follow the user's guide to correctly connect any sensors to avoid damage to lab equipment.
- Remind students to hold the video camera as still as possible. Any movement of the camera will introduce error into their analysis. If using actual camcorders, we recommend using a tripod to hold the camera steady. If students are using a camera on a cell phone or tablet, we recommend using a table to help steady the camera.
- Remind students to place a meterstick in the same field of view as the motion they are capturing with the video camera. Also, the meterstick should be approximately the same distance from the camera as the motion. Most video analysis software requires the user to define a scale in the video (this allows the software to establish distances and, subsequently, other variables dependent on distance and displacement).

LAB 17

Connections to Standards

Table 17.2 highlights how the investigation can be used to address specific performance expectations from the *NGSS;* learning objectives from AP Physics 1 and 2; learning objectives from AP Physics C: Electricity and Magnetism; *Common Core State Standards for English Language Arts (CCSS ELA);* and *Common Core State Standards for Mathematics (CCSS Mathematics).*

TABLE 17.2

Lab 17 alignment with standards

NGSS performance expectation	• HS-PS2-5: Plan and conduct an investigation to provide evidence that an electric current can produce a magnetic field and that a changing magnetic field can produce an electric current.
AP Physics 1 and AP Physics 2 learning objectives	• 1.A.5.2: Construct representations of how the properties of a system are determined by the interactions of its constituent substructures. • 2.D.1.1: Apply mathematical routines to express the force exerted on a moving charged object by a magnetic field. • 3.C.3.2: Plan a data collection strategy appropriate to an investigation of the direction of the force on a moving electrically charged object caused by a current in a wire in the context of a specific set of equipment and instruments, and analyze the resulting data to arrive at a conclusion. • 4.E.1.1: Use representations and models to qualitatively describe the magnetic properties of some materials that can be affected by magnetic properties of other objects in the system.
AP Physics C: Electricity and Magnetism learning objectives	• CNV-8.B.a: Derive the expression for the magnitude of magnetic field on the axis of a circular loop of current or a segment of a circular loop. • CNV-8.C.a: Explain Ampère's Law and justify the use of the appropriate Amperian loop for current-carrying conductors of different shapes such as straight wires, closed circular loops, conductive slabs, or solenoids. • ACT-4.A.a: Determine if a net force or net torque exists on a conductive loop in a region of changing magnetic field.
Literacy connections (CCSS ELA)	• *Reading*: Key ideas and details, craft and structure, integration of knowledge and ideas • *Writing*: Text types and purposes, production and distribution of writing, research to build and present knowledge, range of writing • *Speaking and listening:* Comprehension and collaboration, presentation of knowledge and ideas

Continued

Table 17.2 (continued)

Mathematics connections (*CCSS Mathematics*)	• *Mathematical practices*: Make sense of problems and persevere in solving them, reason abstractly and quantitatively, construct viable arguments and critique the reasoning of others, model with mathematics, use appropriate tools strategically, attend to precision • *Number and quantity*: Reason quantitatively and use units to solve problems, represent and model with vector quantities, perform operations on vectors • *Algebra*: Interpret the structure of expressions, understand solving equations as a process of reasoning and explain the reasoning, solve equations and inequalities in one variable, represent and solve equations and inequalities graphically • *Functions*: Understand the concept of a function and use function notation; interpret functions that arise in applications in terms of the context; analyze functions using different representations; construct and compare linear, quadratic, and exponential models and solve problems; interpret expressions for functions in terms of the situation they model • *Statistics and probability*: Summarize, represent, and interpret data on two categorical and quantitative variables; interpret linear models; make inferences and justify conclusions from sample surveys, experiments, and observational studies

LAB 17

Lab Handout

Lab 17. Electromagnetism: Why Does the Battery-and-Magnet "Car" Roll When It Is Placed on a Sheet of Aluminum Foil?

Introduction

Children of all ages often like to play with magnets. Simple combinations of magnets can lead to complex behavior—for example, pushing the north poles of two magnets together and letting go causes the two magnets to move away from each other. Your teacher is going to demonstrate another complex phenomenon where two magnets are connected to a battery and then placed on a sheet of aluminum foil. Figure L17.1 shows the setup of the magnets and battery.

FIGURE L17.1
A battery-and-magnet "car"

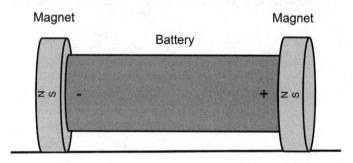

Electricity and magnetism were viewed as two different things prior to the 19th century. The work of scientists including André-Marie Ampère (1775–1836), Hans Christian Ørsted (1777–1851), and Michael Faraday (1791–1867), among others, led to the eventual development of a *unified theory of electromagnetism* (Giancoli 2005). One of the fundamental postulates of the unified theory of electromagnetism is that an electric current will produce a magnetic field surrounding the current-carrying object (often a wire, but not always). Another important idea of the unified theory of electromagnetism is that a change in the magnetic field near an electrical conductor will cause a current to flow through the conductor. Other findings related to the unified theory of electromagnetism have shown that a moving point charge (such as an electron) produces a magnetic field and that a magnetic field exerts a force on a charged object moving through the magnetic field. These findings are important for the working of electrical infrastructure and many modern electronic devices, such as magnetic resonance imaging (MRI) machines.

Electromagnetism
Why Does the Battery-and-Magnet "Car" Roll When It Is Placed on a Sheet of Aluminum Foil?

Besides underlying the working of many of our modern technologies, knowledge of electromagnetism can also help us explain many other observed phenomena and inform the way we design tools. Our understanding of magnetic fields helps explain why two north poles will push each other apart. And our understanding of electric currents informs the design of power strips, leading to power strips being wired in parallel and not in series.

Your Task
Use what you know about electromagnetism, forces, rotational motion, systems and system models, and stability and change to design and carry out an investigation to develop a model that explains the movement of the battery-and-magnets "car." Your model should allow you to make predictions about variables such as the total mass of the car, the voltage of the battery, and the strength of the magnets. There may be other variables that influence the movement of the car that your model will want to account for as well.

The guiding question of this investigation is, *Why does the battery-and-magnet "car" roll when it is placed on a sheet of aluminum foil?*

Materials
You may use any of the following materials during your investigation:

Consumables
- AA batteries
- C batteries
- D batteries
- Duct tape
- Aluminum foil
- Wax paper
- Butcher paper

Equipment
- Safety glasses with side shields or goggles (required)
- Neodymium magnets
- Ceramic magnets
- Voltmeter
- Ammeter
- Multimeter
- Meterstick
- Stopwatch
- Electronic or triple beam balance

If you have access to the following equipment, you may also consider using a video camera, a digital magnetic field sensor, and a digital current sensor and/or digital voltage sensor with an accompanying interface and a computer or tablet. Also, your teacher may give you an electronic pole identifier, which will allow you to determine the north and south poles of your magnets if they are not already labeled.

Safety Precautions
Follow all normal lab safety rules. In addition, take the following safety precautions:

1. Wear sanitized safety glasses with side shields or safety goggles during lab setup, hands-on activity, and takedown.

2. Never put consumables in your mouth.

LAB 17

3. Wire and other metals with electric current flowing through them may get hot. Use caution when handling components of a closed circuit.

4. Us caution in working with sharp objects (e.g., wires) because they can cut or puncture skin.

5. Neodymium magnets should be at least 30 cm away from sensitive electronic and storage devices. These strong magnets could affect the functioning of pacemakers and implanted heart defibrillators.

6. Big magnets have a very strong attractive force. Unsafe handling could cause jamming of fingers or skin in between magnets. This may lead to contusions and bruises.

7. Neodymium magnets are brittle. Colliding magnets could crack, and sharp splinters could be catapulted away for several meters and injure eyes.

8. Sharp splinters could be catapulted away for several meters and injure eyes.

9. Wash your hands with soap and water when you are done collecting the data.

Investigation Proposal Required? ☐ Yes ☐ No

Getting Started

To answer the guiding question, you will need to design and carry out an investigation to determine the mechanisms underlying the movement of the car. You will need to develop a conceptual model that allows you to describe the motion of the car and to predict if a particular arrangement of the battery and magnets will result in the car moving when placed on a certain surface. Furthermore, your model should also allow you to predict how a change in the arrangement of the battery and magnets will result in a change to the motion of the car. Before you can design your investigation, however, you must determine what type of data you need to collect, how you will collect it, and how you will analyze it.

To determine *what type of data you need to collect*, think about the following questions:

- What are the boundaries and components of the system?
- How do the components of the system interact with each other?
- When is this system stable and under which conditions does it change?
- How could you keep track of changes in this system quantitatively?
- What forces, if any, are acting on the objects in the system?
- Which factor(s) might control rates of change in this system?

To determine *how you will collect the data*, think about the following questions:

- What scale or scales should you use when you take your measurements?

Electromagnetism
Why Does the Battery-and-Magnet "Car" Roll When It Is Placed on a Sheet of Aluminum Foil?

- How will you make sure that your data are of high quality (i.e., how will you reduce error)?
- How will you keep track of and organize the data you collect?
- What are the boundaries of this phenomenon or system?
- What are the components of this phenomenon or system and how do they interact?
- How will you measure change over time during your investigation?

To determine *how you will analyze the data*, think about the following questions:

- What type of calculations, if any, will you need to make?
- What types of patterns might you look for as you analyze your data?
- What type of table or graph could you create to help make sense of your data?
- How could you use mathematics to describe a change over time?

Connections to the Nature of Scientific Knowledge and Scientific Inquiry

As you work through your investigation, you may want to consider

- how scientific knowledge changes over time, and
- how scientists use different methods to answer different types of questions.

Initial Argument

Once your group has finished collecting and analyzing your data, your group will need to develop an initial argument. Your initial argument needs to include a claim, evidence to support your claim, and a justification of the evidence. The *claim* is your group's answer to the guiding question. The *evidence* is an analysis and interpretation of your data. Finally, the *justification* of the evidence is why your group thinks the evidence matters. The justification of the evidence is important because scientists can use different kinds of evidence to support their claims. Your group will create your initial argument on a whiteboard. Your whiteboard should include all the information shown in Figure L17.2.

FIGURE L17.2
Argument presentation on a whiteboard

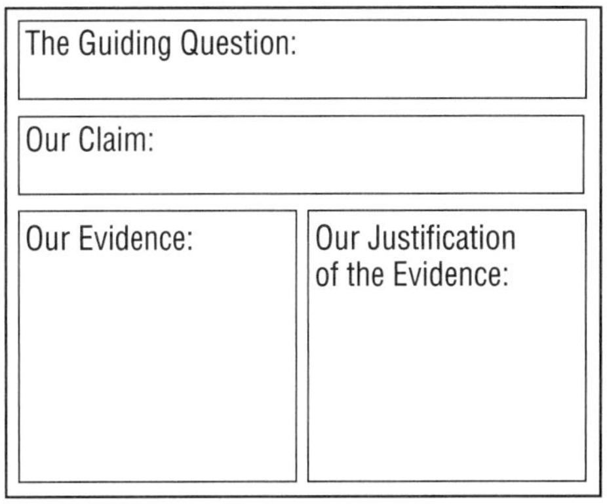

The Guiding Question:	
Our Claim:	
Our Evidence:	Our Justification of the Evidence:

Argumentation Session

The argumentation session allows all of the groups to share their arguments. One or two members of each group will stay at the lab station to share that group's argument, while the other members of the group go to the other lab stations to listen to and critique the other arguments. This is similar to what scientists do when they propose, support, evaluate, and refine new ideas during a poster session at a conference. If you are presenting your group's argument, your goal is to share your ideas and answer questions. You should also keep a record of the critiques and suggestions made by your classmates so you can use this feedback to make your initial argument stronger. You can keep track of specific critiques and suggestions for improvement that your classmates mention in the space below.

Critiques about our initial argument and suggestions for improvement:

If you are critiquing your classmates' arguments, your goal is to look for mistakes in their arguments and offer suggestions for improvement so these mistakes can be fixed. You should look for ways to make your initial argument stronger by looking for things that the other groups did well. You can keep track of interesting ideas that you see and hear during the argumentation in the space below. You can also use this space to keep track of any questions that you will need to discuss with your team.

Interesting ideas from other groups or questions to take back to my group:

Once the argumentation session is complete, you will have a chance to meet with your group and revise your initial argument. Your group might need to gather more data or design a way to test one or more alternative claims as part of this process. Remember, your goal at this stage of the investigation is to develop the best argument possible.

Report

Once you have completed your research, you will need to prepare an investigation report that consists of three sections. Each section should provide an answer to the following questions:

1. What question were you trying to answer and why?
2. What did you do to answer your question and why?
3. What is your argument?

Your report should answer these questions in two pages or less. This report must be typed, and any diagrams, figures, or tables should be embedded into the document. Be sure to write in a persuasive style; you are trying to convince others that your claim is acceptable or valid!

Reference

Giancoli, D. G. 2005. *Physics: Principles with applications.* 6th ed. Upper Saddle River, NJ: Pearson.

LAB 17

Checkout Questions

Lab 17. Electromagnetism: Why Does the Battery-and-Magnet "Car" Roll When It Is Placed on a Sheet of Aluminum Foil?

Use the picture below to answer questions 1–3.

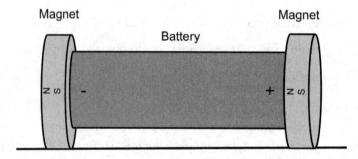

1. The magnet car is set up so that two magnets of equal strength **B** have the north pole facing to the left. When placed on a piece of aluminum foil, they begin rolling. If the two magnets were replaced with magnets of equal mass but with a magnet field strength of 3**B** with both north poles facing to the left, how would this affect the motion of the car?

    a. The car would roll slower.

    b. The car would roll faster.

    c. The car would not roll at all.

    d. The car would roll the same.

    How do you know?

2. Assume now that the magnet on the right is flipped, so that its north pole faces to the right. How would this affect the motion of the car?

   a. The car would roll slower.
   b. The car would roll faster.
   c. The car would not roll at all.
   d. The car would roll the same.

   How do you know?

3. Assume now that the magnets both have the north poles facing to the left. However, the magnet on the left has a strength of **B** while the magnet on the right has a strength of 3**B**. How would this affect the movement of the car?

   How do you know?

LAB 17

4. Scientific knowledge does not change—that is why we still learn about Newton's laws over 300 years after he published them.

    a. I agree with this statement.
    b. I disagree with this statement.

    Explain your answer, using examples from this investigation and at least one other investigation you have conducted.

5. Scientists have always used the same method for investigating questions regarding the interaction between electricity and magnetism.

    a. I agree with this statement.
    b. I disagree with this statement.

    Explain your answer, using examples from this investigation and at least one other investigation you have conducted.

6. Why is it useful to understand factors that influence rates of change in systems? In your answer, be sure to use examples from this investigation and at least one other investigation you have conducted.

7. Why is it useful to assume that you are studying a closed system during an investigation? In your answer, be sure to include examples from your investigation about the battery-and-magnet car and one other investigation you have carried out.

SECTION 5
Appendixes

APPENDIX 1

Standards Alignment Matrixes

Standards Matrix A. Alignment of the Argument-Driven Inquiry Lab Investigations With the Scientific Practices, Crosscutting Concepts, and Disciplinary Core Ideas in *A Framework for K–12 Science Education* (NRC 2012)

	Lab 1. Coulomb's Law	Lab 2. Electric Fields and Electric Potential	Lab 3. Electric Fields in Biotechnology	Lab 4. Capacitance, Potential Difference, and Charge	Lab 5. Resistors in Series and Parallel	Lab 6. Series and Parallel Circuits	Lab 7. Resistance of a Wire	Lab 8. Power, Voltage, and Resistance in a Circuit	Lab 9. Unknown Resistors in a Circuit	Lab 10. Magnetic Field Around a Permanent Magnet	Lab 11. Magnetic Forces	Lab 12. Magnetic Fields Around Current-Carrying Wires	Lab 13. Electromagnets	Lab 14. Wire in a Magnetic Field	Lab 15. Electromagnetic Induction	Lab 16. Lenz's Law	Lab 17. Electromagnetism
NRC *Framework* scientific practices																	
Asking Questions and Defining Problems	■	■	■	■	■	■	■	■	■	■	■	■	■	■	■	■	■
Developing and Using Models	□	□	■	■	□	□	■	■	■	■	■	■	■	■	■	■	■
Planning and Carrying Out Investigations	■	■	■	■	■	■	■	■	■	■	■	■	■	■	■	■	■
Analyzing and Interpreting Data	■	■	■	■	■	■	■	■	■	■	■	■	■	■	■	■	■
Using Mathematics and Computational Thinking	■	■	■	■	■	■	■	■	■	■	■	■	■	■	■	■	■
Constructing Explanations and Designing Solutions	■	■	■	■	■	■	■	■	■	■	■	■	■	■	■	■	■
Engaging in Argument From Evidence	■	■	■	■	■	■	■	■	■	■	■	■	■	■	■	■	■
Obtaining, Evaluating, and Communicating Information	■	■	■	■	■	■	■	■	■	■	■	■	■	■	■	■	■

Key: ■ = strong alignment; □ = moderate alignment.

APPENDIX 1

	Lab Investigations																
	Lab 1. Coulomb's Law	Lab 2. Electric Fields and Electric Potential	Lab 3. Electric Fields in Biotechnology	Lab 4. Capacitance, Potential Difference, and Charge	Lab 5. Resistors in Series and Parallel	Lab 6. Series and Parallel Circuits	Lab 7. Resistance of a Wire	Lab 8. Power, Voltage, and Resistance in a Circuit	Lab 9. Unknown Resistors in a Circuit	Lab 10. Magnetic Field Around a Permanent Magnet	Lab 11. Magnetic Forces	Lab 12. Magnetic Fields Around Current-Carrying Wires	Lab 13. Electromagnets	Lab 14. Wire in a Magnetic Field	Lab 15. Electromagnetic Induction	Lab 16. Lenz's Law	Lab 17. Electromagnetism
NRC *Framework* crosscutting concepts																	
Patterns	■				■					■	■						
Cause and Effect: Mechanism and Explanation			■	■							■	■	■				
Scale, Proportion, and Quantity								■	■				■			■	
Systems and System Models		■		■									■			■	■
Energy and Matter: Flows, Cycles, and Conservation		■			■	■		■	■					■			
Structure and Function	■						■	■		■				■			
Stability and Change			■								■			■	■		■
NRC *Framework* disciplinary core ideas																	
PS1: Matter and Its Interactions								■									
PS2: Motion and Stability: Forces and Interactions	■	■	■							■	■	■	■	■	■	■	■
PS3: Energy		□		■	■	■		■	■					■			

Key: ■ = strong alignment; □ = moderate alignment.

APPENDIX 1

Standards Matrix B. Alignment of the Argument-Driven Inquiry Lab Investigations With the *NGSS* (NGSS Lead States 2013) Performance Expectations for High School Physical Science

NGSS performance expectations	Lab 1. Coulomb's Law	Lab 2. Electric Fields and Electric Potential	Lab 3. Electric Fields in Biotechnology	Lab 4. Capacitance, Potential Difference, and Charge	Lab 5. Resistors in Series and Parallel	Lab 6. Series and Parallel Circuits	Lab 7. Resistance of a Wire	Lab 8. Power, Voltage, and Resistance in a Circuit	Lab 9. Unknown Resistors in a Circuit	Lab 10. Magnetic Field Around a Permanent Magnet	Lab 11. Magnetic Forces	Lab 12. Magnetic Fields Around Current-Carrying Wires	Lab 13. Electromagnets	Lab 14. Wire in a Magnetic Field	Lab 15. Electromagnetic Induction	Lab 16. Lenz's Law	Lab 17. Electromagnetism
HS-PS2-1: Analyze data to support the claim that Newton's second law of motion describes the mathematical relationship among the net force on a macroscopic object, its mass, and its acceleration.			□								■						
HS-PS2-4: Use mathematical representations of Newton's Law of Gravitation and Coulomb's Law to describe and predict the gravitational and electrostatic forces between objects.	■	□	□														
HS-PS2-5: Plan and conduct an investigation to provide evidence that an electric current can produce a magnetic field and that a changing magnetic field can produce an electric current.												■	■	■	■	■	■
HS-PS2-6: Communicate scientific and technical information about why the molecular-level structure is important in the functioning of designed materials.							■										

Key: ■ = strong alignment; □ = moderate alignment.

APPENDIX 1

NGSS performance expectations	Lab Investigations																
	Lab 1. Coulomb's Law	Lab 2. Electric Fields and Electric Potential	Lab 3. Electric Fields in Biotechnology	Lab 4. Capacitance, Potential Difference, and Charge	Lab 5. Resistors in Series and Parallel	Lab 6. Series and Parallel Circuits	Lab 7. Resistance of a Wire	Lab 8. Power, Voltage, and Resistance in a Circuit	Lab 9. Unknown Resistors in a Circuit	Lab 10. Magnetic Field Around a Permanent Magnet	Lab 11. Magnetic Forces	Lab 12. Magnetic Fields Around Current-Carrying Wires	Lab 13. Electromagnets	Lab 14. Wire in a Magnetic Field	Lab 15. Electromagnetic Induction	Lab 16. Lenz's Law	Lab 17. Electromagnetism
HS-PS3-1: Create a computational model to calculate the change in the energy of one component in a system when the change in energy of the other components(s) and energy flows in and out of the system are known.				■				■									
HS-PS3-5. Develop and use a model of two objects interacting through electric or magnetic fields to illustrate the forces between objects and the changes in energy of the objects due to the interaction.	■	■											☐	☐	■	■	☐

Key: ■ = strong alignment; ☐ = moderate alignment.

APPENDIX 1

Standards Matrix C. Alignment of the Argument-Driven Inquiry Lab Investigations With the Nature of Scientific Knowledge (NOSK) and the Nature of Scientific Inquiry (NOSI) Concepts*

	Lab 1. Coulomb's Law	Lab 2. Electric Fields and Electric Potential	Lab 3. Electric Fields in Biotechnology	Lab 4. Capacitance, Potential Difference, and Charge	Lab 5. Resistors in Series and Parallel	Lab 6. Series and Parallel Circuits	Lab 7. Resistance of a Wire	Lab 8. Power, Voltage, and Resistance in a Circuit	Lab 9. Unknown Resistors in a Circuit	Lab 10. Magnetic Field Around a Permanent Magnet	Lab 11. Magnetic Forces	Lab 12. Magnetic Fields Around Current-Carrying Wires	Lab 13. Electromagnets	Lab 14. Wire in a Magnetic Field	Lab 15. Electromagnetic Induction	Lab 16. Lenz's Law	Lab 17. Electromagnetism
NOSK concepts																	
The difference between observations and inferences in science				■			■	■			■			■			
How scientific knowledge changes over time		■							■								■
The difference between laws and theories in science						■	■							■		■	
The difference between data and evidence in science	■				■						■		■		■		
NOSI concepts																	
How the culture of science, societal needs, and current events influence the work of scientists	■		■						■	■							
How scientists use different methods to answer different types of questions		■	■										■				■
The assumptions made by scientists about order and consistency in nature						■	■				■			■		■	
How scientists investigate questions about the natural or material world			■											■			
The nature and role of experiments in science											■				■		

Key: ■ = strong alignment; □ = moderate alignment.

*The NOS/NOSI concepts listed in this matrix are based on the work of Abd-El-Khalick and Lederman 2000; Akerson, Abd-El-Khalick, and Lederman 2000; Lederman et al. 2002, 2014; NGSS Lead States 2013; and Schwartz, Lederman, and Crawford 2004.

APPENDIX 1

Standards Matrix D. Alignment of the Argument-Driven Inquiry Lab Investigations With the Science Practices and Big Ideas in AP Physics 1 and 2

	Lab Investigations																
	Lab 1. Coulomb's Law	Lab 2. Electric Fields and Electric Potential	Lab 3. Electric Fields in Biotechnology	Lab 4. Capacitance, Potential Difference, and Charge	Lab 5. Resistors in Series and Parallel	Lab 6. Series and Parallel Circuits	Lab 7. Resistance of a Wire	Lab 8. Power, Voltage, and Resistance in a Circuit	Lab 9. Unknown Resistors in a Circuit	Lab 10. Magnetic Field Around a Permanent Magnet	Lab 11. Magnetic Forces	Lab 12. Magnetic Fields Around Current-Carrying Wires	Lab 13. Electromagnets	Lab 14. Wire in a Magnetic Field	Lab 15. Electromagnetic Induction	Lab 16. Lenz's Law	Lab 17. Electromagnetism
AP Physics 1 and 2 science practices																	
Modeling	■	■	■	■	■	■	■	■	■	■	■	■	■	■	■	■	■
Mathematical Routines	■	■	■	■	■	■	■	■	■	■	■	■	■	■	■	■	■
Scientific Questioning	■	■	■	■	■	■	■	■	■	■	■	■	■	■	■	■	■
Experimental Methods	■	■	■	■	■	■	■	■	■	■	■	■	■	■	■	■	■
Data Analysis	■	■	■	■	■	■	■	■	■	■	■	■	■	■	■	■	■
Argumentation	■	■	■	■	■	■	■	■	■	■	■	■	■	■	■	■	■
Making Connections	■	■	■	■	■	■	■	■	■	■	■	■	■	■	■	■	■
AP Physics 1 and 2 big ideas																	
Systems: Objects and systems have properties such as mass and charge. Systems may have internal structure.	■	■	■	■	■	■	■										
Fields: Fields existing in space can be used to explain interactions.	■	■	■							■	■	■	■	■	■	■	■
Force Interactions: The interactions of an object with other objects can be described by forces.	■	■	■								■	■		■			■
Change: Interactions between systems can result in changes in those systems.				■				■		■				■		■	
Conservation: Changes that occur as a result of interactions are constrained by conservation laws.					■	■		■	■						■		

Key: ■ = strong alignment; □ = moderate alignment.

APPENDIX 1

Standards Matrix E. Alignment of the Argument-Driven Inquiry Lab Investigations With the Course Content and Science Practices and Skills in AP Physics C: Electricity and Magnetism*

	Lab 1. Coulomb's Law	Lab 2. Electric Fields and Electric Potential	Lab 3. Electric Fields in Biotechnology	Lab 4. Capacitance, Potential Difference, and Charge	Lab 5. Resistors in Series and Parallel	Lab 6. Series and Parallel Circuits	Lab 7. Resistance of a Wire	Lab 8. Power, Voltage, and Resistance in a Circuit	Lab 9. Unknown Resistors in a Circuit	Lab 10. Magnetic Field Around a Permanent Magnet	Lab 11. Magnetic Forces	Lab 12. Magnetic Fields Around Current-Carrying Wires	Lab 13. Electromagnets	Lab 14. Wire in a Magnetic Field	Lab 15. Electromagnetic Induction	Lab 16. Lenz's Law	Lab 17. Electromagnetism
AP Physics C: Electricity and Magnetism course content units																	
Electrostatics	■	■	■														
Conductors, Capacitors, Dielectrics				■													
Electric Circuits					■	■	■	■	■								
Magnetic Fields										■	■						
Electromagnetism												■	■	■	■	■	■
AP Physics C: Electricity and Magnetism science practices and skills																	
Visual Representations	□	□	□	□	□	□	□	□	□	□	□	□	□	□	□	□	□
Question and Method	■	■	■	■	■	■	■	■	■	■	■	■	■	■	■	■	■
Representing Data and Phenomena	■	■	■	■	■	■	■	■	■	■	■	■	■	■	■	■	■
Data Analysis†	□	□	□	□	□	□	□	□	□	□	□	□	□	□	□	□	□
Theoretical Relationships	■	■	■	■	■	■	■	■	■	■	■	■	■	■	■	■	■
Mathematical Routines	■	□	□	■	■	■	■	■	□	■	□	■	■	■	■	■	□
Argumentation	■	■	■	■	■	■	■	■	■	■	■	■	■	■	■	■	■

Key: ■ = strong alignment; □ = moderate alignment.

* The AP Physics C: Electricity and Magnetism course description also includes the big ideas of Fields, Force Interactions, Change, and Conservation. For alignment with the big ideas, see Standards Alignment Matrix D in this appendix.

† The AP Physics C: Electricity and Magnetism practices define *Data Analysis* as "Analyze quantitative data represented in graphs." This is a more constrained definition than we have used in this book. We find moderate alignment for data analysis for many of the labs because students can choose to use a graph for displaying their evidence in support of their claim, but are not required to use a graph; in addition to graphs, there are multiple options for students to display their evidence.

APPENDIX 1

Standards Matrix F. Alignment of the Argument-Driven Inquiry Lab Investigations With the *Common Core State Standards for English Language Arts (CCSS ELA;* NGAC and CCSSO 2010)

Grades 6–12 literacy in science and technical subjects	Lab 1. Coulomb's Law	Lab 2. Electric Fields and Electric Potential	Lab 3. Electric Fields in Biotechnology	Lab 4. Capacitance, Potential Difference, and Charge	Lab 5. Resistors in Series and Parallel	Lab 6. Series and Parallel Circuits	Lab 7. Resistance of a Wire	Lab 8. Power, Voltage, and Resistance in a Circuit	Lab 9. Unknown Resistors in a Circuit	Lab 10. Magnetic Field Around a Permanent Magnet	Lab 11. Magnetic Forces	Lab 12. Magnetic Fields Around Current-Carrying Wires	Lab 13. Electromagnets	Lab 14. Wire in a Magnetic Field	Lab 15. Electromagnetic Induction	Lab 16. Lenz's Law	Lab 17. Electromagnetism
Reading																	
Key ideas and details	■	■	■	■	■	■	■	■	■	■	■	■	■	■	■	■	■
Craft and structure	■	■	■	■	■	■	■	■	■	■	■	■	■	■	■	■	■
Integration of knowledge and ideas	■	■	■	■	■	■	■	■	■	■	■	■	■	■	■	■	■
Writing																	
Text types and purposes	■	■	■	■	■	■	■	■	■	■	■	■	■	■	■	■	■
Production and distribution of writing	■	■	■	■	■	■	■	■	■	■	■	■	■	■	■	■	■
Research to build and present knowledge	■	■	■	■	■	■	■	■	■	■	■	■	■	■	■	■	■
Range of writing	■	■	■	■	■	■	■	■	■	■	■	■	■	■	■	■	■
Speaking and listening																	
Comprehension and collaboration	■	■	■	■	■	■	■	■	■	■	■	■	■	■	■	■	■
Presentation of knowledge and ideas	■	■	■	■	■	■	■	■	■	■	■	■	■	■	■	■	■

Key: ■ = strong alignment; ☐ = moderate alignment.

APPENDIX 1

Standards Matrix G. Alignment of the Argument-Driven Inquiry Lab Investigations With the *Common Core State Standards for Mathematics (CCSS Mathematics;* NGAC and CCSSO 2010)

High school mathematics standards	Lab 1. Coulomb's Law	Lab 2. Electric Fields and Electric Potential	Lab 3. Electric Fields in Biotechnology	Lab 4. Capacitance, Potential Difference, and Charge	Lab 5. Resistors in Series and Parallel	Lab 6. Series and Parallel Circuits	Lab 7. Resistance of a Wire	Lab 8. Power, Voltage, and Resistance in a Circuit	Lab 9. Unknown Resistors in a Circuit	Lab 10. Magnetic Field Around a Permanent Magnet	Lab 11. Magnetic Forces	Lab 12. Magnetic Fields Around Current-Carrying Wires	Lab 13. Electromagnets	Lab 14. Wire in a Magnetic Field	Lab 15. Electromagnetic Induction	Lab 16. Lenz's Law	Lab 17. Electromagnetism
Mathematical practices																	
Make sense of problems and persevere in solving them	■	■	■	■	■	■	■	■	■	■	■	■	■	■	■	■	■
Reason abstractly and quantitatively	■	■	■	■	■	■	■	■	■	■	■	■	■	■	■	■	■
Construct viable arguments and critique the reasoning of others	■	■	■	■	■	■	■	■	■	■	■	■	■	■	■	■	■
Model with mathematics	■	■	■	■	■	■	■	■	■	■	■	■	■	■	■	■	■
Use appropriate tools strategically	■	■	■	■	■	■	■	■	■	■	■	■	■	■	■	■	■
Attend to precision	■	■	■	■	■	■	■	■	■	■	■	■	■	■	■	■	■
Look for and make use of structure																	
Look for and express regularity in repeated reasoning																	
Number and quantity																	
Extend the properties of exponents to rational exponents																	
Use properties of rational and irrational numbers																	
Reason quantitatively and use units to solve problems	■	■	■	■	■	■	■	■	■	■	■	■	■	■	■	■	■
Perform arithmetic operations with complex numbers																	
Represent complex numbers and their operations on the complex plane																	

Key: ■ = strong alignment; □ = moderate alignment.

APPENDIX 1

High school mathematics standards	Lab 1. Coulomb's Law	Lab 2. Electric Fields and Electric Potential	Lab 3. Electric Fields in Biotechnology	Lab 4. Capacitance, Potential Difference, and Charge	Lab 5. Resistors in Series and Parallel	Lab 6. Series and Parallel Circuits	Lab 7. Resistance of a Wire	Lab 8. Power, Voltage, and Resistance in a Circuit	Lab 9. Unknown Resistors in a Circuit	Lab 10. Magnetic Field Around a Permanent Magnet	Lab 11. Magnetic Forces	Lab 12. Magnetic Fields Around Current-Carrying Wires	Lab 13. Electromagnets	Lab 14. Wire in a Magnetic Field	Lab 15. Electromagnetic Induction	Lab 16. Lenz's Law	Lab 17. Electromagnetism
Use complex numbers in polynomial identities and equations																	
Represent and model with vector quantities	■	■	■	■	■	■	■			■	■	■	■	■	■	■	■
Perform operations on vectors	■	■	■	■	■	■				■	■	■	■	■	■	■	■
Perform operations on matrices and use matrices in applications																	
Algebra																	
Interpret the structure of expressions	■	■	■	■	■	■	■	■	■	■	■	■	■	■	■	■	■
Write expressions in equivalent forms to solve problems																	
Perform arithmetic operations on polynomials																	
Understand the relationship between zeros and factors of polynomials																	
Use polynomial identities to solve problems																	
Rewrite rational expressions																	
Create equations that describe numbers or relationships	■	■		■	■	■	■	■			■			■	■	■	
Understand solving equations as a process of reasoning and explain the reasoning	■	■	■	■	■	■	■	■	■	■	■	■	■	■	■	■	■
Solve equations and inequalities in one variable	■	■	■	■	■	■	■	■	■	■	■	■	■	■	■	■	■
Solve systems of equations																	

Key: ■ = strong alignment; □ = moderate alignment.

APPENDIX 1

High school mathematics standards	Lab 1. Coulomb's Law	Lab 2. Electric Fields and Electric Potential	Lab 3. Electric Fields in Biotechnology	Lab 4. Capacitance, Potential Difference, and Charge	Lab 5. Resistors in Series and Parallel	Lab 6. Series and Parallel Circuits	Lab 7. Resistance of a Wire	Lab 8. Power, Voltage, and Resistance in a Circuit	Lab 9. Unknown Resistors in a Circuit	Lab 10. Magnetic Field Around a Permanent Magnet	Lab 11. Magnetic Forces	Lab 12. Magnetic Fields Around Current-Carrying Wires	Lab 13. Electromagnets	Lab 14. Wire in a Magnetic Field	Lab 15. Electromagnetic Induction	Lab 16. Lenz's Law	Lab 17. Electromagnetism
Represent and solve equations and inequalities graphically	■	■	■	■	■	■	■	■	■	■	■	■	■	■	■	■	■
Functions																	
Understand the concept of a function and use function notation	■	■	■	■	■	■	■	■	■	■	■		■	■	■	□	■
Interpret functions that arise in applications in terms of the context	■	■	■	■	■	■	■	■	■	■	■		■	■	■	■	■
Analyze functions using different representations	■	■	■	■	■	■	■	■	■	■	■		■	■	■		■
Build a function that models a relationship between two quantities	■	■			■	■	■	■	■			■		■	■		
Build new functions from existing functions																	
Construct and compare linear, quadratic, and exponential models and solve problems	■	■	■	■	■	■	■	■	■	■	■	■	■	■	■	■	■
Interpret expressions for functions in terms of the situation they model	■	■	■	■	■	■	■	■	■	■	■	■	■	■	■	■	■
Extend the domain of trigonometric functions using the unit circle																	
Model periodic phenomena with trigonometric functions																	
Prove and apply trigonometric identities																	
Statistics and probability																	

Key: ■ = strong alignment; □ = moderate alignment.

APPENDIX 1

High school mathematics standards	Lab 1. Coulomb's Law	Lab 2. Electric Fields and Electric Potential	Lab 3. Electric Fields in Biotechnology	Lab 4. Capacitance, Potential Difference, and Charge	Lab 5. Resistors in Series and Parallel	Lab 6. Series and Parallel Circuits	Lab 7. Resistance of a Wire	Lab 8. Power, Voltage, and Resistance in a Circuit	Lab 9. Unknown Resistors in a Circuit	Lab 10. Magnetic Field Around a Permanent Magnet	Lab 11. Magnetic Forces	Lab 12. Magnetic Fields Around Current-Carrying Wires	Lab 13. Electromagnets	Lab 14. Wire in a Magnetic Field	Lab 15. Electromagnetic Induction	Lab 16. Lenz's Law	Lab 17. Electromagnetism
Summarize, represent, and interpret data on a single count or measurement variable																	
Summarize, represent, and interpret data on two categorical and quantitative variables	■	■	■	■	■	■	■	■	■	■	■	■	■	■	■	■	■
Interpret linear models	■	■	■	■	■	■	■	■	■	■	■	■	■	■	■	■	■
Understand and evaluate random processes underlying statistical experiments																	
Make inferences and justify conclusions from sample surveys, experiments, and observational studies	■	■	■	■	■	■	■	■	■	■	■	■	■	■	■	■	■
Understand independence and conditional probability and use them to interpret data																	
Use the rules of probability to compute probabilities of compound events																	
Calculate expected values and use them to solve problems																	
Use probability to evaluate outcomes of decisions																	

Key: ■ = strong alignment; □ = moderate alignment.

References

Abd-El-Khalick, F., and N. G. Lederman. 2000. Improving science teachers' conceptions of nature of science: A critical review of the literature. *International Journal of Science Education* 22: 665–701.

Akerson, V., F. Abd-El-Khalick, and N. Lederman. 2000. Influence of a reflective explicit activity-based approach on elementary teachers' conception of nature of science. *Journal of Research in Science Teaching* 37 (4): 295–317.

College Board. 2019. *AP Physics 1: Course and exam description. https://apcentral.collegeboard.org/pdf/ap-physics-1-course-and-exam-description.pdf?course=ap-physics-1-algebra-based.*

College Board. 2019. *AP Physics 2: Course and exam description. https://apcentral.collegeboard.org/pdf/ap-physics-2-course-and-exam-description.pdf?course=ap-physics-2-algebra-based.*

College Board. 2019. *AP Physics C: Electricity and Magnetism: Course and exam description. https://apcentral.collegeboard.org/pdf/ap-physics-c-electricity-and-magnetism-course-and-exam-description.pdf?course=ap-physics-c-electricity-and-magnetism.*

Lederman, N. G., F. Abd-El-Khalick, R. L. Bell, and R. S. Schwartz. 2002. Views of nature of science questionnaire: Toward a valid and meaningful assessment of learners' conceptions of nature of science. *Journal of Research in Science Teaching* 39 (6): 497–521.

Lederman, J., N. Lederman, S. Bartos, S. Bartels, A. Meyer, and R. Schwartz. 2014. Meaningful assessment of learners' understanding about scientific inquiry: The Views About Scientific Inquiry (VASI) questionnaire. *Journal of Research in Science Teaching* 51 (1): 65–83.

National Governors Association Center for Best Practices and Council of Chief State School Officers (NGAC and CCSSO). 2010. *Common core state standards*. Washington, DC: NGAC and CCSSO.

NGSS Lead States. 2013. *Next Generation Science Standards: For states, by states*. Washington, DC: National Academies Press. www.nextgenscience.org/next-generation-science-standards.

National Research Council (NRC). 2012. *A framework for K–12 science education: Practices, crosscutting concepts, and core ideas*. Washington, DC: National Academies Press.

Schwartz, R. S., N. Lederman, and B. Crawford. 2004. Developing views of nature of science in an authentic context: An explicit approach to bridging the gap between nature of science and scientific inquiry. *Science Education* 88: 610–645.

APPENDIX 2

OVERVIEW OF THE *NGSS* CROSSCUTTING CONCEPTS

Patterns
Scientists look for patterns in nature and attempt to understand the underlying cause of these patterns. Scientists, for example, often collect data and then look for patterns to identify a relationship between two variables, a trend over time, or a difference between groups.

Cause and Effect: Mechanism and Explanation
Natural phenomena have causes, and uncovering causal relationships (e.g., how changes in x affect y) is a major activity of science.

Scale, Proportion, and Quantity
It is critical for scientists to be able to recognize what is relevant at different sizes, times, and scales. Scientists must also be able to recognize proportional relationships between categories, groups, or quantities. In physics, quantity takes on additional importance as physicists differentiate between vector quantities (quantities with magnitude and direction) and scalar quantities (quantities with only magnitude).

Systems and System Models
To develop a better understanding of natural phenomena in science, scientists often need to define a system under study and make a model of it.

Energy and Matter: Flows, Cycles, and Conservation
It is important to track how energy and matter move into, out of, and within systems.

Structure and Function
In nature, the way an object or a material is structured or shaped determines how it functions and places limits on what it can and cannot do.

Stability and Change
It is critical to understand what makes a system stable or unstable and what controls rates of change in a system.

OVERVIEW OF NATURE OF SCIENTIFIC KNOWLEDGE AND SCIENTIFIC INQUIRY CONCEPTS

Nature of Scientific Knowledge Concepts

The difference between observations and inferences in science

An *observation* is a descriptive statement about a natural phenomenon, whereas an *inference* is an interpretation of an observation. Students should also understand that current scientific knowledge and the perspectives of individual scientists guide both observations and inferences. Thus, different scientists can have different but equally valid interpretations of the same observations due to differences in their perspectives and background knowledge.

How scientific knowledge changes over time

A person can have confidence in the validity of scientific knowledge but must also accept that scientific knowledge may be abandoned or modified in light of new evidence or because existing evidence has been reconceptualized by scientists. There are many examples in the history of science of both *evolutionary changes* (i.e., the slow or gradual refinement of ideas) and *revolutionary changes* (i.e., the rapid abandonment of a well-established idea) in scientific knowledge.

The difference between laws and theories in science

A *scientific law* describes the behavior of a natural phenomenon or a generalized relationship under certain conditions; a *scientific theory* is a well-substantiated explanation of some aspect of the natural world. Theories do not become laws even with additional evidence; they explain laws. However, not all scientific laws have an accompanying explanatory theory. It is also important for students to understand that scientists do not discover laws or theories; the scientific community develops them over time.

The difference between data and evidence in science

Data are measurements, observations, and findings from other studies that are collected as part of an investigation. *Evidence*, in contrast, is analyzed data and an interpretation of the analysis.

Nature of Scientific Inquiry Concepts

How the culture of science, societal needs, and current events influence the work of scientists

Scientists share a set of values, norms, and commitments that shape what counts as knowing, how to represent or communicate information, and how to interact with other scientists. The culture of science affects who gets to do science, what scientists choose to investigate, how investigations are conducted, how research findings are interpreted, and what people see as implications. People also view some research as being more important than other research because of cultural values and current events.

How scientists investigate questions about the natural or material world

Not all questions can be answered by science. Science and technology may raise ethical issues for which science, by itself, does not provide answers and solutions. Scientists attempt to answer question about what can happen in natural systems, why things happen, or how things happen. Scientists do not attempt to answer questions about what should happen. To answer questions about what should happen requires consideration of issues related to ethics, morals, values, politics, and economics.

How scientists use different methods to answer different types of questions

Examples of methods include experiments, systematic observations of a phenomenon, literature reviews, and analysis of existing data sets; the choice of method depends on the objectives of the research. There is no universal step-by step scientific method that all scientists follow; rather, different scientific disciplines (e.g., chemistry vs. physics) and fields within a discipline (e.g., organic vs. physical chemistry) use different types of methods, use different core theories, and rely on different standards to develop scientific knowledge.

The nature and role of experiments in science

Scientists use experiments to test the validity of a hypothesis (i.e., a tentative explanation) for an observed phenomenon. Experiments include a test and the formulation of predictions (expected results) if the test is conducted and the hypothesis is valid. The experiment is then carried out and the predictions are compared with the actual results of the experiment. If the predictions match the actual results, then the hypothesis is supported. If the actual results do not match the predicted results, then the hypothesis is not supported. A signature feature of an experiment is the control of variables to help eliminate alternative explanations for the results.

The assumptions made by scientists about order and consistency in nature

Scientific investigations are designed based on the assumption that natural laws operate today as they did in the past and they will continue to do so in the future. Scientists also assume that the universe is a vast single system in which basic laws are consistent.

APPENDIX 3
Timeline Options for Implementing ADI Lab Investigations

Option A: 6 days (280 minutes), no homework

Day	Stage	Time
1	1: Introduce the task and the guiding question	20 minutes
	2: Design a method	30 minutes
2	2: Collect data	50 minutes
3	3: Develop an initial argument	20 minutes
	4: Argumentation session (and revise initial argument)	30 minutes
4	5: Explicit and reflective discussion	20 minutes
	6: Write investigation report (draft)	30 minutes
5	7: Double-blind peer review	50 minutes
6	8: Revise and submit the investigation report	30 minutes

Option B: 5 days (220 minutes), writing done as homework

Day	Stage	Time
1	1: Introduce the task and the guiding question	20 minutes
	2: Design a method	30 minutes
2	2: Collect data	50 minutes
3	3: Develop an initial argument	20 minutes
	4: Argumentation session (and revise initial argument)	30 minutes
4	5: Explicit and reflective discussion	20 minutes
	6: Write investigation report (draft)	Homework
5	7: Double-blind peer review	50 minutes
	8: Revise and submit the investigation report	Homework

Option C: 5 days (230 minutes), no homework

Day	Stage	Time
1	1: Introduce the task and the guiding question	20 minutes
	2: Design a method and collect data	30 minutes
2	3: Develop an initial argument	20 minutes
	4: Argumentation session (and revise initial argument)	30 minutes
3	5: Explicit and reflective discussion	20 minutes
	6: Write investigation report (draft)	30 minutes
4	7: Double-blind peer review	50 minutes
5	8: Revise and submit the investigation report	30 minutes

APPENDIX 3

Option D: 4 days (170 minutes), writing done as homework

Day	Stage	Time
1	1: Introduce the task and the guiding question	20 minutes
	2: Design a method and collect data	30 minutes
2	3: Develop an initial argument	20 minutes
	4: Argumentation session (and revise initial argument)	30 minutes
3	5: Explicit and reflective discussion	20 minutes
	6: Write investigation report (draft)	Homework
4	7: Double-blind peer review	50 minutes
	8: Revise and submit the investigation report	Homework

Option E: 6 days (280 minutes), no homework

Day	Stage	Time
1	1: Introduce the task and the guiding question	20 minutes
	2: Design a method	30 minutes
2	2: Collect data	30 minutes
	3: Develop an initial argument	20 minutes
3	4: Argumentation session (and revise initial argument)	30 minutes
	5: Explicit and reflective discussion	20 minutes
4	6: Write investigation report (draft)	50 minutes
5	7: Double-blind peer review	50 minutes
6	8: Revise and submit the investigation report	30 minutes

Option F: 4 days (200 minutes), writing done as homework

Day	Stage	Time
1	1: Introduce the task and the guiding question	20 minutes
	2: Design a method	30 minutes
2	2: Collect data	30 minutes
	3: Develop an initial argument	20 minutes
3	4: Argumentation session (and revise initial argument)	30 minutes
	5: Explicit and reflective discussion	20 minutes
	6: Write investigation report (draft)	Homework
4	7: Double-blind peer review	50 minutes
	8: Revise and submit the investigation report	Homework

APPENDIX 4
Investigation Proposal Options

This appendix presents six investigation proposals (long and short versions of three different types of proposals) that may be used in most labs. Investigation Proposal A is appropriate for descriptive studies, whereas Investigation Proposal B and Investigation Proposal C are appropriate for comparative or experimental studies. The development of these proposals was supported by the Institute of Education Sciences, U.S. Department of Education, through grant R305A100909 to Florida State University.

The format of investigation proposals B and C is modeled after a hypothetical deductive-reasoning guide described in *Exploring the Living World* (Lawson 1995) and modified from an investigation guide described in an article by Maguire, Myerowitz, and Sampson (2010).

References

Lawson, A. E. 1995. *Exploring the living world: A laboratory manual for biology*. McGraw-Hill College.

Maguire, L., L. Myerowitz, and V. Sampson. 2010. Diffusion and osmosis in cells: A guided inquiry activity. *The Science Teacher* 77 (8): 55–60.

APPENDIX 4

Investigation Proposal A: Descriptive Study

The Guiding Question…

⬇

What data will you collect?

⬇

How will you collect your data?

Your Procedure

What safety precautions will you follow?

⬇

How will you analyze your data?

I approve of this investigation. _____ _____

Instructor's Signature Date

Your actual data

↓

Your analysis of the data

↓

The claim you will make

APPENDIX 4

Investigation Proposal A: Descriptive Study (Short Form)

The Guiding Question...

↓

What data will you collect?

↓

How will you collect your data?

Your Procedure   What safety precautions will you follow?

↓

How will you analyze your data?

↓

Your actual data

I approve of this investigation. _____ _____
                                         Instructor's Signature   Date

National Science Teaching Association

Investigation Proposal B: Comparative or Experimental Study

The Guiding Question...

Hypothesis 1
IF...

Hypothesis 2
IF...

The Test

AND...
Procedure

What data will you collect?

How will you analyze the data?

What safety precautions will you follow?

Predicted Result if hypothesis 1 is valid
THEN...

Predicted Result if hypothesis 2 is valid
THEN...

I approve of this investigation. _____ _____
Instructor's Signature   Date

APPENDIX 4

Your actual data

AND...

↓

Your analysis of the data

↓

The claim you will make

APPENDIX 4

Investigation Proposal B: Comparative or Experimental Study (Short Form)

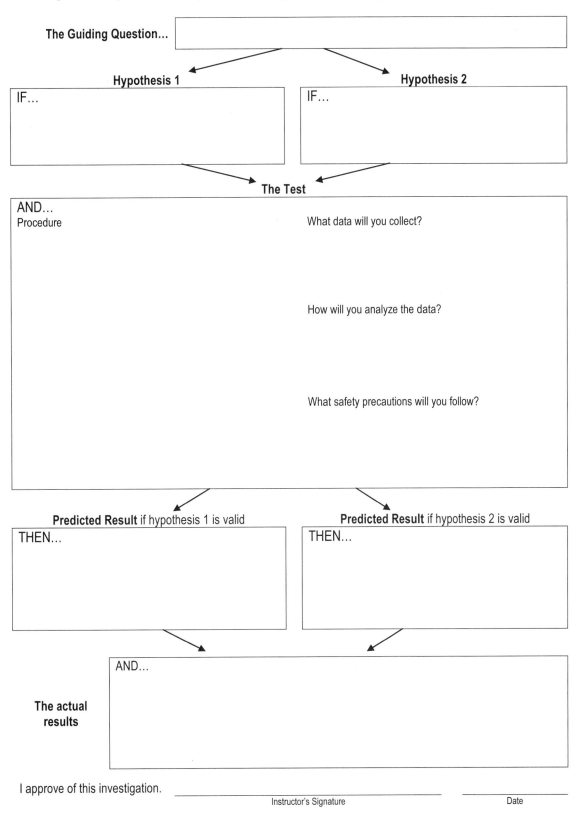

I approve of this investigation. _____ _____
                                                 Instructor's Signature   Date

Argument-Driven Inquiry in Physics, Volume 2: Electricity and Magnetism Lab Investigations for Grades 9–12

APPENDIX 4

Investigation Proposal C: Comparative or Experimental Study

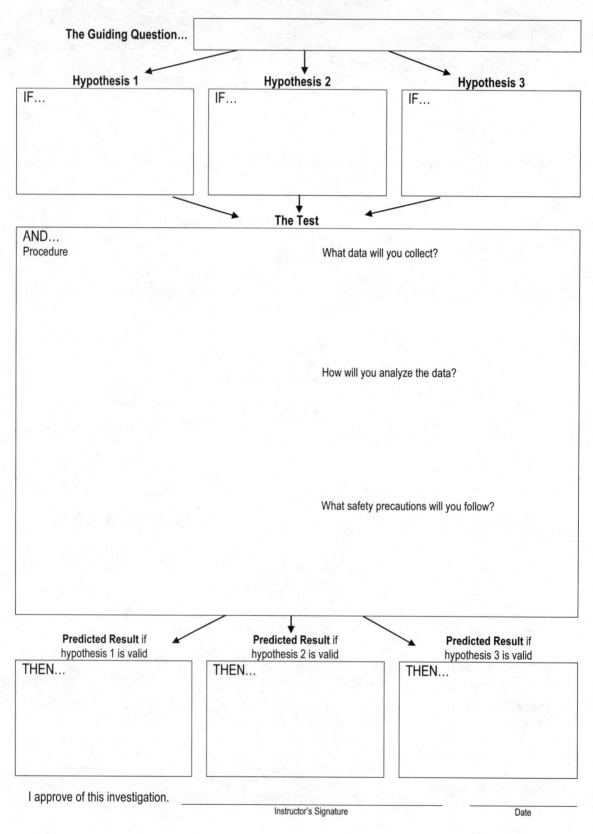

APPENDIX 4

Your actual data

AND…

↓

Your analysis of the data

↓

The claim you will make

APPENDIX 4

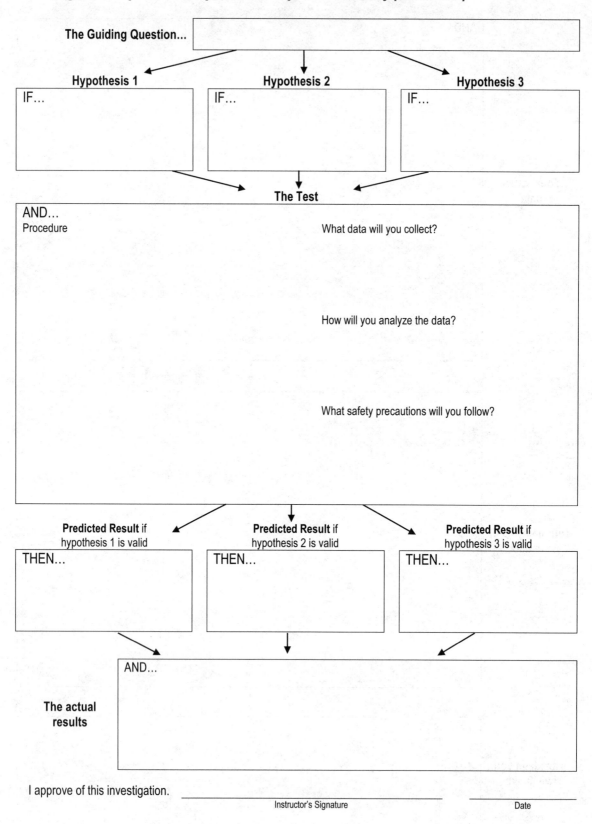

APPENDIX 5
Investigation Report Peer-Review Guide: High School Version

Report By: _____ Author: Did the reviewers do a good job? 1 2 3 4 5
                ID Number   Rate the overall quality of the peer review

Reviewed By: _____   _____   _____   _____
                ID Number   ID Number   ID Number   ID Number

Section 1: Introduction and Guiding Question	Reviewer Rating			Teacher Score
1. Did the author **provide a context** for the story?	☐ No	☐ Partially	☐ Yes	0 1 2
2. Did the author provide enough **background information** about the phenomenon being studied?	☐ No	☐ Partially	☐ Yes	0 1 2
3. Is the background information **accurate**?	☐ No	☐ Partially	☐ Yes	0 1 2
4. Did the author make the **guiding question** explicit and explain how the guiding question is related to the background information?	☐ No	☐ Partially	☐ Yes	0 1 2

Reviewers: If your group made any "No" or "Partially" marks in this section, please **explain how the author could improve** this part of his or her report.

Author: What revisions did you make in your report? Is there anything you decided to keep the same even though the reviewers suggested otherwise? Be sure to explain why.

Section 2: Method	Reviewer Rating			Teacher Score
1. Did the author describe **the procedure** he or she used to gather data and then explain why he or she used this procedure?	☐ No	☐ Partially	☐ Yes	0 1 2
2. Did the author explain **what data** were collected (or used) during the investigation and why they were collected (or used)?	☐ No	☐ Partially	☐ Yes	0 1 2
3. Did the author describe **how he or she analyzed the data** and explain why the analysis helped him or her answer the guiding question?	☐ No	☐ Partially	☐ Yes	0 1 2

APPENDIX 5

Section 2: Method *(continued)*	Reviewer Rating			Teacher Score
4. Did the author use the **correct term** to describe his or her investigation (e.g., experiment, observations, interpretation of a data set)?	☐ No	☐ Partially	☐ Yes	0 1 2
Reviewers: If your group made any "No" or "Partially" marks in this section, please **explain how the author could improve** this part of his or her report.	**Author:** What revisions did you make in your report? Is there anything you decided to keep the same even though the reviewers suggested otherwise? Be sure to explain why.			

Section 3: The Argument	Reviewer Rating			Teacher Score
1. Did the author provide a **claim** that answers the guiding question?	☐ No	☐ Partially	☐ Yes	0 1 2
2. Did the author include **high-quality evidence** in his/her argument? <ul><li>Were the data collected in an appropriate manner?</li><li>Is the analysis of the data appropriate and free from errors?</li><li>Is the author's interpretation of the analysis (what it means) valid?</li></ul>	☐ No ☐ No ☐ No	☐ Partially ☐ Partially ☐ Partially	☐ Yes ☐ Yes ☐ Yes	0 1 2 0 1 2 0 1 2
3. Did the author **present the evidence** in an appropriate manner by <ul><li>using a correctly formatted and labeled graph (or table);</li><li>including correct metric units (e.g., m/s, g, ml); and</li><li>referencing the graph or table in the body of the text?</li></ul>	☐ No ☐ No ☐ No	☐ Partially ☐ Partially ☐ Partially	☐ Yes ☐ Yes ☐ Yes	0 1 2 0 1 2 0 1 2
4. Is the claim **consistent with the evidence**?	☐ No	☐ Partially	☐ Yes	0 1 2
5. Did the author include a **justification of the evidence** that <ul><li>explains why the evidence is important (why it matters) and</li><li>defends the inclusion of the evidence with a specific science concept or by discussing his/her underlying assumptions?</li></ul>	☐ No ☐ No	☐ Partially ☐ Partially	☐ Yes ☐ Yes	0 1 2 0 1 2
6. Is the **justification of the evidence** acceptable?	☐ No	☐ Partially	☐ Yes	0 1 2
7. Did the author discuss **how well his/her claim agrees with the claims made by other groups** and explain any disagreements?	☐ No	☐ Partially	☐ Yes	0 1 2
8. Did the author **use scientific terms correctly** (e.g., *hypothesis* vs. *prediction*, *data* vs. *evidence*) and **reference the evidence in an appropriate manner** (e.g., *supports* or *suggests* vs. *proves*)?	☐ No	☐ Partially	☐ Yes	0 1 2

APPENDIX 5

Section 3: The Argument *(continued)*	Reviewer Rating	Teacher Score
Reviewers: If your group made any "No" or "Partially" marks in this section, please ***explain how the author could improve*** this part of his or her report.	**Author:** What revisions did you make in your report? Is there anything you decided to keep the same even though the reviewers suggested otherwise? Be sure to explain why.	

Mechanics	Reviewer Rating			Teacher Score
1. **Organization:** Is each section easy to follow? Do paragraphs include multiple sentences? Do paragraphs begin with a topic sentence?	☐ No	☐ Partially	☐ Yes	0 1 2
2. **Grammar:** Are the sentences complete? Is there proper subject-verb agreement in each sentence? Are there no run-on sentences?	☐ No	☐ Partially	☐ Yes	0 1 2
3. **Conventions:** Did the author use appropriate spelling, punctuation, and capitalization?	☐ No	☐ Partially	☐ Yes	0 1 2
4. **Word Choice:** Did the author use the appropriate word (e.g., *there* vs. *their*, *to* vs. *too*, *than* vs. *then*)?	☐ No	☐ Partially	☐ Yes	0 1 2

Teacher Comments:

Total: _____ /50

APPENDIX 5

Investigation Report Peer-Review Guide: Advanced Placement Version

Report By: _____ Author: Did the reviewers do a good job?   1 2 3 4 5
　　　　　　　　ID Number　　　　　　　　　　　　　　　　　　　　　　　　　　　　　　Rate the overall quality of the peer review

Reviewed By: _____ _____ _____ _____
　　　　　　　　　ID Number　　　　　　　　ID Number　　　　　　　ID Number　　　　　　　ID Number

Section 1: Introduction and Guiding Question	Reviewer Rating			Teacher Score
1. Did the author **provide a context** for the story?	☐ No	☐ Partially	☐ Yes	0　1　2
2. Did the author provide enough **background information** about the phenomenon being studied?	☐ No	☐ Partially	☐ Yes	0　1　2
3. Is the background information **accurate**?	☐ No	☐ Partially	☐ Yes	0　1　2
4. Did the author make the **guiding question** explicit and explain how the guiding question is related to the background information?	☐ No	☐ Partially	☐ Yes	0　1　2

Reviewers: If your group made any "No" or "Partially" marks in this section, please **explain how the author could improve** this part of his or her report.

Author: What revisions did you make in your report? Is there anything you decided to keep the same even though the reviewers suggested otherwise? Be sure to explain why.

Section 2: Method	Reviewer Rating			Teacher Score
1. Did the author describe **the procedure** he or she used to gather data and then explain why he or she used this procedure?	☐ No	☐ Partially	☐ Yes	0　1　2
2. Did the author explain **what data** were collected (or used) during the investigation and why they were collected (or used)?	☐ No	☐ Partially	☐ Yes	0　1　2
3. Did the author describe **how he or she analyzed the data** and explain why the analysis helped him or her answer the guiding question?	☐ No	☐ Partially	☐ Yes	0　1　2

National Science Teaching Association

APPENDIX 5

Section 2: Method *(continued)*	Reviewer Rating			Teacher Score
4. Did the author use the **correct term** to describe his or her investigation (e.g., experiment, observations, interpretation of a data set)?	☐ No	☐ Partially	☐ Yes	0 1 2
Reviewers: If your group made any "No" or "Partially" marks in this section, please **explain how the author could improve** this part of his or her report.	**Author:** What revisions did you make in your report? Is there anything you decided to keep the same even though the reviewers suggested otherwise? Be sure to explain why.			

Section 3: The Argument	Reviewer Rating			Teacher Score
1. Did the author provide a **claim** that answers the guiding question?	☐ No	☐ Partially	☐ Yes	0 1 2
2. Did the author include **high-quality evidence** in his/her argument? • Were the data collected in an appropriate manner? • Is the analysis of the data appropriate and free from errors? • Is the author's interpretation of the analysis (what it means) valid?	☐ No ☐ No ☐ No	☐ Partially ☐ Partially ☐ Partially	☐ Yes ☐ Yes ☐ Yes	0 1 2 0 1 2 0 1 2
3. Did the author **present the evidence** in an appropriate manner by • using a correctly formatted and labeled graph (or table); • including correct metric units (e.g., m/s, g, ml); and • referencing the graph or table in the body of the text?	☐ No ☐ No ☐ No	☐ Partially ☐ Partially ☐ Partially	☐ Yes ☐ Yes ☐ Yes	0 1 2 0 1 2 0 1 2
4. Is the claim **consistent with the evidence**?	☐ No	☐ Partially	☐ Yes	0 1 2
5. Did the author include a **justification of the evidence** that • explains why the evidence is important (why it matters) and • defends the inclusion of the evidence with a specific science concept or by discussing his/her underlying assumptions?	☐ No ☐ No	☐ Partially ☐ Partially	☐ Yes ☐ Yes	0 1 2 0 1 2
6. Is the **justification of the evidence** acceptable?	☐ No	☐ Partially	☐ Yes	0 1 2
7. Did the author discuss **how well his/her claim agrees with the claims made by other groups** and explain any disagreements?	☐ No	☐ Partially	☐ Yes	0 1 2
8. Did the author **use scientific terms correctly** (e.g., *hypothesis* vs. *prediction*, *data* vs. *evidence*) and **reference the evidence in an appropriate manner** (e.g., *supports* or *suggests* vs. *proves*)?	☐ No	☐ Partially	☐ Yes	0 1 2

APPENDIX 5

Section 3: The Argument *(continued)*		
Reviewers: If your group made any "No" or "Partially" marks in this section, please *explain how the author could improve* this part of his or her report.	**Author:** What revisions did you make in your report? Is there anything you decided to keep the same even though the reviewers suggested otherwise? Be sure to explain why.	

Section 4: Limitations and Implications	Reviewer Rating			Teacher Score
1. Did the author discuss the **limitations of the study** and what he or she could have done in order to **increase the rigor** of the investigation?	☐ No	☐ Partially	☐ Yes	0 1 2
2. Did the author discuss **sources of error** that were unavoidable in the collection of the data?	☐ No	☐ Partially	☐ Yes	0 1 2
3. Did the author discuss **new questions** to explore?	☐ No	☐ Partially	☐ Yes	0 1 2
Reviewers: If your group made any "No" or "Partially" marks in this section, please *explain how the author could improve* this part of his or her report.	**Author:** What revisions did you make in your report? Is there anything you decided to keep the same even though the reviewers suggested otherwise? Be sure to explain why.			

Mechanics	Reviewer Rating			Teacher Score
1. **Organization:** Is each section easy to follow? Do paragraphs include multiple sentences? Do paragraphs begin with a topic sentence?	☐ No	☐ Partially	☐ Yes	0 1 2
2. **Grammar:** Are the sentences complete? Is there proper subject-verb agreement in each sentence? Are there no run-on sentences?	☐ No	☐ Partially	☐ Yes	0 1 2
3. **Conventions:** Did the author use appropriate spelling, punctuation, and capitalization?	☐ No	☐ Partially	☐ Yes	0 1 2
4. **Word Choice:** Did the author use the appropriate word (e.g., *there* vs. *their, to* vs. *too, than* vs. *then*)?	☐ No	☐ Partially	☐ Yes	0 1 2

Teacher Comments:

Total: _____ /56

IMAGE CREDITS

Lab 1
Figure L1.1: Abasaa, Wikimedia Commons, Public domain. *https://commons.wikimedia.org/wiki/File:Measurement_Station_of_the_real-time_radiation_level_measurement_system.JPG*

Lab 2
Figure L2.1a: Geek3, Wikimedia Commons, CC BY-SA 3.0, *https://commons.wikimedia.org/wiki/Category:Electric_field#/media/File:VFPt_plus_thumb.svg*

Figure L2.1b: Geek3, Wikimedia Commons, CC BY-SA 3.0, *https://commons.wikimedia.org/wiki/Category:Electric_field#/media/File:VFPt_minus_thumb.svg*

Lab 4
Figure L4.1: Walter Larden, Wikimedia Commons, Public domain. *https://commons.wikimedia.org/wiki/File:Leyden_jar_cutaway.png*

Lab 5
Figure L5.1: Modified from Carlosar, Wikimedia Commons, CC BY-SA 3.0, *https://upload.wikimedia.org/wikipedia/commons/b/b9/Computer_inside.jpg*

Lab 6
Figure L6.1: Jonathan Cardy, Wikimedia Commons, CC BY-SA 3.0, *https://upload.wikimedia.org/wikipedia/commons/2/21/1902_Ohm_Resistor.jpg*

Figure L6.2: Dmitry G, Wikimedia Commons, Public domain. *https://commons.wikimedia.org/wiki/Category:USSR_resistors#/media/File:Resistors-Old_USSR_Resistor-1W-1100ohm.JPG*

Lab 7
Figure L7.1: Trogain, Wikimedia Commons, CC BY-SA 4.0, *https://commons.wikimedia.org/wiki/File:Piece_of_first_transatlantic_telegraph_cable_at_the_Rupriikki_Media_Museum.jpg*

Lab 8
Figure L8.1: Modified from Koefbac, Wikimedia Commons, CC BY-SA 4.0, *https://commons.wikimedia.org/wiki/File:%C5%BBar%C3%B3wka_60W_produkcji_POLAMP-2.jpg*

Lab 9
Figure L9.1: Alkivar, Wikimedia Commons, CC BY-SA 3.0, *https://upload.wikimedia.org/wikipedia/commons/7/76/Edison_bulb.jpg*

Lab 10
Figure L10.1: Typo~commonswiki, Wikimedia Commons, CC BY-SA 3.0, *https://commons.wikimedia.org/wiki/File:Model_Si_Nan_of_Han_Dynasty.jpg*

Lab 11
Figures 11.2 and L11.1: Geek3, Wikimedia Commons, CC BY-SA 3.0, *https://commons.wikimedia.org/wiki/File:VFPt_cylindrical_magnets_attracting.svg*

Figures 11.3 and L11.2: Geek3, Wikimedia Commons, CC BY-SA 3.0, *https://commons.wikimedia.org/wiki/File:VFPt_cylindrical_magnets_repelling.svg*

Lab 12
Figure L12.1: Paris Wellcome, Wikimedia Commons, CC BY-SA 4.0, *https://commons.wikimedia.org/wiki/Maria_Sk%C5%82odowska-Curie#/media/File:Marie_and_Pierre_Curie_(centre)_in_their_laboratory,_Paris_Wellcome_V0030700.jpg*

Figure L12.2: U.S. Department of Energy, Wikimedia Commons, Public domain. *https://commons.wikimedia.org/w/index.php?search=fermi+lab&title=Special:Search&go=Go&searchToken=2ta16u1mg6ib6d7zd8mui1rpr#/media/File:U.S._Department_of_Energy_-_Science_-_270_006_002_(16187282680).jpg*

Lab 13
Figure 13.2: Modified from Zureks, Wikimedia Commons, Public domain. *https://upload.wikimedia.org/wikipedia/commons/0/0d/Solenoid-1_%28vertical%29.png*

Figure L13.1: From patent #769203, U. S. Patent and Trademark Office, Public domain. *http://pdfpiw.uspto.gov/.piw?docid=00769203&PageNum=1&IDKey=6AE17CF04AC2&HomeUrl=http://patft.uspto.gov/netahtml/PTO/patimg.htm*

IMAGE CREDITS

Lab 14

Figure 14.2: Modified from Acdx, Wikimedia Commons, CC BY-SA 3.0, *https://upload.wikimedia.org/wikipedia/commons/d/d2/Right_hand_rule_cross_product.svg*

Figure L14.1: From patent US1707570A, U. S. Patent and Trademark Office, Public domain. *https://patents.google.com/patent/US1707570*

Figure L14.2: JGHowes, Wikimedia Commons, CC BY-SA 4.0, *https://commons.wikimedia.org/wiki/Category:Speaker_units#/media/File:Johannus_Vivaldi_370_speakers.jpg*

Figure L14.3: Modified from Svjo, Wikimedia Commons, CC BY-SA 3.0, *https://commons.wikimedia.org/wiki/File:Loudspeaker-bass.png*

Lab 15

Figure L15.1: Émile Alglave, Wikimedia Commons, Public domain. *https://commons.wikimedia.org/wiki/File:Faraday_disk_generator.jpg*

INDEX

Page numbers printed in **boldface** type refer to figures or tables.

A

A Framework for K–12 Science Education
    alignment of lab investigations with, 23–24, **403–404**
    importance of science education, ix, xvii
    role of argumentation, xxiii
Ampère, André-Marie, 390
Ampère's law, 268–269, 290–291
AP Physics learning objectives
    alignment of lab investigations with, 23–24, **408–409**
    alignment of standards with, xxvii
    Capacitance, Potential Difference, and Charge lab, 100, **108**
    Circuits in Series and Parallel lab, 138, **147–148**
    Coulomb's Law lab, 32, **42**
    Electric Fields and Electric Potential lab, 54, **63–64**
    Electromagnetic Induction lab, 332, **342–343**
    Electromagnetism lab, 378, **388**
    Electromagnets lab, 290, **299**
    Gel Electrophoresis lab, 76, **85**
    Lenz's Law lab, 354, **365–366**
    Magnetic Field Around a Permanent Magnet lab, 230, **237–238**
    Magnetic Fields Around Current-Carrying Wires lab, 268, **278**
    Magnetic Forces lab, 248, **257**
    performance expectations, 26
    Power, Voltage, and Resistance in a Circuit lab, 182, **191–192**
    Resistors in Series and Parallel lab, 118, **126–127**
    Unknown Resistors in a Circuit lab, 204, **215**
    Wire in a Magnetic Field lab, 310, **320**
    Wire Resistance lab, 160, **170**
Apple Keynote presentations, 9
Application labs, xxv, 25
Argumentation
    defined, xxiii
    role in science, xxiii, 10
    and science education, xvii
    scientific argument
        components, 6, **6**
        evaluation criteria, 6–8, **6**
        layout on a whiteboard, 8–9, **9**
    See also Initial argument
Argumentation session
    described, **4**, 10–12
    gallery walk format for, 11–12, **11**
    role of teacher in, 11–12, 17–18
    See also specific labs
Argument-driven inquiry (ADI) model
    about, x–xi, xxi–xxii, xxiii, 3–17
    role of teacher in, 18, **19–20**
    stages of, 3–17, **4**
        stage 1: identification of task and guiding question, 3–5, **4**
        stage 2: designing a method and collecting data, **4**, 5
        stage 3: data analysis and development of an initial argument, 5–10
        stage 4: argumentation session, 10–12
        stage 5: explicit and reflective discussion, 13–14
        stage 6: writing the investigation report, 14–15
        stage 7: double-blind group peer review, 15–17, **16**
        stage 8: revision and submission of investigation report, 17
Assessment
    checkout questions for, xxvi, 27
    peer-review guide for, 16, **16**, 27
Attractive fields, 248–250, **249**, **250**, 258–259, **258**
Attractive force, 45

B

Batteries. *See* Power, Voltage, and Resistance in a Circuit lab; Unknown Resistors in a Circuit lab
Biotechnology. *See* Gel Electrophoresis lab
Biot-Savart law, xxvi, 318, 323, 354–355, 356, 367, 378
    See also Electromagnetism lab; Lenz's Law lab

C

Capacitance, Potential Difference, and Charge lab, 100–117
    checkout questions, 116–117
    lab handouts
        argumentation session, 114–115
        getting started, 112–113, **112**
        initial argument, 114, **114**
        introduction, 110–111, **110**
        investigation proposal, 112
        materials, 111
        NOSK/NOSI connections, 113
        report, 115
        safety precautions, 112
        task and guiding question, 111
    teacher notes
        content, 100–102
        explicit and reflective discussion topics, 104–106
        implementation hints, 107–108
        materials and preparation, 102–104, **103**
        purpose, 100
        safety precautions, 104
        standards connections, 108, **108–109**
        timeline, 102
Careers in science, ix, xvii
Charge. *See* Capacitance, Potential Difference, and Charge lab
Checkout questions, xxvi, 27
    See also specific labs
Chernobyl, 44
Circuits, electric
    about, 128–129, **128**
    See also Capacitance, Potential Difference, and Charge lab; Circuits in Series and Parallel lab; Power, Voltage, and Resistance in a Circuit lab; Resistors in Series and Parallel lab; Unknown

Index

Resistors in a Circuit lab; Wire Resistance lab
Circuits in Series and Parallel lab, 138–159
    checkout questions, 156–159
    lab handouts
        argumentation session, 153–155
        getting started, 152–153
        initial argument, 153, **153**
        introduction, 150–151, **150**
        investigation proposal, 152
        materials, 151
        NOSK/NOSI connections, 153
        report, 155
        safety precautions, 151–152
        task and guiding question, 151
    teacher notes
        content, 138–141, **139**, **140**, **141**
        explicit and reflective discussion topics, 143–145
        implementation hints, 146–147
        materials and preparation, 141–142, **142**
        purpose, 138
        safety precautions, 143
        standards connections, 147, **147–149**
        timeline, 141
Claim, 6–9, **6**, **9**
Common Core State Standards for English language arts *(CCSS ELA)*, xi, 15, 17, 24, 26, **410**
    See also Literacy connections
Common Core State Standards for Mathematics *(CCSS Mathematics)*, xi, 24, 26, **411–414**
    See also Mathematics connections
Communication skills
    argumentation session, **4**, 10–12, **10**, **11**
    importance of, xvii, 14–15
    investigation report, 14–15
    See also Explicit and reflective discussion; Literacy connections
Computers, 128, **128**
Computer simulations, 4
Conservation laws, 45, **63**, 100, 118, **126**, 137, 138, 143, **147–148**, 151, 182, 184, 188, 202
Content of lab, 24–25
    See also specific labs
Coulomb's Law lab, 32–53
    checkout questions, 50–53
    lab handouts
        argumentation session, 47–49
        getting started, 46–47
        initial argument, 47, **47**
        introduction, 44–45, **44**
        investigation proposal, 46
        materials, 45
        NOSK/NOSI connections, 47
        report, 49
        safety precautions, 46
        task and guiding question, 45
    teacher notes
        content, 32–34, **34**
        explicit and reflective discussion topics, 37–39
        implementation hints, 40–41
        materials and preparation, 35–36, **35–36**
        purpose, 32
        safety precautions, 37
        standards connections, 41, **41–43**
        timeline, 34–35

Criteria for evaluation of scientific argument, 6–9, **6**
Critical-thinking skills, 10
Crosscutting concepts (CCs)
    about, ix–xi, **x**, 417–419
    and ADI model design, 3
    alignment of lab investigations with, 13, 23, **404**
    Capacitance, Potential Difference, and Charge lab, 100, 104–105
    Circuits in Series and Parallel lab, 138, 143–144
    Coulomb's Law lab, 32, 37–38
    Electric Fields and Electric Potential lab, 54, 59–61
    Electromagnetic Induction lab, 332, 338–339
    Electromagnetism lab, 378, 383–385
    Electromagnets lab, 290, 295–296
    Gel Electrophoresis lab, 76, 81–82
    Lenz's Law lab, 354, 361–362
    Magnetic Field Around a Permanent Magnet lab, 230, 233–235
    Magnetic Fields Around Current-Carrying Wires lab, 268, 274–275
    Magnetic Forces lab, 248, 253–254
    Power, Voltage, and Resistance in a Circuit lab, 182, 187–188
    Resistors in Series and Parallel lab, 118, 122–123
    Unknown Resistors in a Circuit lab, 204, 210–211
    Wire in a Magnetic Field lab, 310, 316–317
    Wire Resistance lab, 160, 165–166
Curie, Marie, 280, **280**
Curie, Pierre, 280, **280**
Current, electric, 129, 150–151, 172–173
    See also Circuits in Series and Parallel lab; Electromagnetism lab; Magnetic Fields Around Current-Carrying Wires lab; Power, Voltage, and Resistance in a Circuit lab; Resistors in Series and Parallel lab; Wire Resistance lab

D
Data
    analysis, 4, 5, 6–9, **19**
    collection, **4**, 5, **19**
    described, 418
    designing a method and collecting data, **4**, 5, **19**
    digital interface measurement equipment and analysis, 63, 107–108, 125–126, 146–147, 169, 214, 277, 299, 319–320, 342, 387
Descartes, René, 65
Disciplinary core ideas (DCIs)
    about, ix–xix, **x**, 3
    alignment of lab investigations with, xi, **xii**, **404**
    argumentation session, 11
    Capacitance, Potential Difference, and Charge lab, 100, 104–105
    Circuits in Series and Parallel lab, 138, 143–144
    Coulomb's Law lab, 32, 37–38
    Electric Fields and Electric Potential lab, 54, 59–61
    Electromagnetic Induction lab, 332, 338–339
    Electromagnetism lab, 378, 383–385
    Electromagnets lab, 290, 295–296
    Gel Electrophoresis lab, 76, 81–82
    Lenz's Law lab, 354, 361–362
    Magnetic Field Around a Permanent Magnet lab, 233–235, 239
    Magnetic Fields Around Current-Carrying Wires lab, 268, 274–275

Index

Magnetic Forces lab, 248, 253–254
Power, Voltage, and Resistance in a Circuit lab, 182, 187–188
Resistors in Series and Parallel lab, 118, 122–123
Unknown Resistors in a Circuit lab, 204, 210–211
Wire in a Magnetic Field lab, 310, 316–317
Wire Resistance lab, 160, 165–166
Discipline-specific norms and criteria, **6**, 8
DNA profiles, 87–88, **87**
Doorbells, 301, **301**
Dosimeters, 44–45, **44**
Duschl, R. A., xvii

E

Earth's magnetic field, 239–240, **239**
Edison, Thomas, 217
E-folding constant, 101
Electrical energy, 44–45
Electric current, 129, 150–151, 172–173
    See also Circuits in Series and Parallel lab; Electromagnetism lab; Magnetic Fields Around Current-Carrying Wires lab; Power, Voltage, and Resistance in a Circuit lab; Resistors in Series and Parallel lab; Wire Resistance lab
Electric fields, 54–57, 65–66, **66**, 68, **68**, 230–231, 240
    See also Electric Fields and Electric Potential lab; Gel Electrophoresis lab
Electric Fields and Electric Potential lab, 54–74
    checkout questions, 72–74
    lab handouts
        argumentation session, 70–71
        getting started, 68–69, **68**
        initial argument, 69, **69**
        introduction, 65–66, **66**
        investigation proposal, 68
        materials, 67
        NOSK/NOSI connections, 69
        report, 71
        safety precautions, 67
        task and guiding question, 67
    teacher notes
        content, 54–57
        explicit and reflective discussion topics, 59–62
        implementation hints, 62–63
        materials and preparation, 57–59, **58**
        purpose, 54
        safety precautions, 59
        standards connections, 63, **63–64**
        timeline, 57
Electric force, 45
Electric generator, 344, **344**
Electrification, 217
Electromagnetic Induction lab, 332–352
    checkout questions, 350–352
    lab handouts
        argumentation session, 348–349
        getting started, 346–347, **346**
        initial argument, 347–348, **348**
        introduction, 344–345, **344**
        investigation proposal, 346
        materials, 345
        NOSK/NOSI connections, 347
        report, 349
        safety precautions, 345–346
        task and guiding question, 345
    teacher notes
        content, 332–335, **333–335**
        explicit and reflective discussion topics, 338–340
        implementation hints, 341–342
        materials and preparation, 336–337, **336–337**
        purpose, 332
        safety precautions, 338
        standards connections, 342, **342–343**
        timeline, 336
Electromagnetism lab, 378–399
    checkout questions, 396–399
    lab handouts
        argumentation session, 393–395
        getting started, 392–393
        initial argument, 393, **393**
        introduction, 390–391, **390**
        investigation proposal, 392
        materials, 391
        NOSK/NOSI connections, 393
        report, 395
        safety precautions, 391–392
        task and guiding question, 391
    teacher notes
        content, 378–380, **378**, **379**
        explicit and reflective discussion topics, 383–386
        implementation hints, 386–387
        materials and preparation, 381–383, **381–382**
        purpose, 378
        safety precautions, 383
        standards connections, 388, **388–389**
        timeline, 380–381
Electromagnets, 280–281, **281**, 301–302, 303, **303**
Electromagnets lab, 290–309
    checkout questions, 307–309
    lab handouts
        argumentation session, 305–306
        getting started, 303–304, **303**
        initial argument, 304, **304**
        introduction, 301–302, **301**
        investigation proposal, 303
        materials, 302
        NOSK/NOSI connections, 304
        report, 306
        safety precautions, 303
        task and guiding question, 302
    teacher notes
        content, 290–292, **290**, **291**
        explicit and reflective discussion topics, 295–297
        implementation hints, 297–299
        materials and preparation, 293–294, **293–294**
        purpose, 290
        safety precautions, 294–295
        standards connections, 299, **299–300**
        timeline, 293
Electromotive force, 332–333, 355
Electrostatics. See Coulomb's Law lab; Electric Fields and Electric Potential lab; Gel Electrophoresis lab
Empirical criteria, **6**, 8
Enlightenment, 110
Equivalent resistance (R_{eq}), 118–120, **118**, **119**
Evidence, 6–10, **6**, **9**, 13, 418
Evolutionary changes, 418
Explicit and reflective discussion

Index

ADI instructional model, 13–14, **20**
Capacitance, Potential Difference, and Charge lab, 104–106
Circuits in Series and Parallel lab, 143–145
communication skills, **4**, 13–14
Coulomb's Law lab, 37–39
Electric Fields and Electric Potential lab, 59–62
Electromagnetic Induction lab, 338–340
Electromagnetism lab, 383–386
Electromagnets lab, 295–297
Gel Electrophoresis lab, 81–84
Lenz's Law lab, 361–363
Magnetic Field Around a Permanent Magnet lab, 233–236
Magnetic Fields Around Current-Carrying Wires lab, 274–276
Magnetic Forces lab, 253–255
Power, Voltage, and Resistance in a Circuit lab, 187–190
Resistors in Series and Parallel lab, 122–124
Unknown Resistors in a Circuit lab, 210–212
Wire in a Magnetic Field lab, 316–318
Wire Resistance lab, 165–168

F

Faraday disk, 344, **344**
Faraday, Michael, 344, 390
Faraday's law of induction, 332, 344–345, 354–355, 356, 367
Fermi National Accelerator Laboratory, 280, **281**
Fields, electric, 54–57, 65–66, **66**, 68, **68**, 230–231, 240
See also Electric Fields and Electric Potential lab; Gel Electrophoresis lab
Fields, magnetic. See Magnetic Field Around a Permanent Magnet lab; Magnetic Fields Around Current-Carrying Wires lab
Force and particle accelerators, 281
Free-body diagrams for attractive and repulsive fields, 249–250, **249**, **250**
Fukushima, 44

G

Gel Electrophoresis lab, 76–95
    checkout questions, 94–95
    lab handouts
        argumentation session, 92–93
        getting started, 89–91, **90**
        initial argument, 91, **91**
        introduction, 87–88, **87**
        investigation proposal, 89
        materials, 88
        NOSK/NOSI connections, 91
        report, 93
        safety precautions, 89
        task and guiding question, 88
    teacher notes
        content, 76–78
        explicit and reflective discussion topics, 81–84
        implementation hints, 84
        materials and preparation, 79–80, **79**
        purpose, 76
        safety precautions, 80–81
        standards connections, 84, **85–86**
        timeline, 78

Gravitational fields, 230–231, 240
Group peer review of investigation report
    about, **4**, 15–17, **16**
    peer-review guide for, 16, 27
    revisions based on, 17
    role of teacher in, **20**
Guiding question
    components of initial argument for, 6–7
    designing a method and collecting data for investigation of, 5
    See also specific labs

H

Han-era Si Nan, 239, **239**

I

Identification of task, 3–5, **19**
Inference, 418
Initial argument
    argumentation session on, **4**, 10–12, **19**
    choice of medium for, 8–9
    development of, 5–10, **19**
    example of, 9, **9**
    goal of, 9–10
    layout on a whiteboard, 8–9, **9**
    modification of, 8–9
    role of teacher in, 9–10, **19**
    See also specific labs
Inquiry, science and engineering practices (SEPs), xix
Introduction labs, xxv, 23–28
Investigation design
    Capacitance, Potential Difference, and Charge lab, 105–106
    Circuits in Series and Parallel lab, 144–145
    Coulomb's Law lab, 38–39
    Electric Fields and Electric Potential lab, 61
    Electromagnetic Induction lab, 339–340
    Electromagnetism lab, 385
    Electromagnets lab, 296
    Gel Electrophoresis lab, 82–83
    Lenz's Law lab, 362
    Magnetic Field Around a Permanent Magnet lab, 235
    Magnetic Fields Around Current-Carrying Wires lab, 275
    Magnetic Forces lab, 254
    Power, Voltage, and Resistance in a Circuit lab, 188–189
    Resistors in Series and Parallel lab, 123–124
    Unknown Resistors in a Circuit lab, 211
    Wire in a Magnetic Field lab, 317–318
    Wire Resistance lab, 167
Investigation proposals
    described, 5, 28, 423
    Proposal A (Descriptive Study), 5, 28, **424–426**
    Proposal B (Comparative or Experimental Study), 5, 28, **427–429**
    Proposal C (Comparative or Experimental Study), 5, 28, **430–432**
Investigation report
    double-blind peer group review of, **4**, 15–17, **16**
    format and length of, 15
    peer-review guides for, 16, **16**, 27, **433–439**
    revision and submission of, **4**, 17
    role of teacher in, 15, **20**

Index

writing of, **4**, 14–15
See also specific labs

J
Junction rule, 138–141, 143
Justification of evidence, 6, **6**, 7

K
Kellogg, Edward W., 322
Kirchhoff, Gustav, 138

L
Lab handouts, xxvi, 3–4, 13, 25, 27
See also specific labs
Lab investigations
    alignment with standards, 23–24, **405–406**
    allowing students to fail during, 13, 18
    application labs, xxv, 25
    authenticity of, xxiii, 3
    changing the focus of instruction for, xxi–xxiii
    content of, 24–25
    implementation hints, 27
    instructional materials for, 26–28
        checkout questions, 27
        investigation proposal, 28, 423
        lab handouts, 27
        peer-review guide, 27
    introduction labs, xxv, 24–25
    lab equipment, 4
    limitations of standard format for, xx–xxi
    materials and preparation for, 25
    purpose of, 24–25
    resources for, 25
    role of teacher in, 18, **19–20**
    safety precautions, xxvii–xxviii, 4, 26
    and science proficiency, xx–xxiv
    standards connections, 26
    teacher notes for, 24–26
    timelines for, 25, **421–422**
    See also Investigation report
Learning and science education, xvii–xviii
Lenz's law, 332, 355, 367
Lenz's Law lab, 354–376
    checkout questions, 373–376
    lab handouts
        argumentation session, 371–372
        getting started, 369–370
        initial argument, 370, **370**
        introduction, 367–368, **368**
        investigation proposal, 369
        materials, 369
        NOSK/NOSI connections, 370
        report, 372
        safety precautions, 369
        task and guiding question, 368–369
    teacher notes
        content, 354–358, **356–358**
        explicit and reflective discussion topics, 361–363
        implementation hints, 363–365
        materials and preparation, 359–360, **359–360**
        purpose, 354
        safety precautions, 360
        standards connections, 365, **365–366**
        timeline, 358–359

Leyden jar, 110, **110**
Lightbulbs, 193–194, **193**, 217
Literacy connections
    about, 15, 24, 27, **410**
    Capacitance, Potential Difference, and Charge lab, **109**
    Circuits in Series and Parallel lab, **148**
    Coulomb's Law lab, **42**
    Electric Fields and Electric Potential lab, **64**
    Electromagnetic Induction lab, **343**
    Electromagnetism lab, **388**
    Electromagnets lab, **299**
    Gel Electrophoresis lab, **86**
    Lenz's Law lab, **366**
    Magnetic Field Around a Permanent Magnet lab, **238**
    Magnetic Fields Around Current-Carrying Wires lab, **278**
    Magnetic Forces lab, **257**
    Power, Voltage, and Resistance in a Circuit lab, **192**
    Resistors in Series and Parallel lab, **127**
    Unknown Resistors in a Circuit lab, **216**
    Wire in a Magnetic Field lab, **321**
    Wire Resistance lab, **170**
Lodestones, 239, **239**
Loop rule, 138–141, 143–144, 158
Loudspeaker systems, 322–323, **322**, **323**

M
Magnetic Field Around a Permanent Magnet lab, 230–247
    checkout questions, 244–247
    lab handouts
        argumentation session, 242–243
        getting started, 241
        initial argument, 242, **242**
        introduction, 239–240, **239**
        investigation proposal, 241
        materials, 240
        NOSK/NOSI connections, 242
        report, 243
        safety precautions, 241
        task and guiding question, 240
    teacher notes
        content, 230–231, **231**
        explicit and reflective discussion topics, 233–236
        implementation hints, 236–237
        materials and preparation, 232–233, **232–233**
        purpose, 230
        safety precautions, 233
        standards connections, 237, **237–238**
        timeline, 231–232
Magnetic Fields Around Current-Carrying Wires lab, 268–289
    checkout questions, 286–289
    lab handouts
        argumentation session, 284–285
        getting started, 282–283
        initial argument, 284, **284**
        introduction, 280–281, **280**, **281**
        investigation proposal, 282
        materials, 282
        NOSK/NOSI connections, 283
        report, 285
        safety precautions, 282
        task and guiding question, 282
    teacher notes
        content, 268–271, **268–271**

Index

explicit and reflective discussion topics, 274–276
implementation hints, 276–277
materials and preparation, 272–273, **272–273**
purpose, 268
safety precautions, 273–274
standards connections, 277, **278–279**
timeline, 272
Magnetic flux, 333–335, **333**, 344, 367, **368**
See also Lenz's Law lab
Magnetic Forces lab, 248–267
checkout questions, 263–267
lab handouts
argumentation session, 261–262
getting started, 260–261, **260**
initial argument, 261, **261**
introduction, 258–259, **258**
investigation proposal, 260
materials, 259
NOSK/NOSI connections, 261
report, 262
safety precautions, 259–260
task and guiding question, 259
teacher notes
content, 248–250, **248**, **249**, **250**
explicit and reflective discussion topics, 253–255
implementation hints, 255–256
materials and preparation, 251–252, **251–252**
purpose, 248
safety precautions, 252
standards connections, 256, **257**
timeline, 251
Magnetic resonance imaging (MRI) machine, 281
Magnitude of electric force, 45
Materials and preparation for labs, 25
See also specific labs
Mathematics connections
about, 24, 27, **279**, **411–414**
Capacitance, Potential Difference, and Charge lab, **109**
Circuits in Series and Parallel lab, **148–149**
Coulomb's Law lab, **42**
Electric Fields and Electric Potential lab, **64**
Electromagnetic Induction lab, **343**
Electromagnetism lab, **389**
Electromagnets lab, **300**
Gel Electrophoresis lab, **86**
Lenz's Law lab, **366**
Magnetic Field Around a Permanent Magnet lab, **238**
Magnetic Forces lab, **257**
Power, Voltage, and Resistance in a Circuit lab, **192**
Resistors in Series and Parallel lab, **127**
Unknown Resistors in a Circuit lab, **216**
Wire in a Magnetic Field lab, **321**
Wire Resistance lab, **171**
Metal resistance, 6–8, 160–162, **161**, **162**
See also Power, Voltage, and Resistance in a Circuit lab; Wire Resistance lab
Methods used in scientific investigations, 419
Motors. *See* Power, Voltage, and Resistance in a Circuit lab

N

National Research Council (NRC), x–xi, xvii, xxi
National Science Education Standards, xxi–xxii
National Science Teaching Association, xxviii
Nature of scientific knowledge (NOSK) and nature of scientific inquiry (NOSI)
about, xxvii, **418–419**
alignment of lab investigations with, 14, 23, **407**
Capacitance, Potential Difference, and Charge lab, 106, 113
Circuits in Series and Parallel lab, 145, 153
Coulomb's Law lab, 39, 47
Electric Fields and Electric Potential lab, 61–62, 69
Electromagnetic Induction lab, 340, 347
Electromagnetism lab, 385–386, 393
Electromagnets lab, 297, 304
Gel Electrophoresis lab, 83–84, 91
Lenz's Law lab, 362–363, 370
Magnetic Field Around a Permanent Magnet lab, 235–236, 242
Magnetic Fields Around Current-Carrying Wires lab, 275–276, 283
Magnetic Forces lab, 254–255, 261
Power, Voltage, and Resistance in a Circuit lab, 189–190, 196
Resistors in Series and Parallel lab, 124, 131
Unknown Resistors in a Circuit lab, 211–212, 220
Wire in a Magnetic Field lab, 318, 325
Wire Resistance lab, 167–168, 175
Newton, Isaac, 65
Newton's second law, 34, 76–77, 248–249, 357
Newton's third law, 55, 270
Next Generation Science Standards (NGSS)
alignment of lab investigations with, 23–24, **405–406**
Capacitance, Potential Difference, and Charge lab, **108**
Coulomb's Law lab, **41**
Electric Fields and Electric Potential lab, **63**
Electromagnetic Induction lab, **342**
Electromagnetism lab, **388**
Electromagnets lab, **299**
Gel Electrophoresis lab, **85**
Lenz's Law lab, **365**
Magnetic Fields Around Current-Carrying Wires lab, **278**
Magnetic Forces lab, **257**
performance expectations, xxiv
Power, Voltage, and Resistance in a Circuit lab, **191**
and science proficiency, ix, xvii
and three-dimensional instruction, ix–xi, **x**, xix
Wire in a Magnetic Field lab, **320**
Wire Resistance lab, **170**
Normal of the surface, 333
Nuclear power, 44–45

O

Observation, 418
Ohm, Georg Simon, 217
Ohm's law, 7, 118, 160, 182, 193, 204, 217
Ørsted, Hans Christian, 390

P

Peer-review guide/teacher scoring rubric (PRG/TSR)
AP version, **436–439**
double-blind peer group review, 15–17, **16**, **17**
high school version, **433–435**
revision and submission of investigation report, 17
writing the investigation report, 15, **15**
Peer review of investigation report
ADI instructional model, **4**

peer-review guide for, xxvii, 16, 17, 27
revisions based on, 17
role of teacher in, **20**
Performance expectations, **405–406**
Personal protective equipment (PPE), xxvii–xxviii
Philosophiae Naturalis Principia Mathematica (Newton), 65
Physics. *See* AP Physics learning objectives
Potential difference. *See* Capacitance, Potential Difference, and Charge lab
PowerPoint presentations, 9
Power, Voltage, and Resistance in a Circuit lab, 182–202
    checkout questions, 199–202
    lab handouts
        argumentation session, 196–198
        getting started, 195, **195**
        initial argument, 196, **196**
        introduction, 193–194, **193**
        investigation proposal, 195
        materials, 194
        NOSK/NOSI connections, 196
        report, 198
        safety precautions, 194
        task and guiding question, 194
    teacher notes
        content, 182–185, **184**
        explicit and reflective discussion topics, 187–190
        implementation hints, 190–191
        materials and preparation, 185–187, **186**
        purpose, 182
        safety precautions, 187
        standards connections, 191, **191–192**
        timeline, 185
Preparation for labs, 26
    See also specific labs
Purpose of lab, 24–25
    See also specific labs

R

RC circuits, 110–111, 112–113, **112**, 116
RC time constant, 101
Reading skills. *See* literacy connections
Repulsive fields, 248–250, **249**, 258–259, **258**
Repulsive force, 45
*R*eq (equivalent resistance), 118–120, **118**, **119**
Resistance, metal, 6–8, 129, 160–162, **161**, **162**
    See also Power, Voltage, and Resistance in a Circuit lab; Wire Resistance lab
Resistors in Series and Parallel lab, 118–137
    checkout questions, 134–137
    lab handouts
        argumentation session, 132–133
        getting started, 130–131
        initial argument, 131, **131**
        introduction, 128–129, **128**
        investigation proposal, 130
        materials, 129–130
        NOSK/NOSI connections, 131
        report, 133
        safety precautions, 130
        task and guiding question, 129
    teacher notes
        content, 118–120, **118**, **119**
        explicit and reflective discussion topics, 122–124
        implementation hints, 124–126
        materials and preparation, 120–121, **121**
        purpose, 118
        safety precautions, 122
        standards connections, 126, **126–127**
        timeline, 120
Resistors. *See* Circuits in Series and Parallel lab; Resistors in Series and Parallel lab; Unknown Resistors in a Circuit lab; Wire Resistance lab
Revision and submission of investigation report, **4**, 17
Revolutionary changes, 418
Rice, Chester W., 322
Right-hand rule, 291, **291**, 312, **313**
Rigor and science education, xvii–xviii

S

Safety precautions, xxvii–xxviii, 4, 26
    See also specific labs
Schweingruber, H. A., xvii
Science and engineering practices (SEPs), ix–xi, **x**, xix, 3
Science proficiency
    definition and components of, xvii
    importance of helping students to develop, xvii–xviii
    instructional approaches for development of, ix
    labs to foster development of, xx–xxiv
    and teaching that emphasizes breadth over depth, xvii
    and three-dimensional instruction, xviii–xxiv
Scientific argumentation, xxiii, 3
Scientific explanations, 5
Scientific habits of mind, 10
Scientific laws and theories, 6, 8, 12, 14, 418
    See also specific laws
Scientific practices
    alignment of lab investigations with, 3, 13, 14, **403**, **408–409**
    NGSS standards, ix–xi, **x**
Second Industrial Revolution, 217
Shouse, A. W., xvii
Solenoids, 291–292, **291**, 301–302, 307
    See also Electromagnets lab
Speaker systems, 322–323, **322**
Speaking and listening skills. *See* Literacy connections
Standards connections for labs, 26
    See also specific labs
Sturgeon, William, 301

T

Teacher notes for labs
    content, 24–25
    explicit and reflective discussion, 26
    hints for implementing lab, 27
    materials and preparation, 25
    purpose, 24–25
    safety precautions, 26
    standards connections, 26
    timeline, 25
    See also specific labs
Teacher's roles in argument-driven inquiry, 18, **19–20**
Theoretical criteria, **6**, 8
Three-dimensional instruction, xix–xxiv
Three Mile Island, 44
Timeline for labs, 25, **421–424**
    See also specific labs
"Tool talk," 4–5, **19**
Transatlantic telegraph cable, 172, **172**

Index

U

Unified theory of electromagnetism, 390
Unknown Resistors in a Circuit lab, 204–225
    checkout questions, 223–225
    lab handouts
        argumentation session, 220–222
        getting started, 219–220
        initial argument, 220, **221**
        introduction, 217–218, **217**
        investigation proposal, 219
        materials, 219
        NOSK/NOSI connections, 220
        report, 222
        safety precautions, 219
        task and guiding question, 218
    teacher notes
        content, 204–207, **205**, **206**
        explicit and reflective discussion topics, 210–212
        implementation hints, 212–214, **213**
        materials and preparation, 208–209, **208–209**
        purpose, 204
        safety precautions, 209
        standards connections, 214, **215–216**
        timeline, 207–208

V

Van de Graaff generator, 33, **35**, 37, 40, 46
Video analysis software, 41, 190–191, 256, 364–365
Volta, Alessandro, 217
Voltage, 129, 150
    See also Circuits in Series and Parallel lab; Power, Voltage, and Resistance in a Circuit lab; Resistors in Series and Parallel lab

W

Waste disposal for labs, 26
    See also specific labs
Wheeler, W. T., 301
Wire in a Magnetic Field lab, 310–331
    checkout questions, 328–331
    lab handouts
        argumentation session, 326–327
        getting started, 324–325
        initial argument, 325, **326**
        introduction, 322–323, **322**, **323**
        investigation proposal, 324
        materials, 324
        NOSK/NOSI connections, 325
        report, 327
        safety precautions, 324
        task and guiding question, 323
    teacher notes
        content, 310–313, **312**, **313**
        explicit and reflective discussion topics, 316–318
        implementation hints, 318–320
        materials and preparation, 314–315, **314–315**
        purpose, 310
        safety precautions, 315–316
        standards connections, 320, **320–321**
        timeline, 313
Wire Resistance lab, 160–179
    checkout questions, 178–179
    lab handouts
        argumentation session, 175–177
        getting started, 174–175
        initial argument, 175, **175**
        introduction, 172–173, **172**
        investigation proposal, 174
        materials, 173
        NOSK/NOSI connections, 175
        report, 177
        safety precautions, 173–174
        task and guiding question, 173
    teacher notes
        content, 160–163, **161**, **162**
        explicit and reflective discussion topics, 165–168
        implementation hints, 168–169
        materials and preparation, 163–164, **163–164**
        purpose, 160
        safety precautions, 165
        standards connections, 169, **170–171**
        timeline, 163
Writing skills. *See* Literacy connections
Writing the investigation report, **4**, 14–15
    role of teacher in, 15, **20**